PBX Systems for IP Telephony

Allan Sulkin

McGraw-Hill
New York   Chicago   San Francisco   Lisbon
London   Madrid   Mexico City   Milan   New Delhi
San Juan   Seoul   Singapore   Sydney   Toronto

McGraw-Hill
A Division of The McGraw·Hill Companies

Copyright © 2002 by McGraw-Hill Companies, Inc. All rights reserved. Printed in the United States of America. Except as permitted under the United States Copyright Act of 1976, no part of this publication may be reproduced or distributed in any form or by any means, or stored in a data base or retrieval system, without the prior written permission of the publisher.

1 2 3 4 5 6 7 8 9 0   AGM/AGM   0 9 8 7 6 5 4 3 2

ISBN 0-07-137568-6

The sponsoring editor for this book was Marjorie Spencer, the editing supervisor was Steven Melvin, and the production supervisor was Sherri Souffrance. It was set in New Century Schoolbook by Patricia Wallenburg.

Printed and bound by Quebecor Martinsburg.

A Word about Product Names
From time to time solutions providers change the names of their products or bundle several products as a new offering. A new name does not imply that the original product is no longer available or is no longer supported by the manufacturer/developer. Since this book went to print Avaya has announced a number of name changes reflecting the evolution of their technologies, but will continue to use both old and new names for the convenience of long-time customers. If you have any questions about products mentioned in this book, please contact your Avaya sales representative.

McGraw-Hill books are available at special quantity discounts to use as premiums and sales promotions, or for use in corporate training programs. For more information, please write to the Director of Special Sales, Professional Publishing, McGraw-Hill, Two Penn Plaza, New York, NY 10121-2298. Or contact your local bookstore.

Information contained in this book has been obtained by The McGraw-Hill Companies, Inc., ("McGraw-Hill") from sources believed to be reliable. However, neither McGraw-Hill nor its authors guarantee the accuracy or completeness of any information published herein, and neither McGraw-Hill nor its authors shall be responsible for any errors, omissions, or damages arising out of use of this information. This work is published with the understanding that McGraw-Hill and its authors are supplying information, but are not attempting to render engineering or other professional services. If such services are required, the assistance of an appropriate professional should be sought.

 This book is printed on recycled, acid-free paper containing a minimum of 50 percent recycled, de-inked fiber.

CONTENTS

Chapter 1 Enterprise Communications Systems Today — 1
The Fundamental Enterprise Communications Systems — 5
    KTS — 5
    PBX — 6
    Hybrid System — 7
    ACD Systems — 8
    Voice Messaging System — 9
    Interactive Voice Response System — 11
    Convergence of KTS/Hybrid and PBX Systems — 11

Chapter 2 Evolution of the Digital PBX, 1975–2000 — 15
Digital Switching/Transmission — 17
Computer Stored Program Control — 20
Digital Desktop — 22
Modular System Design — 24
Feature/Function Enhancements — 28
    Basic Voice Call Station/System Features — 28
    Data Communications — 31
    Video Communications — 33
    Networking — 34
    ACD-based Call Centers — 36
    Computer Telephony Integration (CTI) — 38
    Mobile Communications — 41
    Messaging — 42

Chapter 3 Legacy PBX Call Processing Design — 45
Fundamentals of PBX Circuit Switching — 47
    Time Division Multiplexing — 47
    Pulse Code Modulation — 48
    TDM Bus Bandwidth and Capacity — 49
    Port-to-Port Communications over a Single TDM Bus — 51
    Multiple TDM Bus Design — 52
    PBX Circuit Switching Design — 54
    Center Stage Switch Complex — 55
    Broadband TDM Bus — 56
    Single-Stage Circuit Switch Matrix — 57
    Multistage Circuit Switch Matrix — 58
    ATM Center Stage — 59
Local Switching Network Design: TDM Buses and Highway Buses — 60
PBX Switch Network Topologies — 64
    Centralized — 65
    Distributed and Dispersed Switch Network Designs — 67
PBX Switch Network Issues — 71

Switch Network Redundancy	71
Time Slot Access and Segmentation	73
Time Slot Availability: Blocking or Nonblocking	75

Chapter 4 Legacy PBX Switch Network Design — 79

Nonblocking/Blocking PBX Systems	80
PBX Grade of Service (GoS)	82
Defining PBX Traffic: CCS Rating	83
Trunk Traffic Engineering	88
Determining Trunk Circuit Requirements	92

Chapter 5 Legacy PBX Traffic Engineering Analysis — 95

Common Control Complex	96
Main System Processor	98
Operating System Platform	98
Basic Call Processing Functions	100
Main System Memory	103
Local Processors	104
Port Circuit Card Microcontrollers	104
Cabinet/Carrier Shelf Local Controllers	105
System Processing Bus	106
PBX Call Processing Design Topologies	107
Centralized Control	108
Dispersed Control	110
Distributed Control	111
Call Processing Redundancy Issues	112
PBX Call Processing Power: BHC Rating	116

Chapter 6 Legacy PBX Common Equipment — 121

System Cabinets	123
Carrier Types	126
Control	126
Switching	126
Port	127
Auxiliary	128
Application	128
Power	129
Cabinet Power System	129
Cabinet Backplane	132
Cabinet/Carrier Expansion Requirements	132
Port Growth	133
Increased Traffic Requirements	133
Increased Call Processing Requirements	133
New Application Requirements	133
Printed Circuit Boards	134
Control Cards	136
Local Loop Interfaces: Maintenance/Diagnostic	137
Service Cards	137
Port Cards	139

Contents

Circuit Card Provisioning Issues	143
TDM Bus Time/Talk Slot Restrictions	143
Call Processing Limitations	144
Power Distribution Limitations	144
Card Slot Requirements	144

Chapter 7 Introduction to IP-PBX Systems — 147

ToIP and IP-PBX Systems	150
IP-PBX SYSTEM: Benefits and Advantages	154
Leverage Existing Investment in LAN/WAN Infrastructure	154
Reduce Capital, Network, and Operating Expenses	154
Simplify System and Network Configuration Upgrades and Expansion	155
Conforms to Standards	156
Support of Applications across the Enterprise	156
Availability of New and Improved Station User Features and Applications	157
The Case for Converged IP-PBX Systems	158
ToIP Requirements	159
Investment Protection	160
Critical Reliability	162
Private Network Compatibility	163
Feature Requirements	165
Pricing	166
The Case for the Client/Server IP-PBX	169
Converged Network	170
Universality of IP Transport	171
Network Bandwidth	171
Simplified Centralized Management and Administration	172
Rapid Deployment of New Technology and Applications	172
Fully Distributed Network Design	172
Scalability	173

Chapter 8 VoIP Standards and Specifications — 175

ITU-T H.323	176
Point-to-Point and Multipoint Conferencing Support	177
Internetwork Interoperability	177
Heterogeneous Client Capabilities	178
Audio and Video Codecs	178
Management and Accounting Support	178
Security	178
Supplementary Services	178
H.323 Benefits	179
H.323 Architecture Components	180
Terminal	181
Gateway	181
Gatekeeper	181
Multipoint Control Unit	182

Border Elements	183
H.323 Architecture Protocols and Procedures	183
Call Signaling and Control	183
H323 Annexes	184
Audio Codecs	184
Video Codecs	184
H.323 Audio Codecs	186
H.323 Control and Signaling Mechanisms	186
H.225.0 RAS	188
Gatekeeper Discovery	188
Endpoint Registration	188
Endpoint Location	188
Other Communications	188
H.225.0 Call Signaling	189
H.245 Media Control	189
Capability Exchange	190
Opening and Closing of Logical Channels	190
Flow Control Messages	190
Other Commands and Messages	190
The Need for RTP/RTCP	190
Real-Time Transport Protocol	192
RTP Payload	192
RTP Packet	193
RTCP Packet	193
Real-Time Transport Control Protocol	194
SIP	197
SIP Call Process	199
SIP Addressing	199
Locating a SIP Server	200
SIP Transaction	200
SIP Invitation	200
H.323 or SIP?	201
Emerging Dominance of IP	201
Signaling Reliability Mechanism	201
Client/Server Design	201
Addresses	202
Complexity and Cost	202
Command/Message Format	202
QoS Management	203
Firewall/Proxy Design and Configuration	203
Extendible and Scalable	203
Other Protocols: MGCP and MEGACO (H.248)	204

Chapter 9 Converged IP-PBX System Design — 207

IP Station Ports	208
Making IP Voice Calls	214
IP Trunk Ports	216
Dispersed Common Equipment over LAN/WAN Infrastructure	220
Upgraded Circuit Switched PBX	220

Contents

      Distributed Modular Design   223

Chapter 10   Client/Server IP-PBX System Design   227
      Telephony Call Server   230
          Server Operating System   232
          Server Redundancy   233
          Multifunction Server Design   239
      Telephony Gateways   242
          Summarizing Client/Server IP-PBX Design Issues   247

Chapter 11   LAN/WAN Design Guidelines for VoIP   251
      Fundamental LAN Planning Guidelines   254
      Factors Affecting QoS: Packet Loss and Latency   260
          Packet Loss   260
          Latency   262
      QoS Controls   268
          802.1p/Q   268
          ToS   269
          Differentiated Services (DiffServ)   270
      QoS Control Summary Points   271
      A Final QoS Factor: Echo   272
      TIA IP Telephony QoS Recommendations   272

Chapter 12   PBX Cabling Guidelines   277
      Cabling System Fundamentals   279
      Cable Interference and Noise Issues   281
      Overview of ANSI/TIA/EIA 568   283
          Entrance Facility   284
          Equipment Room   284
          Backbone Cabling   284
          Telecommunications Closet   285
          Horizontal Cabling   286
          Telecommunications Outlet   289
          Work Area   290
      Overview of ANSI/TIA/EIA 569   291

Chapter 13   PBX Voice Terminals   295
      Voice Terminal Categories   297
          Analog Telephones   298
          Digital Telephones   299
          IP Telephone Design Basics   303
          Distinct IP Telephony Features/Functions   305
          Desktop Voice Terminal Attributes   317
          Power Requirements   324

Chapter 14   PBX Networking   325
      Public Networking   326
          Automatic Route Selection (ARS)   326
          ISDN Features   328

ANI/SID	329
CBCSS	330
Private Networking	332
Single System PBX Network: On-net Multilocation Support	332
Distributed PBX Common Equipment Design	335
Remote Port Cabinet Options	337
Multiple System Private Networking	341
Fundamentals of PBX Private Networks	343
Uniform Dial Plan (UDP)	344
Automatic Alternate Routing (AAR)	345
FRLs and Network Class of Service (NCOS)	345
Automatic Circuit Assurance (ACA)	346
Virtual Private Networks (VPNs)	346
Intelligent Feature Transparent Network (IFTN)	349
IFTN Features and Functions	351
Qsig	354
Qsig Architecture	356
Qsig Supplementary Services and ANFs	356

Chapter 15 PBX Systems Management and Administration 359

System Administration	361
Administration Sequence	361
Performance Management	365
Trunk Usage and Traffic	365
Attendant Consoles	365
Stations	366
Traffic Distribution	366
Busy Hour Traffic Analysis	366
Processor Occupancy	366
Threshold Alarms	366
Feature Usage	367
VoIP Gateways	367
CDR	367
Directory	369
Inventory	369
Cabling	369
System Diagnostics and Maintenance	369

Appendix A Call Processing Feature/Function Glossary and Definitions 373

Appendix B PBX/IP-PBX Cost and Pricing Issues 399

Appendix C Client/Server IP-PBX RFP Example 407

Appendix D PBX/IP-PBX System Feature and Function Matrices 445

Index 469

CHAPTER 1

Enterprise Communications Systems Today

Today's enterprise communications market is in a considerable flux caused by major ongoing changes in the technology of core products and the network infrastructure. Notably, voice communications systems are migrating from a time- to a packet-based switching and transmission design. The last major market and product shift occurred in the mid-1970s when the first computer stored program control (SPC) and digital switching communications systems were announced and shipped to replace older generation electromechanical systems. Although every generation believes that product upgrades and enhancements occurring in their prime are the most significant ever, telecommunications managers who remember the limited feature and function capabilities available on communications systems 30 years ago may be less impressed with the current market upheaval than industry newcomers who have learned to expect a new generation of products every 18 months from the data networking world.

Today's typical enterprise voice communications network includes many, if not all, of the following ingredients:

1. A core communications *switching system* (Private Branch Exchange [PBX] system, Key Telephone System [KTS], or KTS/Hybrid system) that provides dial tone, call setup and teardown functions, and more call processing features than any one customer is likely to use
2. A *management system* to support fault and configuration operations
3. A *call accounting system* that analyzes and processes call detail records to generate billing and traffic reports
4. A *voice messaging system* that offers a wide array of services far beyond a basic answering system

Other widely used products that support basic voice applications in the enterprise include automated attendants, paging systems, and voice announcers. It is naturally assumed, but sometimes overlooked, that each network system user has some type of desktop or mobile telephone to access the core communications switching system. Other stand-alone desktop equipment scattered around the enterprise is likely to include facsimile (fax) machines and modems for dial-up data network access.

Customers with call center system requirements will install, at a minimum:

1. An Automatic Call Distributor (ACD)
2. A Management Information System (MIS)

As the call center requirements become more sophisticated, subsystems and options will be added to the basic ACD. These might include an Interactive Voice Response (IVR) system, an Automatic Speech Recognition (ASR) system, or a Computer Telephony Integration (CTI) application server. Users now routinely expect that all of these call center system elements will gradually merge with the Web server and e-mail server to form a mixed media e-contact center.

Twenty years ago almost none of these products existed beyond the core communications switching system. Small-line-size customers during the early 1980s with basic voice communications requirements would have a KTS or perhaps one of the recently introduced KTS/Hybrid systems. Intermediate and large-line-size customers with more advanced requirements preferred a PBX system, although what counted as advanced capabilities at the time would include features and functions considered basic today, such as Direct Inward Dialing (DID), Call Detail Recording (CDR), and Automatic Route Selection (ARS). These features were once available on only large, sophisticated, and relatively expensive PBX systems, but they can now be found on KTS products targeted at very small customer locations. The trickle-down theory of KTS/PBX feature and function options says that optionally priced advanced features and functions designed primarily for customers of large PBX systems eventually become standard offerings on entry-level KTSs.

The number of available features on PBX systems has increased exponentially since the first SPC models were introduced in the 1970s. A leading-edge PBX system marketed in 1980 had a software package with about 100 features for station, attendant, and system operations. By 1990 the number of features had more than doubled. Today a typical PBX system boasts more than 500 features, including optional hospitality, networking, and ACD options, and today's typical KTS/Hybrid system offers more performance options than any PBX system in 1980. Despite the significant increase in features designed for desktop access and implementation, the majority of PBX station users (i.e., people with phones on their desks) use fewer than ten features on an everyday basis. Ironically, today's typical station user may use fewer features than he would have used 20 years ago because many once-popular features, such as call pick-up and automatic callback, are rarely implemented. One reason for the decline in use of once common desktop features is the prevalence of voice messaging systems that preclude the use of many manually operated features for call coverage situations.

As a result, today's PBX developers continue to write new feature software programs for the non-typical station user. Studies show that most station users implement about six features on an everyday basis, and features in general use are limited to hold, transfer, conference, and a few others. However, system designers cannot assume that the set of features in general use will be the same for every station user. Many features are used by a small number of system subscribers, but they are no less important than those used by the majority. For example, a feature such as Flexible Night Service may be used only by the system's sole attendant console operator, and the Recent Change History feature may be of value only to the system administrator, but these features are as vital to the few individuals who implement them as Call Forwarding is to a typical desktop telephone station user. Many of the hundreds of PBX features introduced during the past 20 years were developed at the explicit request of customers. When a customer or a small group of customers demanded a new feature, a PBX manufacturer first determined that anticipated demand justified its development. Once offered by a major manufacturer, the new feature soon became available on systems from most competitors.

It's important to note that some perfectly viable features are unique to special categories of customers or station users and may be used by as few as one system subscriber per enterprise. A feature's value is not determined solely by how many individual station users implement the feature, but also by potential cost savings and productivity improvement at a station, system, or network level.

Of course, most customers do not have stringent demands on PBX system design architecture attributes; they're looking for basic growth and redundancy requirements. A station user who doesn't have telecommunications system acquisition or management responsibilities cares nothing about the technical underpinnings of the telephone system he's using. He picks up the telephone handset, listens for dial tone, punches a number or activates a feature, and is satisfied by the experience almost every time. As long as that's true, the station user won't be asking whether the system has analog or digital transmission, circuit-switched or packet-switched connections, a proprietary software operating system, or a standard off-the-shelf Windows solution. People who *should* care about PBX system technology and reliability standards, applications support, and future product direction are the telecommunications manager, voice/data networks director, or CIO. This book is written for the individual who must know more about PBX systems than basic calling procedures, who must be able to configure and reconfigure the system when need demands, and who must know the answers when questions are asked.

The Fundamental Enterprise Communications Systems

Current voice communications systems comprise five fundamental product categories:

- Key Telephone System (KTS)
- Private Branch Exchange (PBX)
- Automatic Call Distributor (ACD)
- Voice Messaging System (VMS)
- Interactive Voice Response (IVR)

The first three categories support call processing and switching functions to enable telephone calls between two or more station users. The latter two product categories are designed to work in conjunction with one of the three core communications switching systems to provide optional services beyond a basic station-to-station call. It's possible to integrate voice messaging or IVR capabilities into a KTS, PBX, or ACD product design, but most companies don't. Instead, customers have traditionally chosen to install external, stand-alone systems for messaging and voice response applications. (Many small KTS products have an integrated voice messaging function, but the messaging features are typically not as robust as stand-alone offerings.)

KTS

A KTS is a customer-premises communications system designed to support basic voice applications and relatively small station user requirements. KTS got its name from its historical use with proprietary telephones, known as key telephones, which interface with a central control cabinet known as a Key Service Unit (KSU). The KSU is equipped with the system's call processing control, port interface cards, and a variety of system/service circuit cards, such as Dual Tone Multi Frequency (DTMF) receivers and Input/Output (I/O) interfaces. The KSU performs central office line connection, intercom functions, paging, and station connections. Its common control elements include a call processor and system memory databases, and its most important function is the provisioning of dial tone. Other basic call processing functions are call answering, dialing, and transaction features such as hold, transfer, call forward, and conference.

Oddly, not all KTS products require a KSU. Instead, the intelligence needed to perform call processing and switching can be built into the circuitry of each key telephone instrument. Such systems are easy to install and maintain, but usually have limited feature and function capabilities, and are acceptable only for customers with modest port-size requirements. The KSU-less system is usually installed to work behind Centrex services offered by local telephone operating companies, which provide the more advanced features and functions through their central office communications switching system.

Common to KTS telephone instruments are designated, programmable key buttons for making and receiving off-premises calls over telephone company line circuits (trunks). The term *KTS* takes its name from the telephone instrument keys (analogous to typewriter keys) integrated into the product design. The line keys for each telephone instrument have direct access to off-premises telephones. Because trunks are distributed over and shared across groups of select telephones, each station user's desktop instrument must be provisioned with multiple-line access keys. *This is the distinguishing characteristic of a KTS: station user selection and access to designated telephone company lines.* A pure KTS product supports only multi-line key telephone instruments and is incapable of supporting industry standard 2500-type single-line analog telephones unless a designated line circuit is dedicated to a single analog telephone instrument. Needless to say, it's unusual for a customer to configure a KTS in this manner. Small system customers who want station users with 2500-type telephones to have shared access to a pool of trunks do have an alternative: a KTS/Hybrid product (see below), designed to support a mix of proprietary multiple-line button phones and nonproprietary single-line analog phones.

PBX

PBXs aren't just big KTSs with more features and functions. The two share architecture elements, but PBX systems are designed for more robust functionality, greater growth capacities for ports, traffic, and call processing, and more levels of redundancy. PBX basic architecture includes the common control carrier/cabinet, port carriers/cabinets, and port circuit cards. Peripheral equipment support includes proprietary telephones, both single and multiple line, and industry standard single-line analog telephones. A critical discussion of PBX system design and feature capabilities is the primary objective of this book, and we'll return to it in subsequent chapters.

First, let's look at the major design and operational difference between a KTS and PBX. PBX stations can place calls over telephone company trunks only with the shared pool access method. The sole exceptions are telephones equipped with an optional private line for inbound and outbound calls. [Note: for reasons too complex to explain, the term *line*, when used in relation to a PBX system, usually refers to all customer premises equipment peripheral endpoints and is not a reference to a trunk circuit, as it would be with a KTS. The terms *station* and *line* are both used for PBX endpoints (telephones, modems, fax terminals) that are not off-premises trunk circuits.]

Hybrid System

Hybrid communications systems also share operating functions common to standard KTS and PBX systems. Original Hybrid systems were designed to more closely resemble KTS rather than PBX systems, although differences between the product categories are diminishing. Like KTS, Hybrid systems are based on a control cabinet similar to a KSU and can support a variety of port circuit boards for interfacing to station and trunk circuit equipment. All Hybrids support multiple-line proprietary key telephones and industry-standard single-line analog telephones. In a Hybrid system, phone access to line circuits is identical to that of a pure KTS, but single-line analog phones access a defined pool (group) of telephone company line circuits. *This latter design capability is what distinguishes a pure KTS from a Hybrid system.* The Hybrid's port-oriented architecture design permits custom configurations to suit specific business applications. The architecture and technology design foundations of current Hybrid systems are more similar to PBXs than to KTSs. In fact, features and functions are sometimes difficult to distinguish from more expensive PBX systems.

Unfortunately confusion reigns when it comes to product category typing a communications system as KTS, Hybrid, or PBX. Some manufacturers call their Hybrid offering a KTS/Hybrid and others may refer to it as a Hybrid/PBX. The naming issue gets interesting in the United States because the Federal Communications Commission classifies customer premises communications systems as either KTS or Hybrid based on how single-line telephones access the central office. If the phone can access only one line as programmed by the system administrator, the system is a KTS; if it can access a pooled group of lines, the system is considered a Hybrid. Some manufacturers may even register a single system as both because the call processing software allows configuration flexibility for either pooled- or single-line access from a single-line telephone. Note that designating the product as a KTS, Hybrid, or PBX system may have financial consequences based on telephone company jurisdiction because trunk tariffs for linking a customer premises communications system to the central office can differ between KTS and PBX. (This was more common 15 years ago than it is today.) Ultimately it is the local telephone company that defines the type of system the customer is seeking to connect to the central office.

ACD Systems

The central component of a customer call center is an ACD. ACD systems were originally developed to handle large volumes of incoming calls and automatically route them to designated answering positions. ACD systems are designed and customer programmed to satisfy higher quality of service standards than PBX systems for the following call processing functions:

- Screening
- Routing
- Queuing
- Answering

Most PBX systems can be programmed to function as ACD systems, but few ACDs can be programmed to function as PBXs and continue to support most of the latter product's standard or optional features and functions. Nevertheless, an ACD system shares many of the architecture and feature capabilities of a PBX system. You can think of it as a PBX designed for a very specific application—to distribute incoming calls equitably to a group or groups of answering stations. We usually call ACD answering stations *agents*, and *this is the fundamental difference between PBX and ACD system service: calls handled by a PBX are routed to a specific station user*, whereas ACD calls are routed to a group of stations, although call analysis programs can be used to route the call to a specific agent in a group.

ACDs exhibit several architecture design and feature standards that are often not adhered to for PBXs. A true ACD system is based on a nonblocking switch network design, has sufficient call processing power to handle a large volume of complex call types, and has software programming features that can screen, route, and distribute calls to agent positions fast and efficiently. Other distinguishing standard ACD system attributes are the ability to support specialized stations, known as supervisor positions, and an integrated MIS reporting system used to monitor, track, and analyze call center operations. All of these features are standard in stand-alone ACD systems, but PBX systems may fail to meet the same standards. Indeed most PBXs must be traffic engineered to support non-blocking switch network access and post-equipped with optional software and external applications servers to all but the most basic ACD feature and MIS capabilities.

Early ACD systems were stand-alone products designed exclusively for a call center environment with large call volumes, such as an airline reservation center. During the early 1980s several PBX systems were designed with optional software that could support basic ACD functions but could not match their performance level. By the 1990s a growing number of PBX systems enhanced with optional ACD software and external MIS reporting systems started to look functionally competitive with stand-alone ACDs. Similar PBX systems with optional ACD packages now control more than 80 percent of the market for call center communications systems. Although stand-alone ACDs hold their own at the high end of the market for large, complex feature and function requirements, PBX and ACD systems totally dominate the call center market segment for systems with fewer than 100 agents. Many KTS/Hybrid systems can also be configured with optional ACD capabilities and are gradually penetrating the very small call center market segment for systems with fewer than 20 agents. Why has the average size of an ACD call center system continually declined over the past 20 years? Customers are realizing that programmable call routing, distribution, and reporting features are beneficial for a variety of nontraditional ACD applications, such as internal help desks and groups of attendant positions. Today's average ACD installation has only about 60 agent positions, and that number is declining.

Voice Messaging System

Most enterprise environments use a VMS as the primary call coverage point for unanswered calls, but you'd be selling the VMS short to think of it only in terms of message record and store. A good VMS mailbox can:

- Be programmed for different greetings
- Offer incoming callers a menu of options for leaving a message or transferring to another station
- Act as an automated attendant position to answer and route calls
- Serve as an automated information system or an outbound call messaging system

VMSs can be interfaced to almost all current generation KTS/Hybrid, PBX, and ACD systems by using a signaling link between the VMS, the switching system, and its voice communications channels. The signaling link is usually referred to as the *voice mail system interface* or, in standards parlance, the Station Message Detail Interface (SMDI). Among other functions, the link activates the message-waiting indication at a user station. Its voice communications channels will be based on 2500-type analog station interfaces, a standard interface supported by all systems.

Here's what happens when a caller leaves a message. VMS coders/decoders (codecs) take the transmitted voice communications signals from the switching system, digitize them, and compress them for storage purposes. The same codecs will convert the digitally compressed messages back to analog format for station user playback. The voice quality of VMS playback largely depends on compression algorithms that may degrade the original message to optimize storage capabilities by using a low sampling rate or bit scheme. Filters and automatic gain control improve sound quality, but if stored messages are forwarded to another station or broadcast to hundreds of stations, further digital compression occurs as messages pass through interim mailboxes. Moreover, when VMS is used behind IP telephony, playback quality weakens if stored message transmission is packetized using IP codecs. This is a technical issue being addressed by designers and developers of the new IP-PBX systems.

Because the market trend is toward Local Area Network (LAN)–connected application servers, the traditional method for setting up signaling between the switch and the external VMS is slowly changing. The new, improved method is to insert an Ethernet TCP/IP link between the switching system and the LAN-connected VMS. At the same time, the newer VMS server design is also driving the market for Unified Messaging systems (UMSs). Many people think that first-generation UMS failed to take hold because they were based largely on traditional VM systems design and user operation. In contrast, second-generation systems are based largely on e-mail server design and user operation and are gaining greater market acceptance. Several new client/server PBX system designs include the VMS or UMS function as an integrated feature capability.

Interactive Voice Response System

An IVR system is a communications system product that typically functions as an intermediary between a PBX system and external computer databases. An IVR can also function as a stand-alone enterprise communications system, without a PBX system, because it can support standard PSTN trunk interfaces. Analog and digital trunk interfaces are supported by most IVR systems to connect to the customer premises PBX system or directly to the PSTN. IVR systems are used for several customer applications, including automated voice response and feedback, automated directory, call routing.

The primary system link for an IVR system is not the switching system, but an external computer system. The IVR mediates between voice callers and computer databases with customer-written scripts and menus prompting callers to respond to prompts with dialpad entries on their phones. The IVR system interprets the DTMF signals from the dialpad and is programmed to respond with another voice prompt, a recorded announcement, or a "spoken" answer to the caller's inquiry. IVR speech is based on a text-to-speech programming algorithm, for which appropriate "answers" can be stored in the IVR database (which usually has limited memory storage capacity) or (more commonly) an external computer database.

Some IVR systems with ASR capabilities don't require DTMF input to respond to caller voice commands and questions. Voice-based interaction with the IVR can significantly speed up the IVR transaction process by bypassing many call prompt levels of the programming tree. There are also many instances when callers find it easier and faster to speak digits or words instead of using the dialpad to enter digits in response to a call prompt command. ASR systems have been famously slow to penetrate the market because of reliability and cost barriers, but recent advances in programming and the declining cost of digital signal processors have made ASR more prevalent. We anticipate that the next major technology advance for automated response systems is *speaker verification*, an important capability to simplify the system interaction process and raise security levels.

Convergence of KTS/Hybrid and PBX Systems

KTS and PBX architectures began their gradual convergence of design and function with the introduction of the first KTS/Hybrid systems. The early 1980s saw major differences in call processing, switching, and port cabinet-design pure KTSs and PBXs. KTSs used analog switching and transmission standards and were based on a common control system design, with limited traffic, processing, and port capacities. Features were necessarily limited, and options such as multiple system networking and ACD were unheard of. Migration between KTS models—even on the same manufacturer's product platform—usually required KSU forklifts. However, most PBX systems of the day employed digital switching and transmission standards, with digital desktop telephone and trunk interface support. PBXs were designed to handle greater traffic, processing, and port expansion requirements and offered numerous advanced features not available on any KTS.

Hybrid systems began to blur these category differences as PBX design technology and feature capabilities were applied to a KTS-like platform, nudging Hybrid switching network design away from the traditional dedicated line access and intercom path. The result more closely resembled the digital Time Division Multiplexing (TDM) bus design used by PBX systems. Hybrid systems could support customer port capacities far in excess of KTSs, some basic networking options, and call center applications.

By the mid-1990s many customers couldn't tell the difference between a Hybrid system and a full-function PBX system because the design platform and feature sets were so similar as to be indistinguishable for all but the most unique customer requirements. High-end Hybrid systems could support customer port requirements up to several user stations. They accommodated networking options such as digital T1-carrier trunk interfaces and ISDN PRI services, ACD options including MIS reporting packages similar to the early PBX/ACD systems, and some integrated voice messaging and wireless communications options. Hybrid system digital telephones were more advanced, often offering customers a greater selection of multiple-line key and display models. It was not unusual for Hybrid telephones to offer the large display fields and softkey feature buttons that are only now becoming a PBX standard.

Today's Hybrids are often larger in capacity than many entry-level PBX systems and can support most, if not all, of the same features and functions for the majority of customer requirements. They can be intelligently networked, and they can satisfy complex call center requirements. Some manufacturers offer the same telephone models for both their Hybrid and PBX models, allowing customers a migration path between the two platforms, and most manufacturers have announced that their recently shipped IP telephones will work behind either systems.

As Hybrid systems expanded, PBX systems grew smaller. Until recently PBX systems were not priced to be competitive against Hybrid systems at the same station sizes. (The target market for competitively priced PBX systems usually began at 40 stations.) This PBX price problem was created by the higher cost of its more robust common control cabinet or carrier. During the past few years the PBX manufacturers have responded by redesigning their common control systems and downsizing the control cabinet/carrier hardware equipment. Although today's PBX systems remain higher priced than a KTS/Hybrid configured for the same number of stations and trunks, the price differential has been continually shrinking. There is currently a 10 to 15 percent price difference between the two system platforms compared with 25 to 50 percent a decade ago. Today there are entry PBX models designed around a single printed circuit board for call processing, memory, and switch network functions. These new models are price competitive but port limited, usually designed for customers with 20 to 40 station size requirements at initial installation, but equipped with the call processing capability to support all of the features and functions of PBX models many times their port capacity.

PBX systems based on client/server designs are targeted primarily at the small/medium enterprise (SME) market, defined as ranging from about 20 stations to about 120 station, and overlaps with the Hybrid system market. A typical Hybrid system installation is about 25 stations. Shipments of PBX systems based on client/server designs are currently averaging about the same station size as Hybrid systems, and few have been installed at customer locations with port requirements above 100 stations. As you might expect, most of the systems being replaced by the client/server PBX systems are Hybrid systems, against which the newer PBXs can offer more bundled applications and support (either current or planned) for IP endpoints. The current generation of installed-base Hybrid systems may someday support IP endpoints but only after costly system upgrades. For the vast majority of users, it will be less expensive to install a new system than to upgrade the old. The first generation of IP telephony systems designed for the very small KTS/Hybrid customer is just beginning to make its way into the market.

Under the circumstances, we can predict that the traditional Hybrid system eventually will be replaced by the new communications system, client/server design, competitive product offerings, or upgrades to the system manufacturer's new design platform. But this picture is also affected by customer size. Almost all market demand forecasts predict that the new PBX designs will be far more successful in smaller rather than larger line size markets, because the latter customer segment is less likely to replace a reliable and upgradeable installed system for a system that has not yet proved as reliable. The cost factors for large system replacement go far beyond the system purchase and installation price. Too much is at stake for large line size customers to risk the lesser reliability level of an OEM server running an operating system, such as Windows NT, not originally developed for real-time industrial applications. Only a few of the new client/server systems are built on more reliable and flexible operating systems, such VX Works, and custom-designed common control elements. Branch offices may sacrifice reliability for a less expensive system, but a large line size corporate or regional headquarters customer would not.

CHAPTER 2

Evolution of the Digital PBX, 1975–2000

There are two distinct generations of PBX systems based on the fundamental transmission and switching platform used to support signaling, control, and communications to and from the station user desktop and the common equipment. The first generation was known collectively as *analog PBXs* and included systems with a variety of internal switching network designs, such as step-by-step and crossbar, for port-to-port connections. The second generation, known as *digital PBXs*, converted analog voice signals into digital bit format using a codec in the desktop telephone terminal or at the port interface circuit card. Time division multiplexed (TDM) transmission buses were used as the core switching network for internal connections between peripheral port interfaces. Second-generation PBX systems used circuit-switched connections based on Pulse Code Modulation (PCM) techniques to establish communications channels between stations and/or trunk ports. The emerging generation of PBXs is based on IP signaling and communications protocols and interface standards commonly used for LAN and Wide Area Network (WAN) data communications but adapted for voice communications applications.

The digital circuit-switched PBX systems being marketed and installed today evolved directly from the first PBXs to use a digital transmission format across the internal switching network introduced in the mid-1970s by several manufacturers within a very short period. Before 1975, the earlier generation of premises communications systems was based entirely on an analog transmission and switching platform for communications between station users and/or trunk circuits. Using a digital transmission format was the first step toward the evolution of the PBX system from a voice-only communications system to the mixed-media communications system currently being marketed and sold. Other significant PBX system design changes that have occurred during the past quarter century include computer-stored program control, evolution of a modular, distributed system design for processing, switching, and port interface operations, and digital transmission between the station user desktop and the common equipment. The same basic design elements of a PBX system remain the same—call processing, switching, port interfacing, and transmission—but the technology and architecture of the system have certainly changed (Figure 2-1).

PBX system features and functions have also evolved since the first digital, SPC systems were introduced. The early digital PBX systems had fewer than 100 total features in support of station user, attendant, and system call processing requirements. Slowly, with each new software feature release, PBX system software options expanded to include

Evolution of the Digital PBX, 1975–2000

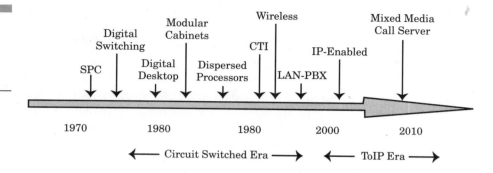

Figure 2-1
PBX evolution timeline: major design developments.

support for multiple system networks, ACD-based call center applications, and integrated voice/data communications. Enhanced system options, such as video communications, computer telephony, mobile communications, and messaging, were continually added to total PBX system offering. Some of the features and functions were based solely on software programming, but many required hardware elements, such as adjunct servers or special signaling interface cards, to implement and operate. Most current communications users are not aware of the significant evolution in system performance capabilities because few station users take advantage of the wide range of features and functions available on their PBX systems.

Herewith is a review and discussion of the major digital PBX system design, feature, and functional changes and enhancements leading to the development of the next generation of IP telephony enterprise communications systems.

Digital Switching/Transmission

PBXs based on digital switching and transmission technology debuted in the mid-1970s. Between 1974 and 1976 several communications system manufacturers claimed to be the first to announce a digital PBX system, including Northern Telecom (currently known as Nortel Networks), Rolm (acquired by IBM and then sold to Siemens), and Digital Telephone Systems (later acquired by Harris Corporation and known as Harris Digital until withdrawing from the market in 2000). The stated driving factor for developing a digital PBX system was to support desktop data communications without a modem, although data communications options would not be widely available until the early 1980s. Other benefits of digital switching/transmission included improved system quality and reliability levels and lower potential manufacturing costs.

There were no established standards for designing a digital PBX system in the 1970s, and the resulting systems reflected each manufacturer's individual design biases. The preferred method of digital transmission used by almost all PBX designers was TDM. TDM is simply described as the sharing of a common transmission bus by many peripheral endpoints. Transmission of digital signals by each endpoint is based on assigned time slots by the PBX common control system. Although TDM was used for transmission of digital signals across the internal PBX switching network, it was possible to use different encoding schemes to convert the original analog signals into a digital format. Although most of the early digital PBXs used an 8-bit word PCM formatting scheme, including Northern Telecom's SL-1 PBX, the first-generation Rolm CBX used a 16-bit word. The typical sampling rate used to convert analog signals to digital format was 8 KHz (a sampling rate double the maximum frequency of a human voice communications signal), but the Rolm CBX used a 12-KHz sampling scheme.

Encoding schemes other than PCM could also be used. In the early 1980s the first-generation Lexar LBX system used a Delta Modulation (DM) sampling/encoding scheme. Some manufacturers evaluated using Adaptive Differential Pulse Code Modulation (ADPCM), based on a 4-bit word encoding format, but no product was ever announced. Although no written industry standard existed, by the early 1980s it became obvious that the 8 KHz sampling using 8-bit word encoding was the preferred digital PBX switching platform. It took Rolm 8 years after its original CBX system made its debut to change its digital switching platform to conform to the 8-KHz, 8-bit word format; Lexar also converted to 8-bit PCM in the late 1980s. By 1990 100 percent of all new PBXs sold in North America were based on digital switching platforms using 8-KHz, 8-bit word TDM/PCM.

The first digital PBX systems digitized the analog voice signal at the port circuit card. Analog voice transmission signals were digitized for transmission across the internal switching network, mostly through the use of a TDM transmission scheme. After being transmitted across the internal switch network, the digitized transmission signal was reconverted back to analog format at the destination port circuit card. Analog station port cards were used to transmit communications to desktop devices, such as telephones or modems, and analog trunk circuit port cards were used to connect to telephone company trunk carrier circuits.

When Intecom introduced the first digital telephone in 1980 for its IBX communications system, the digitization process was performed with a codec in the telephone. Voice signals were digitized and transmitted over the local loop wiring from the telephone to the port circuit card. The first digital telephones used a multiple-channel communications link between the codec and the port circuit card. One channel was used for digitized voice signals and another channel was used for control and signaling functions. A third channel was also available for data communications devices attached to the digital telephone via a data module. Stand-alone data modules for data-only desktops were also available (Figure 2-2).

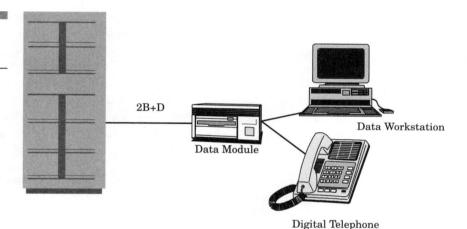

Figure 2-2
Digital PBX data communications.

Desktop-to-desktop digital communications was a major breakthrough for PBX systems. In addition to using the telephony communications network for voice communications, customers could use the PBX system as a local area data communications network. Very expensive modems would no longer be required to convert digital data communications to analog format, and transmission rates up to 64 Kbps could be achieved. Accessing a centralized computer mainframe system would be simplified—no more modems or coaxial cable cluster controllers. LAN technology in 1980 was in its infancy and very expensive. The early Ethernet Network Interface Cards (NICs) were more than double the cost of a digital PBX datastation. A PBX system could support an entire network of data workstations across the entire enterprise when an Ethernet LAN was limited to 50 workstations with major distance limitations. Great things were predicted for the integrated voice/data PBX system because transmission and switching could be all digital. We now know that LAN technology improved; NIC prices rapidly declined; bridges, hubs, Fast Ethernet, and routers were developed; and PBXs as data networking systems never caught on. The irony of the situation is that the digital PBX transmission and switching infrastructure is evolving toward an Ethernet LAN/IP WAN design.

Computer Stored Program Control

Until PBX systems incorporated computer technology into its call processing system design, features and functions were extremely limited. Station user features were restricted to those operations that could be handled by mechanical means. The general availability of computer SPC meant that features could be based on software programming tools, and feature development was limited only by a programmer's imagination. Many PBX functions that are currently viewed as basic telephony capabilities, such as call forwarding and station activated conferencing, were first implemented through computer SPC. Network routing tables and CDR would not be available without computer programming capabilities.

The first SPC PBX system was introduced by Northern Telecom in the early 1970s. Known by a variety of names, including the Pulse and the SG-1, the Northern Telecom system was the first PBX to use a computer software program to perform basic call processing functions, such as provisioning of dial tone, and implement simple station user features, such as hold and transfer. In the United States, Northern Telecom distributed the system through the Pacific Telephone and Telegraph local telephony company, but sales of the new PBX design were limited. It was not until AT&T introduced the Dimension PBX system in 1974 that an SPC communications system was distributed on a large scale through each of the Bell System's local operating companies. Dimension became one of the best-selling PBXs of all time, although AT&T's market share declined throughout the life cycle of the product. After the Dimension PBX announcement, there was a flood of SPC communications systems from AT&T's competitors. Between 1974 and 1980 SPC PBXs went from a 1 to a 95 percent market saturation level for new system installations.

The first computer-based PBXs were based on a centralized processing design. A single computer-based call processing element was used for all system call processing and switching operations. PBX manufacturers of the early digital SPC systems designed and manufactured their own processing hardware and were the designers and developers of the operating system used as a platform for software feature applications. The first-generation digital PBXs were based on call processing designs that closely resembled the minicomputers of the 1970s. Many computer manufacturers became interested in the PBX industry as a new potential market for their products, and a few actually attempted to design a telephony system. Rolm was a manufacturer of military specification computers who successfully entered the PBX market, but most failed. IBM designed, manufactured, and marketed PBXs for the European market but was unable to compete in North America. Digital Equipment Corporation (DEC) was rumored to be developing a PBX based on its VAX minicomputer design, but no product was ever officially announced.

Computer technology in the 1970s was relatively expensive as compared with current prices, and the high cost to design and manufacture a digital PBX was reflected in the enduser price at the time. Common control equipment hardware was priced several times the current cost to customers, even though the features in the 1970s were minimal compared with those of today, and the call processing power of the system was a fraction of today's capacity limits. PBX call processing design evolved significantly during the 1980s when third-party microprocessors were generally available, and prices began their exponential decline. Dispersed and/or distributed call processing designs became the standard architecture platform for PBX systems. The single, centralized, common control element gave way to dedicated processing elements for diagnostics and maintenance operations, localized call processing and switching functions, and systems administration. Basic function electronic telephones with internal processor chips evolved into intelligent digital telephones. Adjunct applications processors provided enhanced functionality behind the core PBX system.

During the 1980s PBXs could be classified into one of three call processing system designs: centralized, distributed, and dispersed. System processor elements expanded from the common control complex to expansion port cabinets and even to individual port circuit cards. The focus of PBX system design was shifting from hardware to software. From the 1970s through the mid-1980s more research and development dollars were spent on hardware upgrades and enhancements, with a focus on digital switching and SPC functions. By the late 1980s most research dollars were being spent on software programming. The emergence in the 1990s of third-party CTI software applications programs running on adjunct servers linked to the PBX officially signaled the beginning of the end of proprietary common control and call processing designs. At the beginning of the twenty-first century, almost 90 percent of PBX research and development dollars were devoted to software applications programming. Little money is spent on core call processing hardware because third-party microprocessors, digital signal processors, and servers, instead of the original self-designed and manufactured computer system, are used.

Today's PBX call processing design is as likely to be based on a customer-provided Windows NT server from Compaq, IBM, or Dell rather than a proprietary common control cabinet from the PBX supplier. Customers may experience lower system reliability levels using third-party servers not designed by their manufacturer for the heavy-duty real-time call processing demands of telephony communications, but the lower price alleviates the risk factor to some extent.

Digital Desktop

In the late 1970s and early 1980s most PBX manufacturers developed an electronic telephone for use behind their systems. The electronic telephone's primary benefit was support of multiple-line appearances. Instead of using KTS equipment behind a PBX to support station user requirements for multiple line appearances, an alternative option was available. The majority of station users at the time used single-line appearance analog telephones, with no feature/function buttons, no speakerphones, and no displays. Only a few lucky station users qualified for multiple-line appearance telephone instruments. Today, of course, the typical PBX telephone instrument looks slightly less complicated than the cockpit of a Boeing 777, with more buttons, bells, and whistles than one knows what to do with.

Like the basic 2500-type analog telephone, voice transmission from the electronic telephone to the port circuit card over the inside wiring was analog, but the built-in intelligence of the telephone instrument provided an array of programmable feature/line buttons and a limited function display. A signaling link between the telephone and the PBX provided the intelligence to identify which line appearance button was being used to place the call or which feature button was depressed for activation. Control signaling between the electronic telephone and the port circuit card was embedded within the instrument's 4-KHz voice transmission channel. The low-frequency signaling stream constrained feature/function development, but was a first step in the evolution of intelligent digital desktops behind the PBX.

The evolutionary step made by electronic telephones was a break from the traditional DTMF signaling techniques for communicating with the PBX common control equipment, as was done with traditional analog telephones. Each PBX manufacturer used a proprietary signaling scheme and dedicated station line circuit cards to support electronic telephones. An industry standard for electronic telephones was not developed for a variety of reasons, although it may have led to more sophisticated desktop terminals. Maintaining a proprietary signaling link meant that electronic telephones could be sold at a high price, with a significant profit margin, if customers required multiple-line appearances. Third-party telephones could not be manufactured unless the signaling scheme specifications were published (which they weren't).

When Intecom introduced the first digital PBX telephone, the product marketing materials emphasized its potential for integrated voice/data communications with an optional data module. Two communications channels were available to the desktop station user, one for voice and the other for data. Little mention was made of the dedicated signaling channel used to link the telephone with the PBX common control equipment. The digital signaling channel was the major breakthrough that would be the distinguishing factor between analog transmission telephones (industry standard, 2500-type, electronic) and digital transmission telephones. The out-of-band signaling channel, operating at transmission rates between 16 and 64 Kbps (based on the individual manufacturer's design specifications), could be used for a variety of new, advanced desktop capabilities.

The primary function of the signaling channel was to alert the PBX common control equipment when the telephone handset was taken off the hook to prepare a voice call. The signaling channel was designed to transmit keypad dialing signals and feature/function activation and implementation signals. Display information, such as calling name and number, was carried over the signaling channel, including call redirection information for forwarded calls. Station users could self-program their telephone instruments with software programs residing in the main PBX control complex but accessible via the signaling channel.

The second communications channel originally developed for desktop data communications applications was rarely used because LANs became the dominant enterprise data communications network. Eventually telephone designers were able to program the PBX to support a second voice channel to the individual station user desktop in support of an adjunct voice terminal. The intelligent signaling channel can distinguish between voice calls placed to different directory numbers and support simultaneous calls to and from discrete desktop devices. Using a special analog line adapter module, a digital telephone can be used to support an adjunct analog telephone, modem, facsimile terminal, or audioconferencing station. The adapter module converts signals from the adjunct analog communications device signals into the proprietary PBX digital format. Other uses of the second communications channel include support of a second digital telephone (using a digital line adapter module) off a single PBX communications port interface and bonding of the two channels for high-speed transmission in support of data or video applications using an ISDN Basic Rate Interface (BRI) type of adapter module.

The most impressive use of the signaling channel is the support of sophisticated display information fields and associated context-sensitive softkey feature/function access. The current generation of digital telephones have large multiple-line display fields that are used to view directories and call logs, access on-line help programs, read text messages, and perform station and/or system management operations.

One of the criticisms of traditional PBXs has been the use of a proprietary control signaling to support digital desktop equipment. The recent development of LAN telephony systems using IP signaling standards someday will eliminate the proprietary signaling link between the call processing system and the telephone, but standards are still in development. Cisco's IP telephone currently does not work on a 3Com IP telephony system, and Avaya's and Nortel's IP telephones interwork do not work on each manufacturer's respective IP-PBX offering. When a high-performance industry standard IP telephone is available to work behind a multitude of IP-PBX systems, it will be possible only because of a standardized signaling link between the call processing server and the desktop, a design specification first developed 20 years ago in the first-generation digital PBXs.

Modular System Design

Until the early 1980s all PBX systems were based on a centralized processing, centralized switching, and centralized cabinet equipment design. Intecom, the developer of the first digital PBX telephone, also broke system design tradition when its IBX system featured distributed port cabinets linked to the main processing/switching complex via fiber optic cabling. Each of the IBX's distributed interface modules (IMs) could be located 10,000 feet from the main equipment room to support campus configuration requirements. Each Intecom IM cabinet had its own local processing unit operating under the control of the centrally located Master Control Unit (MCU).

The distributed cabinet design was dictated by distance limitations imposed by digital signal links to the digital desktop. Intecom was forced to bring the port cabinet closer to the station user. Analog telephones could support cabling loop lengths of 1 or 2 miles, but the Intecom ITE digital telephones were limited to 1,000 feet between wall jack and port circuit card.

The next logical step in a modular system design was to remote port cabinet miles away from the PBX common control complex using telephone company trunk carrier circuits. Northern Telecom was the first to accomplish this when it designed a remote peripheral cabinet for its SL-1 PBX in 1982. Using analog trunk circuits, the remote cabinet depended on the main PBX location for all call processing and switching functions, but at least a customer could support two or more distributed locations with a single PBX system. If the trunk circuit link to the remote location failed, however, the remote location was left without communications service. A spare processing option at the remote location would solve the link failure problem, so Intecom announced such an option about one year after Northern Telecom introduced the first remote cabinet option.

By the mid-1980s several PBXs offered remote cabinet options, but only Intecom has a remote survivable processor option. Other manufacturers offered an alternative solution to the remote cabinet option and in some ways a better PBX system design. A PBX system first announced in the early 1980s, and still working today after many upgrades and enhancements, was the Ericsson MD-110 PBX. Based on its own central office switching system, Ericsson's MD-110 was a fully distributed communications system from a call processing, switching, and cabinet architecture perspective. Each MD-110 Line Interface Module (LIM) contained a common control complex that operated independently yet in coordination with every other LIM cabinet in the system. The LIMs could be geographically dispersed on a campus location or across a telephone network (analog or digital trunks, copper or fiber optic cabling, microwave or satellite transmission). Each LIM had its own switching system backplane and communicated with other LIMs via a centralized group switch complex. PCM links between the LIMs and group switch could be duplicated, as could the group switch (Figure 2-3).

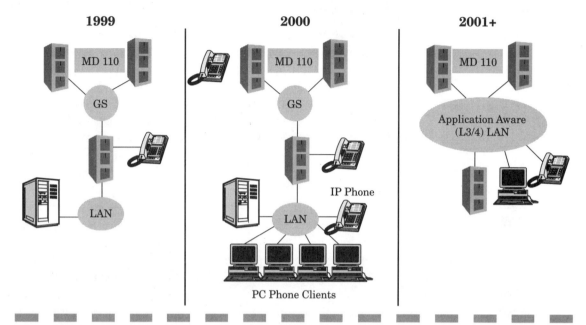

Figure 2-3   MD110 IP evolution.

In the mid-1980s Rolm introduced a PBX design similar to the Ericsson offering. The Rolm CBX II 9000 did not have a centralized group switch but it did have functionally independent control cabinet clusters. The Northern Telecom SL-100, a modified version of the manufacturer's DMS-100 central office switching system, became a popular PBX system for very large (thousands to tens of thousands of user stations) distributed communications configurations requiring an extremely high level of reliability and redundancy. The SL-100 Remote Switch Center (RSC) option could be located hundreds of miles from the main PBX location, support thousands of stations users, and function as a stand-alone system, if necessary, with minimal loss of features if the control link to the main common control complex failed. The growing availability of PBXs capable of supporting multiple common control complexes and port cabinets geographically dispersed across great distances marked a distinct change from the old, monolithic design platform of PBXs before 1980.

For customers with single-location requirements and not interested in remote port cabinet options, the most important PBX cabinet innovation of the early 1980s was the introduction of the stackable cabinet design. PBX control and port cabinets were traditionally based on large, multiple carrier steel frames. Customers would be forced to buy and install an expensive large cabinet capable of supporting several equipment shelves, even if they required expansion for a few stations. The incremental cost to add a few stations was very expensive. When the NEC NEAX2400 was introduced in 1983, it was the first PBX based on a stackable cabinet design, with dedicated single-shelf cabinets for call processing functions and stackable port cabinet shelves. Up to four Port Interface Module (PIM) single-shelf cabinets could be stacked on top of each other, sharing a common switching and processing backplane. Each PIM had a dedicated Port Processor Interface and a dedicated Time Slot Interexchange (TSI). The NEAX2400 offered customers a cost-effective solution for modest growth requirements as compared with PBX systems based solely on large expansion port cabinets costing tens of thousands of dollars even if only a few expansion ports were required.

By the early 1990s almost all PBX systems targeted to customers with small and/or intermediate port requirements were based on modular, stackable port cabinet designs. Many PBX manufacturers offered a remote port cabinet option to customers desiring a single communication system for multiple-location configuration requirements. Distributed processing and switching designs were becoming commonplace. The emergence of CTI in the 1990s allowed manufacturers to offload advanced software options, particularly for call center management, onto adjunct servers dedicated for a specific application. Optional software application programs run on proprietary or customer-provided server equipment reduced the call processing load on the main control complex and offered a more flexible migration and upgrade path to enhance older PBX system platforms that still performed the basic communications functions with little problem. The early CTI hardware solutions required proprietary hardware links between PBX and server, but evolving PBX architecture design led to standardized TCP/IP links over Ethernet LANs.

The development of call processor control signaling over LAN infrastructures simplified the installation of third-party hardware and software solutions behind the core PBX system and kickstarted development activity for IP telephony and the emerging client/server IP-PBX system design. Using the LAN infrastructure (Ethernet switches, multiservice routers) for voice transport and switching between LAN-connected PC client softphones and LAN-connected servers for call processing is the ultimate modular system design because the processing and switching functions are totally distributed across the entire network.

Feature/Function Enhancements

The current number of PBX features, functions, and options appears infinite to older station users who are accustomed to using the same three or four features they had available to them before digital PBX. Although it is true that most station users commonly use fewer than ten features on a frequent basis, the set of features for one station user is different than the set of another. Many features once developed to improve call coverage and handling capabilities, such as call pickup and automatic callback, have fallen into general disuse since the advent of voice messaging systems, but not everyone depends on a machine to answer their calls. E-mail may have overtaken telephone calls as a method of communications among business people, but nothing is better than a real-time voice call for emergency contact between a supplier and a customer.

There are many ways to classify PBX features and functions. To provide an overview of how digital PBX performance levels have continually improved during the past quarter century, the following feature/function categories are briefly reviewed: basic voice call station/system features, data communications, video communications, networking, inbound/outbound call center, CTI, mobile communications, and messaging.

Basic Voice Call Station/System Features

Do station users really need six variations of call forwarding? Do managers still "buzz" their secretaries? Although PBXs have many call forwarding options and still retain the manual signaling (buzz) feature, the most significant station/system feature enhancements during the past two decades have been to improve incoming call coverage, support the needs of the new mobile workforce, and simplify the administration and maintenance operations of the system manager.

An important PBX feature developed in the days before voice messaging systems invaded the workplace was programmed call coverage. Programmed call coverage was a form of enhanced call forwarding, with some important distinctions. First introduced in 1983 by AT&T on the System 85 PBX, call coverage did not receive the market attention it deserved during the 1980s and 1990s, but renewed interest in personalized call screening and routing to improve communications service levels has revitalized the feature. Call coverage capabilities on current-generation PBX systems allow station users to define where incoming calls are directed when they are unable to answer the call and program the coverage path based on who is calling (CLID, Automatic Number Identification [ANI], internal calling number, call prompt), where the call originated (internal or external to the system), how it arrived into the system (trunk group ID), or when a call is placed (time of day, day of week). Building on the concept of call forwarding, personal call coverage programming redirects calls to a defined path of answering stations and will default to the called party's voice mailbox only as a last resort. Calls will not be redirected to the forwarding position or voice mailbox of a station user defined in the call coverage path; the originally called party's coverage path overrides intermediary station user call forwarding commands.

Call coverage tables and station user programming was not possible before the development of digital PBXs. The new CTI-based PBX system designs allows station users to program caller-specific call coverage paths based on identified callers. The personal call coverage function in these new-generation systems is supported at the station user desktop (a PC client softphone), not at the common control call processing system. The objective of personalized call coverage features is to reduce dependency on voice mail systems because a human answering station rather than a noninteractive machine might be preferred by the caller. Voice mailboxes should be the last option in a call coverage environment, not the first or only option.

The new mobile workforce includes station users who are rarely in the office and workers who do not have permanent desk assignments because they are constantly moving or their job function is not desk based. To support these mobile workers, it is necessary to dissociate a station user's telephone directory number from a physical telephone instrument. *Hoteling*, a feature designed to support workers who work at different desks throughout the enterprise, allows station users to log into the system from a telephone and reassign their directory number to their chosen telephone. In addition to their telephone number, the individual's station user profile (service levels, call restriction levels, group assignments) is also assigned to the physical telephone location. Account codes and call records are maintained for the station users for each telephone they use. When done using the telephone, after 1 hour, or 1 week, or 1 month, the station user logs out, freeing the telephone for the next mobile worker. Hoteling is becoming very popular in sales offices. The feature can significantly reduce system costs by optimizing common equipment hardware, telephone instrument, and cabling requirements and, more importantly, minimizing real estate requirements (fewer dedicated desks/telephones, less office space).

Today's mobile workers who are rarely at a fixed telephone location also benefit from recent feature enhancements. The find-me feature allows station users to program their telephone to direct calls to other telephone numbers outside of the PBX system. More than one external number can be programmed. For example, on a no-answer call at the station user desktop, the call can be forwarded to another telephone number after a selected number of rings; if there is no answer at the external number, another telephone number is dialed, and the call is redirected. External telephone numbers likely to be programmed include cellular telephones, home, conference facility, remote office branch, or even a hotel. A relatively recent teleworker option available on some PBXs allows station users to bridge their line appearance to a telephone external to the system. The concept of the PBX as a mobility server can significantly improve call coverage, reduce lost or abandoned calls, and increase the number of successful call attempts between caller and called parties.

Another category of mobile workers consists of station users who require a telephone away from the formal office environment. Known as teleworkers, these station users require their high-performance telephones to function away from the workplace and receive incoming calls redirected to their remote desktop. The original teleworker option was an off-premises extension (OPX) station using highly tariffed telephone trunk circuits to link remote analog station equipment to the main PBX system. Expensive and low-performance analog OPX stations have evolved into affordable and high-performance digital desktops. The same digital telephone supported behind the PBX at the office can be supported remotely with several options, including distance extender modules and analog trunk carrier facilities, ISDN BRI services and equipment, and the recently available IP workstation (hard telephone or PC client softphone).

The most important system feature enhancements during the past decades have been systems administration and maintenance tools. The early PBX management terminals required high-level programming skills and weeks of training. A typical station move, add, or change operation could require at least 15 minutes of keyboard entries. After booting up the systems management terminal the administrator was met by a blank monitor screen waiting for a programming command. There were no menus, on-screen help command, or point, click, and drag. Computer technology evolved during the 1980s and so did PBX management tools. By 1990 a systems management terminal had a basic graphical user interface (GUI), usually a menu selection list and formatted

Evolution of the Digital PBX, 1975–2000

screens. By 2000 PBX management tools were accessed through a web browser via the Internet, and a sophisticated GUI simplified the administration process. Few keyboard entries are now required, and access to a common metadirectory server simplifies the initial station user directory entry.

Data Communications

Digital PBXs were supposed to be the enterprise data communications network backbone, but some things were not meant to be. The first PBX data communications options were introduced in the early 1980s. In 1980 Intecom was the first to offer a high-speed data module capable of transmission rates of up to 57.6 Kbps. This was at the time when most modems were operating below 9.6 Kbps, and 10-Mbps Ethernet was not yet introduced. After the Intecom announcement, most of the older PBX suppliers announced data module options for their systems with maximum transmission rates ranging from 9.6 to 64 Kbps (Figure 2-4).

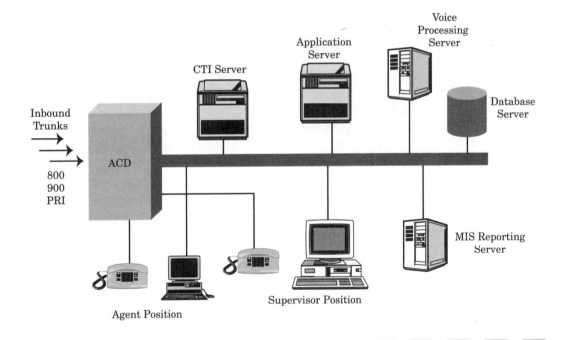

Figure 2-4 Call center configuration.

Data communications options were available for integrated voice/data or stand-alone data ports. The integrated voice/data port option required a data module that attached to a digital telephone and provided a RS-232C or RS-449 interface for an adjunct data terminal. Asynchronous and synchronous interfaces were usually available from each PBX suppler. Stand-alone data modules were also available and may have required a port circuit card dedicated to data-only communications. The early data modules were priced at about $300 to $500 and required the more expensive digital telephones to work. Most PBX systems at the time were not designed to handle long call holding times and required extensive traffic engineering to support significant customer data requirements. When digital trunks were first available in the mid-1980s, the tariffs were very high, and the PBX digital trunk interface cards were expensive compared with analog trunk interfaces. The cost of LAN equipment, at first significantly more expensive than PBX data option pricing, declined rapidly during the 1980s, making it a far more attractive data networking solution than a PBX. PBX data modules, once considered a high-speed option, were viewed as slow when compared with LAN transmission rates. Dreams of the PBX becoming the data networking solution died by the late 1980s when LAN technology matured and network routers first entered the market. Shipment levels of PBX data stations (integrated and stand-alone) never exceeded 3 percent of total annual shipments.

PBXs attempted to make a comeback in the early 1990s by offering a wideband data communications option using an ISDN primary rate interface (PRI) circuit card. By bonding multiple B channels together, a PBX could support transmission rates up to 1.5 Mbps to the desktop and across its digital trunk network. The cost to support the option, however, was seen as excessive because a single wideband port required a dedicated ISDN PRI port circuit card that could cost several thousand dollars. Fujitsu was the first to offer wideband data communications on its F9600 PBX, but customer demand was weak. Other PBX suppliers soon followed the Fujitsu announcement with their own ISDN PRI–based data communications option, but total sales of the option to date have failed to reach 1 percent saturation.

Another PBX system attempt to make a dent in the data communications market came in the mid-1990s when Intecom introduced an Ethernet hub and workstation interface option fully integrated into its port cabinet design. Broadband fiber optic loops between the distributed port cabinets handled intrasystem data traffic and could support Ethernet 10BaseT transmission standards. The broadband data communications option was priced higher than existing LAN interface and switching equipment and failed to find a market.

More than 20 years after the first attempts to position the PBX system as a data communications networking solution, sales of PBX data modules are negligible. The only appreciable data traffic transmitted across a PBX system today is analog-based data communications generated by modems. PBXs are used primarily as a back-up system when LANs are down for service or repair. Ironically, the often unreliable nature of enterprise LANs has made the PBX an invaluable spare data network solution, forcing many voice communications managers to install a significant number of analog ports in support of data modems for use in emergencies. PBX data solutions may not be high speed, but they are reliable.

Video Communications

The older generation of PBX station users may remember their first introduction to the videophone at the 1964 New York World's Fair. Several generations of PBXs have come and gone since 1964, but PBX-based desktop video communications is still a work in progress for most customers. The first major attempts at desktop video communications behind a PBX system occurred during the early 1990s when ISDN BRI options became available. Using both BRI bearer communications channels per video call (128-Kbps transmission rate) provided a fair quality of service, but a killer application for the video option never materialized, and desktop video behind the PBX is rarely implemented today. Desktop video communications today is based primarily on LAN or supported by dedicated trunk circuit facilities.

Lucent Technologies attempted to revive interest in PBX-based video communications in the mid-1990s when it introduced two Definity PBX options designed to support voice calling features on video calls. The MultiMedia Communications Exchange (MMCX) was a server-based system designed to support H.323 mixed-media calling (voice, data, and video). The MMCX provided some basic calling features, including dial plan, conferencing, and call forwarding, to LAN-connected workstations used for video communications applications. The MMCX could also support traffic between PBX ports and LAN peripherals, and a Q signaling (Qsig) link would provide a higher level of PBX system features to the LAN workstations. Another PBX option was called MultiMedia Call Handler (MMCH), which was designed to apply a limited number of voice calling features to ISDN BRI video workstations conforming to H.320 standards. Neither the MMCX nor MMCH offerings gained market acceptance, although the MMCX product was later modified by Lucent and reintroduced as its first Internet telephony gateway product. The IP gateway was further redesigned as an internal IP trunk interface card for the Definity PBX. The concept of the MMCX, a call processing server for LAN workstations, could be considered an early version of today's emerging IP-PBX systems. Although no IP telephones existed when the MMCX was announced, LAN-based voice communications using a video workstation equipped with a microphone and speaker were supported.

Networking

PBX networking has evolved dramatically during the past 25 years. The earliest PBX networking arrangements consisted of two switch nodes linked by a dedicated, private line facility (E&M tie trunk) to save on long distance toll charges. The primary benefit was cost savings. When customers began to use multiple long distance carriers in the late 1970s it became necessary for PBX systems to analyze each placed long distance call to determine which carrier service should handle the call. The preferred carrier service was usually the lowest priced for the call. A new PBX feature was developed and known by several names, including least cost routing (LCR), ARS, and most economical route selection (MERS). Implementing the feature required the system administrator to enter the call destination route and routing pattern data into a database that would price each call based on tariff pricing data. The tariff data was obtained through a service bureau and needed to be updated regularly. The benefit to the customer was to reduce long distance toll expenses. Expensive calling routes were restricted to callers only with permitted network classes of service levels; callers with the lowest service level rating could place calls only when the lowest cost route was available.

Also in the late 1970s, AT&T announced its Electronic Tandem Network (ETN) offering, and PBX networks acquired a greater degree of complexity and functionality. ETN was a private tandem network consisting of a meshed network of private line facilities linking tandem switch PBX nodes, main PBX nodes, and satellite systems. In-band signaling techniques supported a network dialing plan and automatic alternate routing between nodes within the network. In addition to cost-savings benefits using fixed tariff private line carrier facilities, customers enjoyed greater control over network operation and use. All of this was initially done with the use of narrowband analog trunking facilities.

The next step up the evolutionary PBX networking ladder was establishing an intelligent network signaling to support transparent feature/function operations between discrete locations served by independent PBX systems. AT&T's Distributed Communications System (DCS) offering was introduced in 1982 for its Dimension PBX. The first intelligent signaling link required an expensive private data circuit; analog private lines were used to carry voice traffic. The DCS intelligent networking solution allowed customers to use simplified dialing plans (i.e., four- or five-digits) for calls across PBX systems; supported transport of caller name/number display information between telephone desktops working behind different switching nodes; and provided a basic level of transparency within the network for many of the most commonly used station features, such as call transfer, call forwarding, and multiparty conferencing. Shared applications were also supported across a network of PBX systems, such as centralized voice messaging.

The arrival of digital T1-carrier trunk services in the mid-1980s changed the rules for PBX networking because in-band signaling was replaced by out-of-band signaling, and new networking solutions became possible. The same digital trunk circuit used for voice traffic could also be used to support the intelligent signaling link between PBXs. Digital voice carrier services using T1-carrier circuits made out-of-band signaling a more economic and feasible solution for implementing an intelligent network of PBXs. Use of an out-of-band signaling channel allowed PBX systems to communicate with one another at a much higher level than before. The resulting intelligent network configuration could offer customers traditional network transmission costs savings and provide significant productivity gains and additional cost savings through the use of shared application features/functions.

Each PBX manufacturer's intelligent networking solution was proprietary and caused problems for customers with a mix of PBX systems in their networks. An initiative was begun in the late 1980s in Europe by the leading PBX suppliers to create a standardized network signaling protocol to intelligently link dissimilar PBXs. The signaling standard is commonly known as *Qsig*, and was originally developed under the auspices of the ISDN Private Network Systems (IPNS) Forum. PBX systems that support Qsig can interwork intelligently with each other; support basic call set-up and tear-down across the network between dissimilar PBX system platforms; conform to a common dialing plan for limited digit dialing across PBXs; transmit and accept telephone display information, such as calling name and number, and call redirection data between desktops; and support feature-transparent operations for a defined set of features, such as call forwarding, call transfer, conferencing, and network attendant service. Although most leading PBX suppliers support Qsig as part of their networking solutions, the degree of transparency between systems remains limited. Manufacturers must do continual testing of their systems to correct Qsig message and signaling problems.

PBX networking advancements in the late 1980s included support of ISDN PRI services. ISDN PRI service circuits became the preferred trunking solution for implementing an intelligent feature-transparent network because the D channel was a natural communications channel for handling signaling and control data across distributed PBXs. New PBX networking features based on ISDN PRI services included support of incoming ANI, and call-by-call service selection (CBCSS). CBCSS allows a PBX system administrator to define the communications service supported by individual ISDN PRI bearer communications channels. A single T1-carrier trunk circuit, supported by ISDN PRI service, could be used for a variety of services between the PBX system and the network exchange carrier's central office switching system. For example, several bearer channels could be designated incoming DID trunks, others could be designated two-way CO trunks, and others could be designated clear channel data circuits. Using programming tools, the administrator could reconfigure the mix of trunk services on demand or by a schedule, or could even program the channels to reconfigure themselves based on real-time traffic conditions.

Network carrier services in the 1990s designed to support data communications were also supported by PBX systems. Nortel Networks redesigned its Magellan Passport Asynchronous Transfer Mode (ATM) switching system as a Meridian 1 gateway module to support a mix of voice, data, and video communications over broadband trunk carrier circuits. Lucent Technologies, NEC, and Siemens also introduced ATM network interface options for their respective PBX systems. One of the most important PBX networking advancements in the late 1990s was the introduction of external IP telephony gateways, closely followed by integrated IP trunk gateway port circuit cards. Lucent Technologies was the first to market an integrated IP trunk option and was closely followed by other market leaders, including Nortel Networks and Alcatel.

ACD-based Call Centers

The fundamental function of a call center is to direct calls to a group of answering positions, equitably distribute the calls among the group members, and minimize the caller's time in queue waiting for an agent. ACD is the general term used to describe this telephony function and was introduced as a PBX feature in the early 1980s. The early PBX-based ACD software options had limited flexibility in screening, routing, and queuing calls, and the MIS reporting function was minimal. ACD was developed as an enhanced version of the Uniform Call Distribution (UCD) feature, which itself was an enhanced version of Hunting.

ACD systems in the early 1980s were used only in very formal incoming call center environments, and annual shipments were limited to several hundred systems. In the early 1980s the only PBX system with an ACD capability that was competitive with stand-alone ACD systems was the Rockwell Wescom 580. Rockwell was also the leading manufacturer of stand-alone ACD systems at the time; its Galaxy ACD was the leading system in the market in terms of features and functions. The Rolm CBX had a basic ACD package, but most PBXs could offer little more than UCD for distributing calls. By the mid-1980s the market for ACD systems was growing almost 50 percent annually, and many PBX manufacturers looked at the call center application market for potential high profit margins. The first PBX manufacturer to offer a sophisticated ACD option that could compete with the Rockwell Galaxy ACD was AT&T. When AT&T introduced its latest version of System 85 in 1987, the ACD call screening, routing, and queuing functions were based on a new customer programming tool called *Call Vectoring*. Call Vectoring consisted of a series of programming commands that defined call coverage and treatment operations for each incoming call with flow charting methods. A new advanced MIS reporting system, based on an adjunct computer working behind the System 85, offered customers dozens of reports for call center management and monitoring. Supervisor positions, linked to the MIS reporting processing system consisted of GUI workstations displaying real-time call center information. The AT&T announcement marked a major change in the direction of PBX-based call centers.

After the AT&T announcement, other PBX manufacturers began development of similar ACD and MIS reporting options for their communications systems. By the mid-1990s the typical PBX/ACD offering was based on an adjunct application server required to run sophisticated call routing and treatment software, provide advanced MIS reports, and support supervisor PC client positions. Customers with complex call center application requirements would likely link the PBX/ACD system to a CTI application server to provide features such as screen pops, an coordinated voice/data screen transfer. An IVR would likely serve as a front-end system to screen calls and pass call-prompted data to the

PBX/ACD program for analysis. New ACD features that improved call center efficiency and customer satisfaction levels included skills-based routing and agent skill profiling and mapping (Figure 2-4).

By 2000 PBX/ACDs dominated the call center market, accounting for more than 80 percent of total agent shipments. During the next few years the traditional voice-centric call center will migrate toward a mixed-media model that integrates the traditional ACD system with e-mail distribution systems and Internet-based Web sites. The emerging IP-PBX system will be able to support the new-generation mixed-media agent position, which will handle incoming and outgoing voice calls, respond to e-mails, and chat with Web surfers on-line. Despite the changes in technology and the mix of voice calls, e-mail, and on-line chat, the fundamental functions of the new contact center system will be the same as those of the original ACD systems: equitable distribution of calls to agents and minimized response times for customer inquiries.

Computer Telephony Integration (CTI)

CTI had its origins in the early 1980s when AT&T was the first PBX manufacturer to use an adjunct applications processor to provide a message center and directory system function behind its Dimension PBX. Unanswered calls were forwarded to a message center agent who could do a directory lookup for the person being called and enter a message into a computer terminal. The message was then forwarded to a printer close to the called party. The connection between the PBX and the applications processor was a physical link, and no enhanced PBX voice functions were provided. This was CTI in its most rudimentary form.

The concept of enhancing the PBX system with an adjunct applications processor, with physical and logical connectivities between the two independent systems, was not realized until the late 1980s. The first true CTI options were available almost concurrently on two PBX systems. The NEC NEAX2400 and the Intecom IBX offered an Open Applications Interface (OAI) option for third-party application software developers to design applications running on an adjunct client or server to enhance the performance capabilities of the core PBX system. Both suppliers' OAI option included an intelligent signaling link between the PBX and the adjunct applications processing system. The intelligent signaling link provided a communications path for the PBX system to send system status and message packets to the applications processor and provided the means to transmit the software application program commands back to the PBX. The applications programming interface included with the OAI software developer's toolkit was proprietary to each system and required software developers to write different programs for different systems. The PBX manufacturers initiated the first CTI implementations, and customer demand for the NEC and Intecom offerings was limited.

When the first CTI industrywide standards were being developed in the late 1980s and early 1990s, it was the major computer companies, such as DEC and IBM, that took the initiative, not the PBX manufacturers. Most of the early CTI implementations were host-based applications, such as predictive dialing and agent screen pops for outbound calling centers. The PBX manufacturers focused on first-party desktop CTI applications by providing a physical/logical link between their digital telephones and desktop PC clients. The client software applications supported a variety of PC telephony features and functions, including directories, screen pops, PBX feature/function activation, on-screen dialing, and call logs/notes. Client/server CTI applications were initially driven by Novell, which promoted its Telephony Services Application Programming Interface (TSAPI) standard. The CTI standard in Europe was developed by the European Computer Manufacturers Association (ECMA) and was known as Computer Services Telephony Applications (CSTA). Microsoft developed its own desktop CTI standard, Telephony Applications Programming Interface (TAPI), that was later enhanced to support client/server applications, and supplanted Novell's TSAPI as the most popular CTI platform (Figure 2-5).

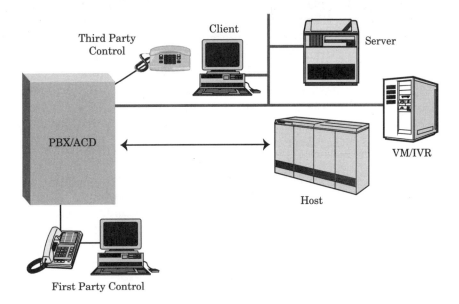

Figure 2-5
CTI system designs.

CTI is used behind a PBX system to provide features and functions not available in the generic communications software package. Many of the CTI applications have limited market potential for which PBX developers cannot afford to expend resources. Desktop and client/server CTI was envisioned by many within the industry has a means to replace the proprietary telephone instrument. Instead of buying an expensive voice terminal with limited feature button and display capabilities, the PC client and monitor would serve as the station user interface to the PBX system for all system operations, including dialing, call screening, call answering, and feature implementation. Despite substantial marketing efforts, customers resisted using CTI PC telephony as a substitute for high-performance telephone instruments. Station users were reluctant to learn new programming tools, the ergonomics were poor, the cost was too high, and the PC reliability factor was unacceptable. Desktop CTI shipments behind PBX systems have been negligible, with annual shipment levels less than 2 percent of total station shipments. The only PBX market segment in which CTI gained a foothold was call centers because the evolving call center process became heavily dependent on computer technology, and ACD agents were already using PC clients to handle the typical caller transaction. About 25 percent of call centers are currently implementing CTI.

In the late 1990s several recent PBX market entrants attempted to design and market enterprise communications systems based on CTI client/server architecture principles: an applications processor served as the call processing manager, and station user desktops were analog telephones logically working with PC telephony clients. The desktop PC telephony applications software was included in the system price, and CTI links were used to provide third-party applications not included in the generic software package. These systems did not support multiple-line appearance digital telephones and found limited market appeal. One manufacturer who first attempted to market a pure client/server design, Vertical Networks, recognized the value of digital, multiline, display telephones, and downplayed the CTI attributes of its system when it belatedly marketed a traditional PBX-like digital telephone.

The new *IP softphones*, available from most longtime and new PBX suppliers, is a proprietary version of desktop CTI, although it is not marketed as such. The PBX system's call processing manager, be it a traditional proprietary common control complex or a third-party server, functions as a server to the PC client desktop. It is forecasted that IP softphones will become more popular throughout the remainder of this decade, although desktop telephone instruments will remain the dominant voice terminal type.

Evolution of the Digital PBX, 1975–2000

Mobile Communications

The first wireless PBX options were announced in 1992. The first to announce was Ericsson, a leading PBX supplier who has also been a dominant competitor in wireless communications, Ericsson introduced a wireless adjunct controller to work behind its MD-110 PBX. A few months later an unknown company called Spectralink introduced its wireless adjunct option designed to work behind any PBX system. Although the Ericsson and Spectralink systems were based on different radio transmission frequencies and used different voice encoding formats, the basic features and functions were similar. Both wireless communications options supported multiple-zone coverage areas and features such as roaming and handoff between coverage zones. In addition to a controller cabinet that linked to the main PBX system, the infrastructure included radio transmission transceivers linked to the controller over standard, unshielded, twisted-pair telephony wiring and provided the interface between the wireless telephone handsets and the PBX system. The adjunct controller served as a mere gateway between PBX and the station users. It was the PBX system that continued to provide all communications services and function to the wireless station users, including dial tone, call processing, and switching functions (Figure 2-6).

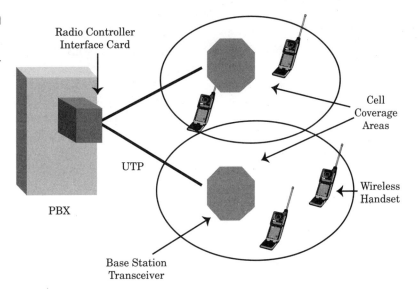

Figure 2-6
Wireless PBX design.

After the initial wireless system announcements from Ericsson and Spectralink, several other leading PBX suppliers developed and introduced their own wireless system options by the mid-1990s. These options were designed to work behind their own communications systems. Among those that developed proprietary wireless communications options were AT&T, Northern Telecom, NEC, Alcatel, and Tadiran. Throughout the 1990s several of the wireless system suppliers upgraded or overhauled their original system designs. The original systems operating in the 800-MHz frequency ranges migrated to 900 MHz and/or 1900 MHz. Ericsson, for example, went through two system redesigns and eventually abandoned its last version to focus on premises cellular systems as an extension of their network-based offerings.

Market demand for wireless PBX system options has been less than originally forecasted. Several reasons have been given for poor shipment levels: the cost to install and operate a wireless station subscriber behind a PBX system could be several times greater than a wired station; there are severe traffic capacity limitations within each coverage zone; and the technology standards are continually changing. For example, the typical price for a proprietary wireless handset is more than $500 at a time when network cellular telephones can be purchased for less than $100.

Evolving mobile communications capabilities behind a PBX system are likely to be based on enterprise mobility servers that link a network carrier's cellular infrastructure with a PBX system. Ericsson's Digital Wireless Office System (DWOS) is an example of an enterprise cellular communications solution that supports network carrier cellular telephones behind a PBX system, with a mobility server as the link between the premises and off-premises communications system. Ericsson, in fact, plans to port its MD-110 PBX features and function to its DWOS mobility server and market a fully integrated PBX/cellular communications system. Other leading PBX suppliers, such as Alcatel and Siemens, offer mobile communications option similar to DWOS and also may integrate PBX functions into the mobility server. The future boundary between premises and network communications systems and services will be blurred when the new mobile communications offerings are generally available.

Messaging

VMSs were introduced to the market in the early 1980s, and originally worked as stand-alone systems behind PBX systems. Although the first VMSs were designed and marketed by third-party suppliers, several of the leading PBX manufacturers eventually entered the market with products of their own design. Rolm and AT&T were among the first PBX manufacturers to enter the voice messaging market with products designed to work behind their own communications systems, although they could also be engineered as stand-alone systems to work behind other suppliers' PBX systems.

Northern Telecom, one of the leading PBX suppliers, came late to the VMS market during the late 1980s, but when it introduced Meridian Mail it became the first messaging system to be fully integrated within the PBX system design. Meridian Mail used the Meridian 1 processing and switching network backplane for supporting PBX station user messaging applications. The Meridian Mail Module was installed as another cabinet stack in the Meridian 1 and tightly integrated within the overall PBX system design. Instead of using analog station interfaces and a dedicated data signaling link between the PBX system and adjunct voice messaging cabinet, Meridian Mail ports appeared to the Meridian 1 switching network as just another station port, and signaling between the Meridian Mail Module and the Meridian 1 common control complex was transmitted over the internal system processor bus. AT&T followed Northern Telecom's example and later redesigned its Audix VMS as a multiple card slot equipment module to be installed within its Definity PBX system. The Definity Audix option offered most of the features and functions available on the larger, stand-alone Audix (later Intuity Audix) system at a reduced price.

During the early 1990s VMSs were redesigned to support integrated messaging applications with e-mail servers. The concept of a UMS designed to support voice and e-mail messaging, with both message mediums sharing a common directory and storage system, was also introduced in the early 1990s. Although demand for the enhanced messaging system designs has been limited to date, there are many productivity and cost benefits attributable to using one mailbox for all types of messages and having a single interface to the mailbox from either a telephone or PC client.

Recognizing the competitive advantage of bundling messaging applications within the PBX system, several recent start-up companies with PBX client/server designs, such as Altigen and NBX (since acquired by 3Com), integrated a UMS application running off the main system server that also provided basic PBX communications features and functions. Recently, Avaya integrated the capabilities of a full-function Intuity Audix system into the main call processing board of its small Definity One PBX system and included the Intuity Integrated Messaging appli-

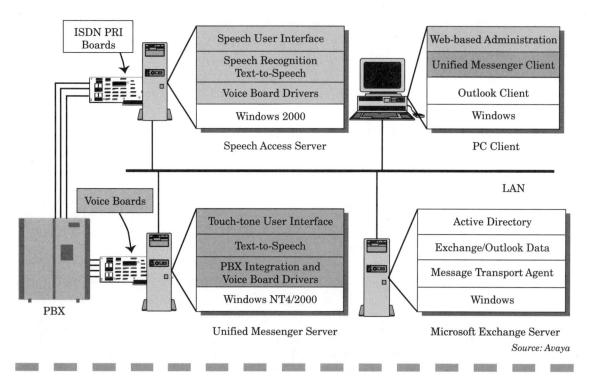

Figure 2-7 Avaya speech access with unified messenger architecture.

cation on the same board. The Altigen, 3Com, and Avaya PBX systems with the bundled messaging capabilities are designed for small/intermediate customer line size requirements, and the message storage capacities and access ports are limited. PBX systems designed for large and very large customer port requirements would not be able to integrate the messaging application into the main common control complex without affecting the basic communications responsibilities of the system. Dedicated messaging application servers will likely be the optimal solution for higher-end PBX customers, even when IP-PBX client/server system designs become standard by the end of this decade.

CHAPTER 3

Legacy PBX Call Processing Design

Telephone communications systems, including PBXs, have always used circuit switched networks to establish connections between network endpoints—the sender and receiver of a voice call. Circuit switching is simply defined as a type of communications in which a dedicated channel, referred to as a circuit, is established for the duration of a transmission between the originating and terminating endpoints. In communications parlance, a channel refers to a communications path between two connected endpoints, for example, two telephones. Circuit switched networks, also known as connection-oriented networks, are ideal for real-time voice communications requirements.

Until the development of digital communications, circuit switched networks used analog switching and transmission techniques to establish connections between calling and called parties. When the first digital PBXs were introduced in the mid-1970s the internal switching networks required a conversion of analog wave signals into a digital transmission format. Voice audio signals from the desktop telephone to the PBX common equipment hardware were transmitted over a 4-KHz communications channel, at which point a codec embedded on the port circuit card converted the analog signal into a digital signal for transmission across the internal circuit switched network. When digital telephones were introduced in 1980, the codec function resided in the desktop instrument itself, and transmission to the PBX common equipment was in digital format, ready for transmission across the internal circuit switched network. Digital switching, as opposed to analog switching, provides improved sound quality and more reliable transmission at a lower cost.

Digital PBXs from different manufacturers use different internal circuit switched network designs. Some use a distributed switching design; others may use a center stage design. The switching and traffic handling capacities of each system are unique to the individual design. There are some design elements and communications standards common to all digital PBXs as they evolved between 1975 and 2002, and I review those in the first two sections of this chapter. The third section defines and describes the different design topologies used by digital circuit switched PBX designers. The fourth section addresses PBX circuit switched network issues that affect reliability and traffic handling capacity. The fifth section provides an overview of PBX traffic analysis covering station and trunk traffic issues.

Fundamentals of PBX Circuit Switching

Time Division Multiplexing

The core design element of a traditional digital PBX is the local transmission bus that connects to a port circuit card. Many port circuit cards may share a common local transmission bus, and a PBX system may have many local buses dedicated to designated port circuit cards housed in different port carrier shelves and/or cabinets. Port circuit cards are used to connect peripheral equipment devices, such as telephones and telephone company trunk circuits, to the internal circuit switched network, where the local transmission bus is the point of entry and exit. Voice signals transmitted from the port circuit card onto the transmission bus are in digital format. The transmission and coding standard used by all current circuit switched PBX systems is known as Time Division Multiplexing/Pulse Code Modulation (TDM/PCM). To fully understand the workings of the PBX circuit switched network, it is necessary to define the basic terminology (Figure 3-1).

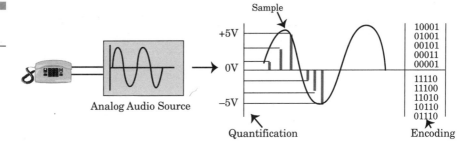

Figure 3-1
TDM/PCM.

Multiplexing is the sharing of a common transmission line (bus) for transport of multiple communications signals. A communications transmission bus is a collection of transmission lines used to transport communications signals between endpoints. TDM is a type of multiplexing that combines multiple digital transmission streams by assigning each stream a different time slot in a set of time slots. TDM repeatedly transmits a fixed sequence of time slots over a single transmission bus. In a PBX system, the transmission bus is usually referred to as the *TDM bus*.

Chapter 3

A PBX TDM bus is used to transport digitized voice signals that originate as continuous (analog format) sinusoidal waveform signals. Digital sampling of a continuous audio signal is a technique used to represent the analog waveform in digital bit format. The sampling technique that has become the accepted standard for circuit switched communications is PCM.

Pulse Code Modulation

PCM is a sampling technique for digitizing the analog voice-originated audio signals. PCM samples the original analog signal 8,000 times a second. This is more commonly referred to as 8-KHz sampling. The sampling rate used to code voice audio signals is based on the frequency range of the original signal. To accurately represent an analog signal in digital format, it is necessary to use a sampling rate twice the maximum analog signal frequency, a calculation based on the Shannon theorem. The maximum frequency of human voice is about 3.1 KHz. This frequency was rounded up to 4 KHz for ease of engineering design, resulting in an 8-KHz (2×4 KHz) sampling rate for digitizing voice audio signals. An 8-KHz sampling rate translates into a one sample every 125 microseconds (8 KHz^{-1}; Figure 3-2).

Figure 3-2
PCM encoding.

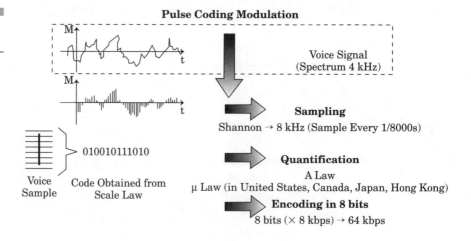

Each digital sample is represented by an 8-bit word ($2^8 = 256$ sample levels) that measures the amplitude of the signal. The amplitude of the signal is based on the power (expressed in units of voltage) of the electrical signal generated by the telephone transmitter/receiver in the handset. This signaling technique has become known as Digital Signaling 0 (DS0), or 64-Kbps (8 bits × 8 KHz) channel transmission format. The term DS0 was defined based on the Digital Signaling 1 (DS1) format used to describe a digital T1-carrier communications circuit supporting 24 64-Kbps communications channels.

The PCM samples generated from each communications system port are transmitted onto the TDM bus in a continuously rotating sequence based on the time slot assignments given to each port circuit interface (see below). Only a single PCM word sample is transmitted at a time; that is the entire electrical transmission line is reserved for use by only one port circuit for transmission of its sample signal. The PBX processing system monitors each port circuit's transmission time assignment in the rotating sequence, controls when the sample is transmitted, and coordinates transmission of the sample between the originating and destination endpoints.

There are two standards for coding the signal sample level. The Mu-Law standard is used in North America and Japan, and the A-Law standard is used in most other countries throughout the world, although each uses the 64-Kbps transmission format. For this reason, PBX systems must be designed and programmed for different geographic markets. Using firmware downloads, system vendors and customers can program their PBX systems to support the local PCM standard. The early digital PBX systems used different hardware equipment based on the location of the installation.

To summarize the fundamentals of PCM:

1. 4-KHz analog voice signals are sampled 8,000 times per second (8-KHz sampling rate)
2. Each sample produces an 8-bit word number (e.g., 11100010)
3. 8-bit samples are transmitted onto the TDM bus at a 64-Kbps transmission rate
4. The samples from each port circuit are transmitted in a continuously rotating sequence

TDM Bus Bandwidth and Capacity

Bandwidth is the amount of data that can be transmitted in a fixed period. For digital transmission, bandwidth is expressed in bits per second; for analog transmission, bandwidth is expressed in cycles per second

(Hertz). The bandwidth of an 8-bit PBX TDM bus is determined by the internal switching system clock rate used to create time slots for each channel's transmission. The faster the clock rate, the more digitized samples per second can be transmitted over the TDM bus. The clock functions merely as a counter; the faster it "counts," the more sampled digital signals within a fixed period (usually defined as 1 second) can be transmitted over the TDM bus. For example, an 8-bit TDM bus operating at 2.048 MHz has a bandwidth of 16 Mbps. If you double the clock rate (double the operating frequency), the bandwidth capacity doubles.

If the operating frequency of the TDM bus is not provided but the number of time slots is known, the bandwidth of a TDM bus can be calculated by multiplying the number of time slots (as determined by the system clock rate) by 64 Kbps (the number of transmitted bits per communications channel). A TDM bus segmented into 32 time slots has a transmission bandwidth of 2.048 Mbps (32 × 64 Kbps). A system with a faster clock rate that is capable of segmenting the TDM bus into 512 time slots would have a bandwidth of 32.64 Mbps (512 time slots × 64 Kbps). It is usually awkward to refer to the TDM bus bandwidth by the exact transmission capacity, so it is common to see the TDM bus bandwidth written, and referred to, as a 32-Mbps TDM bus. The most common PBX TDM bus bandwidths are usually based on exponential multiples of 2 Mbps (2^n Mbps): 2 Mbps, 8 Mbps, 16 Mbps, or 32 Mbps.

Not all of the time slot segments on a TDM bus are designed to handle communications traffic. Most PBX system TDM buses reserve a few time slots for the transmission of control signaling across the internal system processing elements. For example, a control signal time slot is used to alert the main system control complex that a telephone has gone off-hook. The signal is passed from the telephone instrument to the port circuit card and across the internal processing/switching transmission network via the local TDM bus. A variation of this design is to dedicate an entire TDM bus for control signaling. Examples of PBX systems using a dedicated signaling bus for system port control are the Siemens Hicom 300H and Hitachi HCX 5000 products. When a single bus is used for communications and processing functions, the control signaling time slots are not available for port communications requirements and should not be considered in the analysis of the system's traffic handling capabilities. Time slots that can be used for real-time communications applications are sometimes referred to as talk slots. The total number of talk slots and control signaling slots per TDM bus are equal to the number of time slots (Figure 3-3).

Legacy PBX Call Processing Design

Figure 3-3
TDM transmission bus.

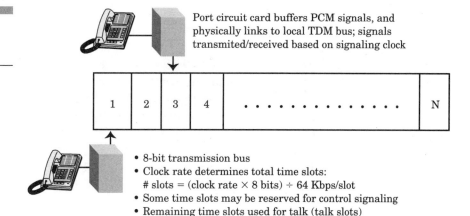

The number of available talk slots limits the number of active PBX ports that can be simultaneously supported by a single, common TDM bus. An active PBX port is simply defined as a port that is transmitting and receiving real-time communications signals—on-line. A port may be customer premises equipment working behind the PBX system, such as a telephone, or an off-premises trunk circuit. For example, a 2.048-Mbps TDM bus with 30 talk slots can support 30 active communications ports (telephones, modems, facsimile terminals, trunk circuits, voice mail ports, etc.).

Port-to-Port Communications over a Single TDM Bus

When a station port is about to become active, the PBX processing system will assign that port a talk slot on the local TDM bus connected to the port's circuit interface card. For the remainder of the call, the port will use the designated talk slot to transmit its digitized voice communications signals across the internal circuit switched network. The port receiving the call may be another station or a port interface connected to a trunk circuit. If the originating station port places an internal system call to another station, the PBX processing system will assign the destination port a designated talk slot on its interface circuit card TDM bus. If the originating station port is making an off-premises call requiring a trunk circuit connection, the processing system will assign the trunk circuit port a talk slot on the TDM bus supporting its interface circuit card.

The same process takes place for incoming trunk calls to user stations: the trunk interface port circuit is assigned a talk slot, as is the destination station interface port circuit. The circuit switching system will use the two designated talk slots to connect the two ports together to transmit and receive communications signals for the duration of the call. The two talks slots will work in tandem for talking and listening between ports, with each port physically linked to both talk slots.

The number of talk slots required per call will depend on two conditions:

1. The number of connected ports per call
2. The number of TDM bus segments required for port connections

A two-party conversation between PBX ports interfacing with the same TDM bus will require two talk slots, but multiparty conference calls will require as many talk slots as conference parties. For example, a four-party conference call would require four TDM bus talk slots. A small PBX system based on a circuit switched network design consisting of a single TDM bus will require only two talk slots per two-party call, but intermediate and large PBX systems designed to support hundreds or thousands of station and trunk ports will have switching network designs based on many interconnected TDM bus segments, and more than two talk slots will be required for an internal two-party call. More than four talk slots will be required for a four-party conference call if the ports are housed in different port equipment cabinets. The answer to the question of how many talk slots are needed per call in an intermediate or large PBX system requires some knowledge of the PBX switch network design.

Multiple TDM Bus Design

Most PBX systems have multiple TDM buses supporting the communications needs of the system ports. The individual TDM bus segments can be linked through a variety of methods based on the topology of the switching network design (see section below). Two user stations connected to port interface circuit cards housed in different port equipment cabinet frames would each be supported by different local TDM buses. A PBX switched network TDM bus may support a few port circuit cards (perhaps half a port carrier shelf), an entire port carrier shelf, or even multiple port carrier shelves within the same cabinet frame. Stackable single carrier cabinet designs sharing a common backplane for process-

Legacy PBX Call Processing Design

ing and switching functions also may be supported by a single TDM bus, but it is more than likely that port circuit interfaces housed in different cabinets will not share the same TDM bus. Based on these TDM bus segment scenarios, it is possible that a call between two user stations housed on the same port carrier shelf would require four talk slots per call and two station users on different port carrier shelves in the same equipment cabinet frame would require only two talk slots. The number of required talk slots will depend on the switch network design.

Whenever two communicating ports do not share a common TDM bus, the PBX processing system will assign the originating port a talk slot on its local TDM bus (the TDM bus directly connected to its port interface card) and a talk slot on the TDM bus that is local to the destination port. The destination port will likewise be assigned two talk slots: one on its local TDM bus and another on the originating port's local TDM bus. This scenario requires at least four available talk slots divided between two TDM buses. Additional communications channels may be required to link the TDM buses, and the PBX processing system makes the necessary assignments per call. Multiparty conference calls across multiple TDM buses will require one talk slot per party per TDM bus used to complete the call connection. For example, a three-party conference call among three internal station users, each supported by a different TDM bus, may require nine talk slots: three talk slots per party (one per TDM bus) × three parties = nine talk slots (Figure 3-4).

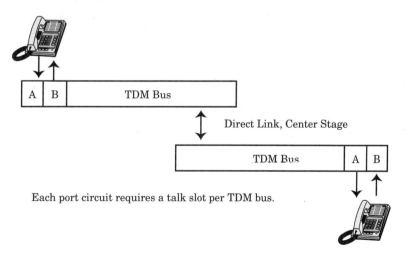

Figure 3-4
Call across multiple TDM buses.

Each port circuit requires a talk slot per TDM bus.

To minimize the number of inter-TDM bus connection requirements and increase the traffic handling capability of the PBX system, it is often recommended that groups of station and trunk circuit ports share a common TDM bus, instead of dedicating different TDM buses to different station or trunk interface port circuits. Station user groups with high intercom traffic requirements also should share common switch networking facilities to minimize inter-TDM bus connection requirements.

In some instances there will not be talk slots available on a local TDM bus when a station call is initiated. Voice-based communications systems traditionally have been designed to support more system ports than available local TDM bus talk slots. In a typical PBX system environment, it is rare that every station or trunk port will be active simultaneously, but there may be a blocked call if there are more provisioned station/trunk ports than total local TDM bus talk slots. PBX systems are designed with traffic engineering calculations to minimize blocked call attempts. Call blocking situations have a low probability of occurring if the system is correctly traffic engineered, but they may occur if there are more potentially active ports than available talk slots required to provide the circuit switched connection between the ports. A more detailed discussion of blocking and nonblocking switch network design and traffic engineering analysis follows at the end of this chapter.

PBX Circuit Switching Design

PBX circuit switched network designs differ between each manufacturer's product portfolio and even among models within a portfolio. Although there are differences in the individual PBX system switch network designs, the main functional elements are the same. All port circuit interface cards transmit and receive communications signals via a directly connected TDM bus, but the time and talk slot capacities are likely to differ between systems. A very small or small PBX system switching network design may consist of a single TDM bus backplane connected to every port interface circuit card, but a larger PBX system with more than one TDM bus must be designed to provide connections between the TDM bus segments. The TDM bus connections may be direct connections or center stage switch connections. The center stage switching system complex may be based on a space switch matrix design using circuit switched connections or a broadband TDM bus interconnecting lower bandwidth TDM buses. Two of the leading suppliers of PBX systems, Avaya and Alcatel, also offer customers of their very large PBX system models a center stage ATM switching option that can also support switched LAN data communications applications.

Center Stage Switch Complex

The primary function of the center stage switching complex is to provide connections between the local TDM buses, which support port carrier interface transmission requirements across the internal switching network. Complex center stage switching systems may be used in PBX systems designed for 100 user stations, although the smaller systems typically have a single TDM bus design or multiple TDM buses with direct link connections between each bus. A center stage switching complex may consist of a single large switching network or interconnected switching networks.

A very small PBX system usually does not require a center stage switching complex because the entire switching network might consist of a single TDM bus. Individual TDM bus switch network designs require a TDM bus with sufficient bandwidth (talk slots) to support the typical communications needs of a fully configured system at maximum port capacity. Most small PBX systems based on a single TDM bus design can provide nonblocking access to the switch network at maximum port capacity levels. If the TDM bus has fewer talk slots than station and trunk ports, the switch network can still support the communications traffic requirements, if properly engineered.

There are a few small and intermediate PBX systems that have multiple TDM buses but no center stage switching complex. For example, an Avaya Definity G3si can support up to 2,400 stations and 400 trunks using three-port equipment cabinets, each with a dedicated TDM bus, but does not use a center stage switching complex to connect the TDM buses. PBX system designs like the Definity G3si use direct cabling connections between each TDM bus for intercabinet connections between ports. This type of design can support a limited number of TDM buses without a center stage switching complex, but more TDM buses require more direct connections between each bus. When the system design includes more than three TDM buses, the switch network connection requirements may become unwieldly and often very costly. During the 1980s the Rolm CBX II 9000 supported up to 15 port equipment nodes that required dedicated fiber optic cabling connections to link each cabinet's TDM bus switching network because it lacked a center stage switching complex. A fully configured system required 105 direct link connections (fiber cabling, fiber interface cards), resulting in a very costly alternative to a center stage switching complex. Every new nodal addition to the system required new fiber optic connections to every existing cabinet node. The advantage of a center stage switching complex in an intermediate/large PBX system design is to simplify switch network connections between endpoints.

There are several center stage switch designs typically used in digital circuit switched PBXs:

1. Broadband (very large bandwidth) TDM bus
2. Single-stage switch matrix
3. Multistage switch matrix

Broadband TDM Bus

Most local TDM buses have limited bandwidths capable of supporting between 32 and 512 time slots. A TDM bus functioning as a center stage switching complex capable of supporting switch connections between many local buses must have a transmission bandwidth equal to or greater than the total bandwidth of the local TDM buses it supports for nonblocking access. For example, a single TDM bus with a bandwidth of 128 Mbps (2,048 time slots) can support switch connections for sixteen 8-Mbps TDM buses or four 32-Mbps TDM buses.

The center stage TDM bus must also support a sufficient number of physical link connections to support all local TDM buses. If the bandwidth of the center stage TDM bus is not sufficient to support switched connections for every local TDM bus time slot, there is a probability of blocking between the local TDM bus and the center stage TDM bus. The number of local TDM bus connections is always limited to ensure non-blocking access to the center stage TDM bus.

Local TDM buses typically interface to the center stage TDM bus through a switch network element known as a Time Slot Interchanger (TSI). The TSI is a switching device embedded on the physical interface circuit card that supports the physical local/center stage bus connection. The primary function of a TSI is to provide time slot connections between two TDM buses with different bandwidths. The simplest definition of a TSI is a portal between the local TDM bus and the center stage bus.

If a single broadband TDM bus cannot support nonblocking connections for all of the installed and configured local TDM buses, it may be necessary to install additional center stage TDM buses. A center stage switching complex based on multiple high bandwidth TDM buses requires connections between each center stage bus, in addition to switched connections to the local buses. Switched connections between any two local TDM buses in the PBX system may require transmission across two center stage buses, which are linked together, because each center stage TDM bus has dedicated connections to a select number of local TDM buses. The bandwidth connections between the high-speed center stage TDM buses must be sufficient to support the port-to-port traffic needs of the local TDM buses. For this reason, system designers use very high-speed optical fiber connections to ensure the switched network traffic requirements.

Single-Stage Circuit Switch Matrix

The most popular center stage switching design is a single-stage circuit switch matrix. A single-stage circuit switch matrix is based on a physical crosspoint switched network matrix design, which supports connections between the originating and destination local TDM buses. A single-stage circuit switch matrix may consist of one or more discrete switch network matrix chips. Most small/intermediate PBX systems use this type of design because of the limited number of local TDM buses needed to support port circuit interface requirements.

The core element of a crosspoint switching matrix is a microelectronic switch matrix chip set. The switch matrix chip sets currently used in PBXs typically support between 512 and 2,048 nonblocking I/O channels. A 1K switch matrix supports 1,024 channels; a 2K switch matrix supports 2,048 channels. Each channel supports a single TDM bus time slot. Larger switch network matrices can be designed with multiple switch matrix chips networked together in an array.

Based on the size of the switch network matrix and the channel capacity of a single chip set, a center stage switching complex may require one or more printed circuit boards with embedded switch matrix chip sets. The number of chips increases exponentially as the channel (time slot) requirements double. For example, if a single 1K switch chip can support 1,024 I/O communications, four interconnected 1K switch chips are required to support 2,048 I/O channels. Doubling the number of channels to 4,096 will require 16 interconnected 1K switch chip sets. Large single-stage switching networks use a square switching matrix array, for example, a 2 × 2 array (four discrete switch matrix chip sets) or a 4 × 4 array (16 discrete switch matrix chip sets).

A 1K switch matrix can support any number of TDM buses with a total channel (time slot) capacity of 1,024, for example, eight 128 time slot TDM buses or four 256 time slot TDM buses. The total bandwidth (time slots) of the networked TDM buses cannot be greater than the switch network capacity of the center stage switch matrix. The physical connection interfaces for the TDM buses are usually embedded on the switching network board, but this is not always the case. The intermediate/large Nortel Networks Meridian 1 models require an intermediary circuit board, known as a Superloop Card, to provide the switch connection between the local TDM buses (Superloops) and the center stage 1K group switch matrix.

Multistage Circuit Switch Matrix

A single-stage circuit switch matrix design is not feasible for the center stage switching system complex of a large or very large PBX system because such a system would have a system traffic requirement for as many as 20,000 time slots. A very large array of switch matrix chip sets would lead to design complications and require several switch network array printed circuit boards. The better switch matrix design solution for a large or very large PBX system is a multistage design. The most common multistage switch network design type is a three-stage network design known as a Time-Space-Time (T-S-T) switch network. A T-S-T switch network connects three layers of switches in a matrix array that is not square (Figure 3-5).

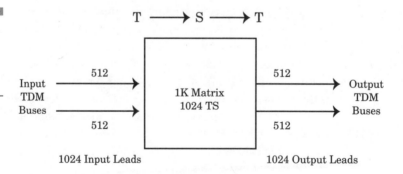

Figure 3-5
TDM bus connections: center stage space switch matrix.

In a T-S-T switching network design, each switch network layer consists of the same number of switch matrix chips. The first switch network layer connects the originating local TDM buses to the second switch network layer; the third switch network layer connects to the second switch network layer and the destination local TDM buses. In this design, the second network switch layer is used to connect the first and third layers only, with no direct connection to the local TDM buses. The term Time-Space-Time was derived from the fact that the first and third switch network layers connect to TDM buses, and the second switch network layer functions solely as a crosspoint space connection switch for the two outer layers.

In a T-S-T switch network configuration, each TDM bus channel entering the first switch network layer has access to each outbound switch connection to the second switch network layer. In turn, each outbound switch connection in the second switch network layer has access to each switch connection in the third switch network layer. Each switch matrix in the first and second layers is connected according to the same pattern.

The T-S-T switch network is contained on a combination of printed circuit boards. Multiple first and third layer switch matrix chip sets may be packaged on a single board, although the usual design is a single switch matrix per board to simplify connections between the local TDM buses and the second switch network layer. Multiple second layer switch matrices are usually packaged on a single board. The total number of boards required for the center stage switching complex will depend on the number of I/O TDM channels configured in the installed system. An 8K switch network will require fewer boards than a 16K switch network.

ATM Center Stage

During the early 1990s, it was believed that traditional circuit switched voice networks would someday be replaced by ATM switch networks. Several PBX manufacturers worked to develop a PBX switch network based on ATM switching and transmission standards. An ATM switching network can provide the same high quality of service as traditional circuit switched networks can for real-time voice communications; it also offers the additional advantage of very high switching and transmission rates. Lucent Technology's enterprise communications system division (now Avaya) and Alcatel developed, announced, marketed, and shipped ATM center stage switching options for its largest PBX models. Implementing the ATM center stage switching option requires a stand-alone ATM switching system equipped with customized interface cards to connect to the PBX processing and switch network subsystems. A gateway interface card is used to link the local TDM buses to the ATM switching complex for intercabinet communications. The gateway interface card converts communications signals from time-based PCM format to ATM packet format.

Shipments of the option have been negligible since its introduction for two important reasons: few customers have installed ATM-based LANs, opting instead to upgrade their IP-based Ethernet LAN infrastructure, and the cost to install the PBX option is greater than the cost of a traditional TDM/PCM center stage switching complex. In addition to the cost of the ATM switching system, there is the cost of high-priced interface cards used to convert TDM/PCM communications signals to ATM format for connecting the local TDM buses to the center stage switch complex. Nortel Networks tested an ATM-based version of its Meridian 1, but canceled development in the late 1990s after determining that the cost to upgrade a customer's installed system was too high.

The Avaya Definity ECS and Alcatel OmniPCX 4400 ATM-based offerings are still being marketed, but too few customers have shown enough interest to make it a viable center stage switching option for the future. Growing market demand for IP-based PBX systems appears to have stunted development of the ATM center stage switching option.

Local Switching Network Design: TDM Buses and Highway Buses

The local switching network in a PBX system can support several basic functions:

1. Port interface circuit card access and egress into the circuit switched network
2. Direct switched connections between port interface circuit cards
3. Switch connections into the center stage switching complex

The primary function of the local switching network is to provide the local communication path for calls between system ports. Small PBX systems without a center stage switching complex depend on the local switching network for all communication paths between station and trunk ports. Much of the communications traffic in many intermediate or large PBX systems is carried exclusively over the local switching network without connections across the center stage switching complex, if the design topology is dispersed or distributed (see next section). When switched connections between endpoints must be made across the center stage switching complex, it is the local switching network that handles most of the call's transmission requirements.

A PBX system's local switching network design may be comprised of the following elements:

Legacy PBX Call Processing Design

1. Local TDM buses
2. Highway TDM buses
3. Switch network interfaces/buffers
4. Time slot interchangers

A traditional PBX switch network local TDM bus is an unbalanced, low characteristic impedance transmission line that directly supports the traffic requirements of port circuit interface cards without intermediary TDM buses. The ends of the TDM bus are usually terminated to ground, with a separate resistor for each bit. Port interface circuit cards typically connect to the TDM bus through a customized bus driver device. A bus driver is a switchable constant current source so that, in the high "output" state during transmission, there is no bus loading to cause reflections.

A Highway TDM bus consolidates traffic from multiple lower bandwidth local TDM buses to facilitate switch network connections between local TDM buses and provide a communications path to the central stage switching complex when needed to connect the originating and destination call endpoints across different local TDM buses and Highway buses.

Although all circuit switched PBXs depend on local TDM buses for transporting communications signals to and from port interface circuit cards, the local switching network design usually differs from one system to another. The local TDM bus in a PBX system may support a few port interface circuit cards, a full port carrier shelf, or an entire port cabinet. The number of port interfaces a TDM bus can adequately support is based on its bandwidth. A limited bandwidth TDM bus that supports 32 time slots may be used to support only a few low-density port circuit cards, whereas a high bandwidth TDM bus that supports 512 time slots can easily support the traffic requirements of a high-density port carrier shelf or a moderate density port cabinet. A few examples illustrate the differences in local TDM bus design:

1. A Fujitsu F9600 16-port card slot Line Trunk Unit (LTU) carrier is supported by eight 2.048 Mbps TDM buses (32 time/talk slots per bus); each local TDM bus supports a maximum of two port interface circuit cards (the number of ports across the two cards must be equal to or less than 32).
2. The switch network architecture of the Avaya Definity PBX family is based on a 32-Mbps TDM bus (512 time slots, 483 talk slots) that can be configured to support a single port carrier shelf or a five-carrier

shelf cabinet. Each Definity G3si/r port carrier shelf supports 20 port interface card slots. The 512 time slot TDM bus can support very high traffic requirements for a single port carrier or moderate traffic requirements across a multiple carrier cabinet if traffic engineering guidelines are used.

3. A Nortel Meridian 1 Option 81C Intelligent Peripheral Equipment Module (IPEM), single port carrier cabinet with 16 port interface card slots, can be configured with one, two, or four Superloops (128 time slots, 120 talk slots per Superloop). A Superloop is the Nortel Networks name for its Meridian 1 local TDM bus. The IPEM port carrier shelf can have access and egress to 120, 240, or 480 talk slots; the number of configured Superloops depends on the traffic capacity requirements of the local ports. Basic traffic requirements can usually be supported by a single Superloop, but nonblocking switch network access requirements may dictate four Superloops per IPEM. A single Superloop can also be configured to support two IPEM stackable cabinets, with a total of 32 port card slots (a maximum of 768 voice ports), if there are very low traffic requirements.

The Fujitsu example illustrates a PBX system with multiple local TDM buses per port carrier shelf, with each local TDM bus supporting only two port cards. The Avaya example illustrates a PBX system with a local TDM bus designed for and capable of supporting a multiple port carrier cabinet capable of housing dozens of port cards and hundreds of ports. The Nortel example illustrates a flexible local TDM bus design that can support low, medium, or high traffic requirements per port carrier shelf by provisioning the appropriate number of local TDM buses.

Even though the Fujitsu, Avaya, and Nortel PBXs use local TDM buses to provide a communications path for port interface circuit cards, the bandwidth of the TDM buses and the number of TDM buses per carrier shelf or cabinet varies among the three systems. There is no standard for local TDM bus bandwidth and provisioning in a PBX system. The concept is the same, but the implementations differ.

In the Fujitsu example, the backplane of the port circuit cards connects directly to the local TDM bus. The Definity port carrier backplane also provides a direct connection to the local TDM bus. In the Nortel example, a Superloop bus supports the communication transmission needs of the port interface circuits in an IPEM cabinet, but there is no direct link between the cards and the Superloop. An interface card is used as a buffer to link the carrier shelf backplane to the electrical transmission wire operating as the Superloop TDM bus. The switch network buffer function is embedded on the IPEM Controller Card, which also provides local processing functions to the port carrier shelf.

Switch network interfaces/buffers are used to consolidate communications signals from multiple port interface circuit cards for access to and egress from the local TDM bus. These specialized interface cards may be dual function interfaces because several PBX switch network interface/buffer cards also have an on-board microprocessor controller used for localized processing functions (see Chapter 4).

The Siemens Hicom 300H has an interface card similar to the Nortel Meridian 1 to support both switching and processing functions as the local port carrier level. The Siemens Line Trunk Unit Controller (LTUC) card provides a link between the main system processor and the port interface circuit card microcontrollers and also serves as a buffer interface between the high-speed 32-Mbps Highway transmission bus (512 time slots) and two segmented TDM buses (256 time slots per TDM bus, 128 time slots per segment) that connect directly to the port circuit interface cards. The Hicom 300H LTUC functions like a TSI because it is multiplexing several moderate bandwidth TDM buses onto a higher bandwidth TDM bus.

From high-level diagrams, it appears that the Nortel and Siemens switch network designs are very similar, but major differences exist. The Meridian 1 Superloop functions as a local TDM bus but requires a buffer interface to link to the port interface carrier; the Hicom 300H has four TDM bus segments directly connected to the LTU port circuit interface cards and requires the LTUC, functioning as a TSI, to link to the Highway bus. The Meridian 1 Superloop and Hicom 300H Highway Bus provide a communications to the central stage switching complex of their respective PBX systems, but the design structures are not identical.

The NEC NEAX2400 IPX also uses a TSI to multiplex local TDM buses onto a higher bandwidth Highway bus. A 384 time slot local TDM bus supports each NEAX2400 PIM single carrier shelf cabinet. Up to four PIMs can be stacked together, and the individual local TDM buses communicate over a common 1,536 time slot Highway bus. A TSI links each PIM's local TDM bus to the Highway bus. Multiple Highway buses across cabinet stacks communicate over a higher bandwidth Highway bus. In the largest NEAX2400 configuration, a Super Highway bus links Highway buses across the entire switch network complex. The broadband Super Highway bus functions as a center stage control complex but only is used for switched connections between local Highway buses. Most system traffic is localized at the PIM cabinet level and uses the Highway and Super Highway buses infrequently if the system is properly engineered. The Highway bus design of the NEAX2400 is not nonblocking for a worse case traffic situation but is essentially nonblocking for most customer requirements.

Highway buses typically operate at very high transmission rates because they are required to provide the communications path across many local TDM buses or between many local TDM buses and the center stage switch complex. The terminology used to describe a PBX switch network system Highway bus varies from system to system, but the function is essentially the same. Avaya calls the optical fiber cable link used to connect Definity Port Network cabinets an Archangel Expansion Link (AEL), and Ericsson calls its MD-110 LIM cabinet communications links FeatureLinks (formerly PCM links), but the two perform the same primary function: linking port cabinets together directly or through a center stage switch complex. The Definity AEL is always a fixed high-bandwidth optical fiber link and provides nonblocking access to the center stage switch complex for each port network cabinet TDM bus. The Ericsson FeatureLink operates at only 2 Mbps and can support only 32 time slots (30 talk slots). Based on customer traffic requirements, up to four FeatureLinks can be equipped per LIM, for a total bandwidth of 8 Mbps, to support a maximum of 120 talk slots. The limited number of talk slots supported by the MD-110's Highway bus would seem to cause switch network access problems, but analysis of customer traffic patterns (inbound trunk calls, outbound trunk calls, intercom calls) indicates that the four FeatureLink capacity is more than sufficient for most customer configurations.

PBX Switch Network Topologies

The design structure of a PBX's switch network architecture is known as its topology. The switch network topology describes how calls are switched and transmitted based on the call origination and destination endpoints. The specific topology category used to classify a PBX's switch network design depends on the required switched connections between the local switch network and the center stage switch complex necessary to establish a call connection.

All PBX switch network topologies can be categorized into three basic designs:

1. Centralized

Legacy PBX Call Processing Design

2. Distributed
3. Dispersed

Centralized

A centralized topology is defined simply as a switch network design that requires all calls, regardless of the origination and destination endpoints, to be connected through the same TDM bus or switch matrix or the center stage switch complex (Figured 3-6).

Figure 3-6
Centralized switching network topology.

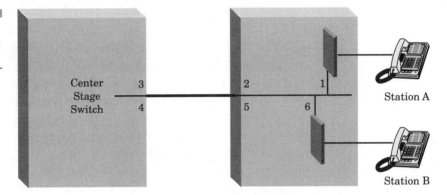

Call Connection: Station A to Station B
1. Station A port circuit card links to local TDM bus.
2. Local TDM bus links to highway bus.
3. Highway bus links to center stage switch.
4. Center stage switch links to highway bus.
5. Highway bus links to local TDM bus.
6. Local TDM bus links to Station B port circuit card.

Any PBX system with a switch network system comprised of a single TDM bus or switch matrix is classified into the centralized switch network design category because all calls are handled through the same "centralized" switch network element. Most small PBX systems have a switch network design based on a single TDM bus because the traffic requirements for equipped systems with fewer than 100 ports (stations and trunks) can easily be supported without multiple TDM bus requirements and/or a center stage switch complex. A single TDM bus design may also be used by PBXs with larger port capacity limits, if the TDM bus bandwidth is sufficient to support the port traffic requirements. For example, the Avaya Definity's switch network design is based on a 32-Mbps TDM bus that can easily support the very small port (20 to 40 stations) traffic requirements of the Definity One model and the larger port (40 to 400 stations) traffic requirements of the Definity ProLogix model.

Many intermediate/large PBX system models have centralized switch network designs because a center stage switch complex handles all call connections regardless of the originating and destination call endpoints. It is easy to see the necessity of using a center stage switch complex to support switch connections between port interface circuit cards housed in different multiple carrier port cabinets because a local TDM bus is not configurable across common cabinets, and the system installation may include many port cabinets. The system switch network architecture is easier to design and program if all calls are connected with a centralized switch network element because the same call processing steps are followed for each and every type of call. It is more difficult to see the necessity of using a center stage switch complex to support switch connections between port interface circuit cards housed on the same port carrier shelf or even between ports on the same port interface circuit card, but a centralized switch network design dictates the same switch network connection protocol (center stage switch complex connection) regardless of originating and destination port interface circuit proximity.

The Nortel Networks Meridian 1 Option 81C, Siemens Hicom 300H, and Fujitsu F9600 XL models are examples of centralized switch network designs. Each of these systems can be installed with multiple port cabinet stacks, with several port carriers per cabinet stack. Each system uses a center stage switch complex to support connections between each port carrier's local switching network (single or multiple local TDM buses), even if the two connected telephones are supported by the same port circuit interface card and connect to the same local TDM bus that connects to the same the Highway bus that connects to the center stage switch complex. It may appear a waste of switch network resources (talk slots, switch connections) to use the center stage switch complex for a call of this type, but that is the way the system is designed and programmed. Figure 3-6 illustrates the call communications path for a Meridian 1 Option 81C between port interface circuits in different cabinet stacks and between port circuit interface circuits on the same port circuit interface card. The call connection protocol is similar, if not identical, for the Siemens and Fujitsu systems.

A centralized switch network design offers no customer benefits, but it can be problematic because a large number of potential switch network elements (local TDM buses, Highway buses, switch network interface/buffer, TSI, center stage switch elements) are required to complete any and all calls. This can affect switch network reliability levels because the probability of switch network element failure or error affecting a call connection is increased. For example, center stage switch complex failures or errors affect all system port connections in the PBX system.

A major disadvantage of the centralized switch network design is when a customer needs to install a remote port cabinet option to support multiple location communications with a single PBX system. A remote port cabinet option requires a digital communications path between it and the main PBX system location. Most remote port cabinet installations are supported with digital T1-carrier trunk services. If the PBX switch network topology is centralized, all calls made or received by station users housed in the remote cabinet must be connected through the center stage switch complex at the main location, even if calls (intercom or trunk) are local to the remote port cabinet. A T1-carrier circuit, with a limited number communications channels, must be used for every remote cabinet call to access the center stage switch complex. Most remote PBX cabinet options require two T1-carrier channels per call connection, thus limiting the number of active simultaneous conversations at the remote location. This may force the customer to install additional T-1 carrier circuits to support the port traffic requirements at the remote location, but there are limits on how many T1-carrier interface circuit interfaces can be supported by the remote port cabinet. The limitations of the centralized switch network design may force a customer to install multiple remote port cabinets at the remote location or a stand-alone PBX system.

Distributed and Dispersed Switch Network Designs

A distributed topology is defined simply as a switch network design comprised of multiple, independent local switching networks that are connected with direct communications links instead of a center stage switch complex. Each local switching network operates independently of the others and supports all of the communications needs of the local port interface circuits it connects to. Communications between user ports housed in different cabinets require a direct communications path between each cabinet's local switch network. There is no center stage switch complex (standard in centralized switch network designs with multiple local switch networks) in a PBX based on a distributed switch network design, which is a potential cost benefit to the customer. Another benefit of a distributed switch network design as opposed to a centralized design is its flexibility in supporting multiple location customer requirements. Without a center stage switch complex, the communications links between remote locations and the main customer site are minimized because most station user traffic is local to the cabinet's switch network. Only intercabinet traffic requires communications link resources (Figure 3-7).

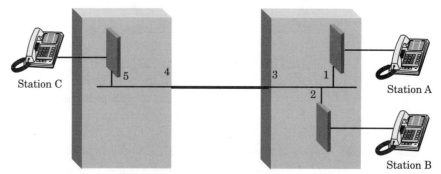

Figure 3-7
Distributed switching network topology.

Call Connection: Station A to Station B
1. Station A port circuit card links to local TDM bus.
2. Local TDM bus links to Station B port circuit card.

Call Connection: Station A to Station C
1. Station A port circuit card links to local TDM bus.
2. Local TDM bus links to highway bus.
3. Highway bus links to local TDM bus.
4. Local TDM bus links to Station C port circuit card.

A distributed switch network design is usually limited to PBX systems with a minimal number of local switching networks supporting two or three port cabinets. Once the number of local switching networks exceeds three, it usually becomes a cumbersome, and expensive, process to upgrade the system because of the necessity of having direct communications links between each cabinet, unless a cabinet can be used as a tandem switching node within the distributed cabinet configuration. The two most popular PBXs based on a distributed switch network design are the Avaya Definity G3si and the Alcatel OmniPCX 4400. A Definity G3si can be installed with up to three port network cabinets (a PPN control cabinet and two EPN expansion port cabinets). Each cabinet has a local switching network based on a 32-Mbps TDM bus and can be equipped with expansion interface circuit boards to connect to an EAL (see above) for intercabinet communications. There is no center stage switch complex, and each port network cabinet TDM bus functions independently.

Legacy PBX Call Processing Design

The Alcatel OmniPCX 4400 is an example of a PBX with a distributed switch network design that can support more than three cabinets, making it the exception that proves the rule. As part of its Alcatel Crystal Technology (ACT) system architecture, a single OmniPCX 4400 system can support up to 19 discrete cabinet clusters (control cabinet and expansion cabinets); each cabinet cluster has a local TDM bus (420 two-way channels) and can be linked to other cabinet clusters over a variety of communications paths based on PCM, ATM, or IP communications standards. A single interface board in the cluster's control cabinet can support up to 28 communications links. The bandwidth of each PCM link is 8 Mbps; the ATM link can operate at transmission rates of up to 622 Mbps. Direct links between any two cabinets can be established, or a control cabinet can function as a tandem switching node to link two or more distributed control cabinets. The availability of very high-speed communications links between cabinet clusters can minimize the number of physical transmission circuits supporting intercabinet cluster communications requirements, and the use of hop-through connections through a tandem switch node allows Alcatel to design large and very large system configurations without a center stage switch complex. Alcatel markets a multiple system version of the OmniPCX 4400, capable of supporting a maximum of 50,000 stations, and can design the network to handle communications traffic between all cabinet clusters across all systems without a center stage switch complex.

The third type of switch network design is dispersed topology. A dispersed switch network combines the design attributes of a distributed design (functionally independent local switch networks) and centralized design (center stage switch complex connecting local switch networks). A dispersed switch network design is comprised of local switch networks that support all of the local communications requirements of its connected port interface circuits and a center stage switch complex that is used only to provide switched connections between local switch networks for calls between ports connected to different local switch networks. For example, a call between two ports in the same cabinet sharing a common switch network would be connected by using only the resources of the cabinet's local switch network, such as a local TDM bus. If a call were placed between ports in different cabinets, the call would be connected through the center stage switch complex, and access to the center stage switch complex would be via the local switching networks (Figure 3-8).

Figure 3-8
Dispersed switching network topology.

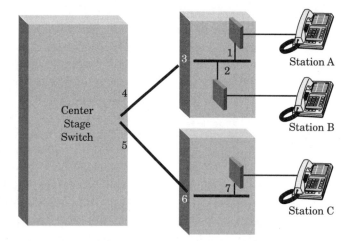

Call Connection: Station A to Station B
1. Station A port circuit card links to local TDM bus.
2. Local TDM bus links to Station B port circuit card.

Call Connection: Station A to Station C
1. Station A port circuit card links to local TDM bus.
2. Local TDM bus links to highway bus.
3. Highway bus links to center stage switch.
4. Center stage bus links to highway bus.
5. Highway bus links to local TDM bus.
6. Local TDM bus links to Station C port circuit card.

For example, the Avaya Definity G3r, a larger version of the Definity G3si, can support up to 40 port network cabinets. The G3r EPN expansion port cabinets are identical to the G3si cabinets; each is designed with a local switching network capable of handling all local communications requirements—calls exclusively between ports (stations and/or trunks) in the same cabinet. Calls between ports in different EPN cabinets are handled through a center stage switching complex in the PPN cabinet (common control cabinet). The Ericsson MD-110 is another example of a dispersed switch network design; communications between LIM cabinets are handled through a centralized group switch network complex, but local communications traffic remains within the LIM. The NEC NEAX2400 IPX also can be considered a dispersed switch network design because communications between ports in the same PIM cabinet (a single carrier shelf cabinet) are supported exclusively on the local TDM bus; intercabinet and intermodule group communications are supported over a hierarchy of Highway buses.

Legacy PBX Call Processing Design

PBX Switch Network Issues

The switch network design in a customer's PBX system is of little interest to most station users and may not even be of interest to the system manager. What is important to stations users and particularly the system manager are system availability, reliability, and survivability. The switch network design issues that affect these system requirements are:

1. Switch network redundancy
2. Time slot access and segmentation
3. Time slot availability: blocking or nonblocking

Switch Network Redundancy

Switch network redundancy is a design criterion that minimizes switch network downtime for one station user, a few station users, or all station users in the system. The term *redundancy* is often confused with the term *duplication*. A switch network design incorporated with duplicated elements is said to be redundant, but a redundant switch network may not necessarily have any duplicated elements that might prevent downtime for some or all station users. Duplication is the highest form of redundancy, but it is not the only type of redundancy, particularly in PBX switch network architectures, as we will shortly see.

If a customer wants a redundant switch network design that is based on duplication of critical design elements, the checklist of duplicated elements may include any or all of the following:

1. Center stage switch complex
2. Local TDM buses
3. Highway buses
4. Switch network interface (including embedded TSI)
5. Intercabinet cabling

Duplication of the center stage switch complex is a vital redundant switch network requirement in a centralized design topology because all calls are connected through this switch network element. Center stage switch complex errors or failure affect every call in the PBX system. Center stage switch problems may be slightly less important in a dispersed design topology, but it would still be highly desirable to have duplication of the switch network element because it is needed to connect all calls between local switch networks. A few PBX systems have a fully duplicated center stage switch complex as a standard design feature, such as the Nortel Networks SL-100, a modified version of the supplier's DMS-100 central office switching system. It is more common that the duplicated center stage switch complex is available as an option, although some intermediate/large PBX systems do not offer it as a standard or optional design element. The Siemens Hicom 300H is available in two models: the large line size Model 80 has an optional duplicated center stage switch complex and the smaller Model 30 does not offer it as standard or optional.

Loss of the local TDM bus will negatively affect the communications capabilities of all ports to which it connects. Redundancy of the local TDM bus can vary between different PBX systems based on the definition of redundancy. The Fujitsu F9600 XL has a fully duplicated local switching network design, including duplication of the local TDM buses. The Avaya Definity PBXs have a redundant local TDM bus design: the local 32-Mbps TDM bus (512 time slots) supporting all of the communications needs within a Port Network cabinet is based on two independent TDM buses, each with a 16-Mbps bandwidth (256 time slots), but operating as a single TDM bus from the viewpoint of the port interface circuit cards. If one of the two 16-Mbps TDM buses fails, all system ports can still connect to the remaining TDM bus. The Siemens Hicom 300H offers a similar redundant design concept for its TDM bus architecture. Two 8-Mbps TDM buses support eight port interface circuit card slots (one half of an LTU carrier shelf), each operating independently and accessible by any of the eight port interface cards. In these Avaya and Siemens models, loss of one TDM bus will place a heavier traffic load on the remaining TDM bus and may increase the number of blocked call attempts due to the reduced number of available time slots. The major difference between the two designs, however, is that a Definity 32 Mbps TDM bus can support a five-carrier shelf cabinet with several hundred stations and associated trunk circuits, and the Siemens 16-Mbps TDM bus design supports only eight port card slots (nominally 192 ports). Failure of a Definity TDM bus segment will have greater traffic handling consequences than failure of a Siemens TDM bus segment. Siemens offers switch network redundancy at a more local level than does Avaya.

Nortel Networks has claimed that the multiple Superloop design in its intermediate/large Meridian 1 models is a form of redundancy because loss of a single Superloop affects only a limited number of the ports in a cabinet stack. The term *limited*, however, can be misleading because a Superloop can support up to 32 port card slots, and each port card slot can support 24 digital telephones. If strategic system ports, such as attendant consoles or trunk circuits, are affected by loss of a Superloop, the redundancy level of the design might not be acceptable.

Highway buses may be fully duplicated, or loss of a TDM bus segment comprising the Highway bus may not affect the remaining bus segments (although traffic handling capacity will be reduced). Highway buses are used for connections between local TDM buses and to provide communications paths to the center stage switch complex. Loss of a Highway bus can be significant in a centralized switch network design.

Switch network interfaces are printed circuit boards connecting local switch networks to each other or the center stage switch complex. It is an electronic switch network design element that can fail and affect communications traffic between port cabinets. A duplicated switch network interface typically links the local switching network element, such as a TDM bus, to a high-speed fiber optic cable communications link. Duplicated center stage switch complex designs usually have duplicated switch network interfaces and duplicated cabling links for intercabinet communications connections.

Time Slot Access and Segmentation

Some switch network designs are based on universal port access to the local switching network; that is all ports in a carrier shelf or cabinet can use the full bandwidth capacity of the local TDM bus. For example, any port interface circuit housed in a Definity PPN cabinet can be assigned any talk slot on the local 32-Mbps TDM bus regardless of its port circuit card slot location. A five-carrier shelf PPN cabinet has 100 port card slots, and the local TDM bus supports every port interface circuit card in the cabinet. The Definity TDM bus is said to be universally accessible to all PPN port circuit terminations. The 512 time slot (483 talk slots) TDM bus supports the traffic needs of hundreds of system ports in the cabinet.

A switch network design is said to be segmented if the local switching network is based on segmented TDM buses supporting a single port carrier shelf or cabinet. For example, the Siemens Hicom 300H LTU carrier connects to the center stage switch complex via a 32-Mbps Highway bus. The local switching network consists of two 16-Mbps segmented local TDM buses, with each local TDM bus supporting different port card slots. Although the total TDM bandwidth at the LTU carrier shelf level is 32 Mbps, the 512 time slots are divided equally between both halves of the shelf (eight port card slots per half). If the segmented TDM bus supporting port card slots 1 to 8 fails, available time slots on the second operational TDM bus are not accessible to port circuit interfaces on the port card housed in slots 1 to 8. The LTU carrier shelf is said to be based on a segmented TDM bus design. This is the downside of a segmented TDM bus design when compared with a universally accessible design. The upside is that the Siemens system can be traffic engineered to a greater degree. The local TDM bus supports only eight port card slots, a fraction of the number the Definity local TDM is required to support (Figure 3-9).

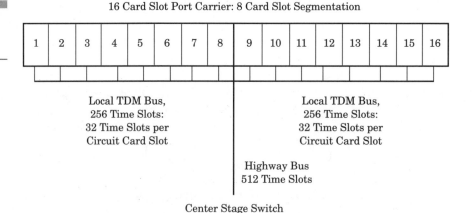

Figure 3-9
Segmented bus design.

The segmented TDM bus design of the F9600 was described earlier. Each port carrier shelf is supported by a 16-Mbps Highway bus that segments into eight 2-Mbps local TDM buses, with each bus supporting two port card slots. Minimizing the number of port card slots supported by a TDM bus is not always a good design objective because there may be less flexibility when configuring the port circuit cards. The F9600 backplane used to access the local TDM bus can support only 32 connections, which is also the maximum number of total port circuit terminations allowed for the two adjacent port cards. If 16 port circuit cards are installed, there is no problem, but problems may occur when a higher-density digital trunk card is installed. A 24-port T1-carrier interface card will limit the flexibility in configuring the adjacent port card slot that shares the 32 connections to the 2-Mbps local TDM bus (32 time/talk slots). Only an eight-port card can be housed in the second port card slot if the configuration rules are followed. If a 16-port card was installed and only eight telephones were installed, thereby limiting the number of configured ports to 32 (24 + 8), the system configuration

Legacy PBX Call Processing Design

guidelines would still prohibit such an installation. The Fujitsu system was programmed for nonblocking switch network access only, and more ports than fixed TDM bus connections/time slots are not allowed. The TDM bus segmentation design can limit port configuration flexibility, if the number of backplane connections per TDM bus is relatively small.

Time Slot Availability: Blocking or Nonblocking

The terms *blocking* and *nonblocking* have been used previously. In PBX terminology, blocking is defined as being denied access to any segment of the internal switch network because there is no available talk slot or communications channel to complete a call connection. Blocked calls are characterized by busy signals. Nonblocking switch network access means that an attempt to access the internal switch network will always be successful because there is a sufficient number of talk slots or communications channels to support simultaneous call attempts by every configured station user in the system.

Blocked calls due to unavailable trunk carrier circuits are not included as part of this discussion because the issue being addressed is blocking and nonblocking access to, and connections across, the internal PBX switch network. Trunk blocking issues are discussed in the next chapter.

There are several connection points in the overall PBX system and switch network design that can cause a call attempt to be blocked (Figure 3-10):

1. Port circuit card
2. Local TDM bus
3. Highway bus
4. Switch network interfaces
5. Center stage switch

Although it may seem strange that a call can be blocked at the port circuit card level, the number of physical communications devices supported by a port card can be greater than the number of communications channels supported by the desktop. For example, an ISDN BRI port circuit card that conforms to passive bus standards can support up to eight BRI telephones, but only two can be active simultaneously, because the BRI desktop communications link is limited to two bearer channels.

Figure 3-10
Traffic handling red flags: blocking points.

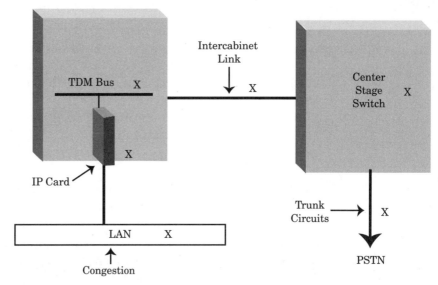

X = Potential Blocking Points

Scenarios also exist where a digital station card can support more desktop communications devices than available desktop communications channels. For example, a Siemens optiSet digital telephone equipped with two adapter modules can be connected concurrently to a second desktop optiSet digital telephone and a desktop analog telephone, with all three communications devices being supported by a single communications link to the Hicom 300H PBX and sharing the wall interface jack, inside telephony wiring, and port circuit card interface. Like the BRI port interface circuit, the optiSet interface can support only two active bearer communications channels, which means that one of the three desktop devices can be blocked from accessing the system.

Several recently introduced IP station cards supporting LAN-connected desktop telephones may also block call attempts because the number of physical telephones supported by the card can be greater than the number of local TDM bus connections supported by the card. For example, the Avaya Definity Media Processing Board for IP Telephony can support 96 IP telephones but can support between 32 and 64 connections to the local TDM bus based on the audio coder standard used for IP:TDM/PCM protocol conversion.

The local TDM bus is usually the most likely switch network element to be the cause of a blocked call. Although a greater number of current PBX systems are designed with a nonblocking switch network architecture, a good percentage of installed and new systems must be traffic engineered because the number of port circuit interfaces is greater than the maximum number of time slots on the local TDM for connecting the call. For example, the local 32-Mbps TDM bus supporting an Avaya Definity port network cabinet can support a maximum of 483 active ports, although the cabinet can physically support several times this number of ports. Avaya typically recommends installing 800 stations per port network cabinet for customers with moderate traffic requirements. It is unlikely that all 800 station users will attempt to place a call at the same time, but if they do only 483 time slots are available, and quite a few station users will hear a busy signal when they make their call attempt. Similarly, a Nortel Meridian 1 Superloop can support 120 active ports at full TDM bus utilization but is typically configured to support at least 200 station users. If properly traffic engineered, based on station user traffic requirements, call blocking should be minimal, but it can occur.

Most Highway buses provide nonblocking switch connections between local TDM buses and have sufficient communications channels for nonblocking access to the center stage switch complex. However, the bandwidth of the Highway bus may be less than the total bandwidth of the local TDM buses it supports and a call may be blocked if based on traffic conditions. Almost all current switch network interfaces and center stage switch complexes are also designed for nonblocking access and transmission, but exceptions do exist. For example, until Nortel recently upgraded the Meridian 1 Option 81C Sub Group Assembly module (a center stage switch complex) with a fiber optic ring design, it was possible, if not highly probable, that calls between switch network groups within the center stage could be blocked.

CHAPTER 4

Legacy PBX Switch Network Design

The most fundamental function of a PBX system is to support switched connections between peripheral endpoints. Stations users are accustomed to picking up their handset, hearing the dial tone, dialing a telephone number, and being connected to the called party. The possibility always exists that the station user receives a busy signal when the dialing process is completed. The most probable reason for a busy signal is that the called party is off-hook and engaged in another call. Infrequently, all telephone company trunk circuits are busy, and the station user hears an announcement to call again at a later time. A busy signal also may be received when all PBX trunk circuits are in use or internal switch network resources are not available. The PBX station user cannot control the availability of the called party or the availability of PSTN trunking facilities but can minimize the probability of busy signals due to blocked access to the internal switch network or local trunk circuits, if the PBX system is properly configured and engineered to meet the expected traffic demands of the customer. PBX traffic analysis and engineering tools are used to achieve acceptable customer service standards for internal switched connections and off-premises trunk calls.

Nonblocking/Blocking PBX Systems

PBX systems can be classified into two switch network design categories based on traffic engineering requirements: nonblocking and blocking. A PBX system is said to be nonblocking where no switch network traffic engineering is required because there will always will be sufficient switch network resources (local TDM bus talk slots, Highway bus communications channels, switch network interfaces, and center stage switch connections) to satisfy worse-case customer traffic demands at maximum system port capacity. Worse-case traffic demand occurs when all equipped ports are simultaneously active; that is, transmitting and receiving across the internal switch network. Although this would be a very unlikely customer situation, because PBXs are never configured at maximum port capacity and the probability is almost infinitesimal that all ports require simultaneous access to the internal switch network for communications applications, the assumption is used to define a nonblocking system.

Station users may have nonblocking access to the internal switch network but receive a busy signal when attempting to place an off-premises trunk call. The PBX system is still classified as a nonblocking PBX system because the term does not apply to access to trunk circuits or other external peripherals that may have limited port capacity, i.e., VMS. Trunk traffic engineering is an independent discipline that will be discussed later in this chapter.

A PBX system is said to be blocking if traffic engineering is required, at maximum port capacity, to satisfy worse-case traffic demand situations. For example, a small/intermediate line size PBX system based on a switch network design consisting of one 16-Mbps (256 talk slots) TDM bus can appear to be nonblocking to customers with requirements of 100 station users and 30 trunk circuits because the total number of potentially active ports is smaller than the total number of available talk slots. If the total port requirements of the customer were to grow beyond 256 ports, e.g., 240 stations and 60 trunk circuits, some ports might be denied access (blocked) to the switch network because the number of active ports may be larger than the number of available talk slots. A PBX system with a blocking switch network design can operationally function as a nonblocking system if two conditions are satisfied:

1. The system is traffic engineered
2. There are sufficient switch network resources to satisfy actual customer traffic requirements

The typical PBX system is usually installed and configured with a number of equipped ports with significantly less than the maximum port capacity. The switch network resources of a blocking PBX system are usually sufficient to provide nonblocking access to the equipped system ports, but as customer port requirements approach maximum port capacity, the probability of blocking increases. The probability of blocked access to the switch network is based on the potential number of active ports (communications sources) and switch network resources required to connect a call.

The most important switch network resource determining the probability of a blocked call placed by a station port is the number of available talks slots on the local TDM bus. The local TDM bus is the most likely switch network element to have insufficient resources; talk slots, because most (if not all) PBX systems are based on switch network designs with sufficient Highway bus traffic capacity to support access to the center stage switch or connect local TDM buses. Most PBXs also are designed with a nonblocking center stage switching system complex, even if local TDM bus traffic capacity is limited. The Definity G3r is a good example of a PBX system that requires traffic engineered local

TDM buses, although the Highway bus/center stage switch complex used to link the local TDM buses supports nonblocked access across the internal switch network. If customer traffic requirements are light to moderate, a G3r local TDM bus (483 talk slots) can adequately support about 800 user stations. A Nortel Networks Meridian 1 Option 81C is typically configured with about 200 stations supported by a single Superloop (120 talk slots). Although the number of equipped ports is larger than the number of available talk slots, the two blocking PBX system designs are usually sufficient to support typical station user traffic demand. Customers with heavy traffic requirements would need to traffic engineer the Definity G3r/Meridian 1 Option 81C because the number of local TDM bus talk slots is smaller than the maximum number of ports, and the number of blocked calls could increase to an unacceptable level.

PBX Grade of Service (GoS)

PBX systems with blocking switch network designs are traffic engineered by the vendor when they are configured and installed based on customer traffic requirements. Customer traffic requirements are based on two parameters: required GoS level and expected traffic load. PBX GoS may be simply defined as the acceptable percentage of calls during a peak calling period that must be completed (connected) by the PBX switch network. Calls that are not completed, because the PBX switch network cannot provide the connection between the originating and destination endpoints, are known as blocked calls. The traditional method of stating a customer GoS level for a PBX system is to use the acceptable level of blocked calls instead of completed calls. The PBX's GoS level is stated with the symbol P, representing a *Poisson distribution*, although it is more commonly referred to as *probability*. For example, P(0.01) represents the probability that one call in 100 will be blocked. This is the same as saying that 99 percent of calls will be completed. P(0.01) is the most common GOS level used for PBX traffic analysis and engineering, although customers with more stringent traffic requirements may require a GoS level of one blocked call in 1,000, or P(0.001).

The GoS level is applied during the peak call period, which is typically 1 hour. In traffic engineering analysis, this peak call period is known as the *Busy Hour*. The Busy Hour for most PBX customers usually occurs during the mid-morning or mid-afternoon hours, although the exact time of day will differ from customer to customer. The GoS at Busy Hour, a

Legacy PBX Switch Network Design

worse-case traffic situation, is a unit of measurement indicating the probability that a call will be blocked during peak traffic demand. There are numerous methods used to find the Busy Hour. A common method is to take the 10 busiest traffic days of the year, sum the traffic on an hourly time basis, and then derive the average traffic per hour.

If a customer does not have access to traffic data over a long period, there is a simple method to estimate Busy Hour traffic load based on daily traffic load. Busy Hour traffic for a typical 8-hour business operation is usually 15 to 17 percent of the total daily traffic. Traffic usually builds up from the early morning to mid-morning, declines as lunch hour approaches, builds up again after lunch hour to mid-afternoon, and then declines toward the end of the business day. Traffic during off-business hours is usually very light, but a 24/7 business is likely to have very different traffic patterns from a business keeping traditional 9 to 5 hours.

The Busy Hour analysis must take into account seasonal variations in customer PBX traffic demand, such as the pre-Christmas holiday period. Although the average hourly PBX traffic load may be significantly less than the Busy Hour, and early Monday morning traffic is usually less than Wednesday mid-afternoon traffic, the worse-case situation is used for traffic engineering purposes. PBX switch network resources cannot be increased and decreased for fluctuations in traffic during the day, week, month, or year.

Defining PBX Traffic: CCS Rating

PBX traffic load is generally measured in 100 call-second units known as Centum Call Seconds (CCS). *Centum* from Latin, signifies 100. The maximum traffic load per station user during the Busy Hour is equal to 36 CCS, which is a shorthand method of stating 3600 seconds. Thirty-six CCS is equivalent to 60 minutes, or 1 hour of traffic load. A station port (telephone, facsimile terminal, modem, etc.) that "talks," or connects, to the switch network for 10 minutes during 1 hour has a traffic rating of 6 CCS (10 minutes = 600 seconds = 600 call-second units). Combining the station user traffic load with an acceptable GoS level results in the following station user traffic requirement: 6 CCS, P(0.01). This notation signifies that a station user with an expected 6 CSS traffic load is willing to accept a 1 percent probability of call blocking when attempting to use the switch network. A 2 percent blocking probability would be expressed as 6 CCS P(0.02); a 0.1 percent blocking probability would be expressed as 6 CCS P(0.001).

A traffic rating of 36 CCS P(0.01) is used for station users who require virtually nonblocking switch network access. A 36 CCS traffic load is a worse-case situation because it is the maximum station user traffic load during the Busy Hour. The usual station user traffic rating requirement is about 6 to 9 CCS, P(0.01). Although a station user might be on a call that lasts for 1 hour or more—a 36 CCS traffic load—there is a very small probability all station users are simultaneously engaged in calls of at least 1 hour during the same Busy Hour. It is far more likely that an individual station user will have a 0 rather than a 36 CCS, traffic load during Busy Hour because that person may be in a meeting, traveling, on vacation, or too busy with paperwork to take or place telephone calls. Even if a station user makes several calls per hour, it is possible that each will be of short duration because many calls today are answered by a VMS with limited available time to leave a message. Most business-to-business calls today are connections between a station user and a VMS, and each of these calls typically last for less than 2 minutes and many last for less than 1 minute. An increasing number of callers no longer leave messages; they disconnect and send an e-mail.

The total PBX station traffic load during Busy Hour is simply the sum of the individual station user traffic requirements. If ten station users are connected to the network for 10 minutes during the same hour, the total traffic load on the switch network would be 60 CCS (10 station users $\times$ 10 minutes/station user, or 10 $\times$ 6 CCS). If the probability of blocking level was 1 percent, the traffic requirement would be noted as 60 CCS P(0.01). The total PBX station traffic load is rarely calculated, however, unless the switch network design is based on a single TDM bus or switch matrix. PBX traffic loads are better calculated for groups of station users sharing access to the same switch network element, assuming station users with similar traffic requirements are grouped together.

For switch network traffic engineering calculations, most customers use an average traffic load estimate to represent all station users instead of segmenting the station user population into like traffic load requirements. It is recommended that a different approach be used to traffic engineer a PBX system. Station users should be segmented into different traffic rating groups to ensure that switch network resources are optimized for each category of station user. In every PBX system there are some station users with very high traffic rating requirements, such as attendant console operators. Other station port types with very high traffic rating requirements include ACD call center agents, group answering positions, voice mail ports, and IVR ports. Each station port typically will have a 24 CCS traffic load, although customers usually prefer these ports to have nonblocking [36 CSS P(0.01)] switch network access and state so in their system requirements. Averaging the high traffic, moderate traffic, and low traffic station ports will result in a traffic engineered system that blocks an unacceptable percentage of calls for attendant positions because a rarely used telephone in the basement is using switch network resources instead of more important user stations.

As an example, a Nortel Networks Meridian 1 Option 81 C, based on a 120 talk slot Superloop local TDM bus design and a port carrier shelf that can typically support 384 ports, should be configured as follows to satisfactorily support the following station user traffic groups:

1. A maximum of 120 very high traffic station users, 36 CCS, P(0.01): stations configured on a single port carrier shelf supported by a dedicated Superloop bus
2. About 250 moderate traffic station users, 9 CCS, P(0.01): stations configured on a single port carrier shelf supported by a dedicated Superloop bus
3. About 500 low traffic station users, 4 CCS, P(0.01): configured across two port carrier shelves supported by a dedicated Superloop bus.

A single Superloop bus can adequately support each traffic group in this example, although the number of station users differs across the group categories. If the maximum number of potential traffic sources, or station users, is no larger than 120, then the Superloop bus is rated at 3,600 CCS, P(0.01). This is the maximum traffic handling capacity of a Superloop bus. The Superloop bus is rated at slightly less than 3,000 CCS, P(0.01), if the port carrier shelf is configured for about 256 station users, according to the original Meridian 1 documentation guide. If the number of potential traffic sources increases, then the traffic handling capacity decreases for a given probability of blocking level. The exact traffic rating for a specific number of station users is available with the use of a computer-based Meridian 1 configurator. Figure 4-1 illustrates CCS traffic handling capabilities of a Meridian 1 Superloop with 120 available talk slots. Customers with very high traffic requirements can configure a single Meridian 1 IPE shelf with up to four SuperLoops. Each SuperLoop is dedicated to four port card slots. Figure 4-2 illustrates how a port carrier shelf can be segmented.

Figure 4-1
Meridian 1 Superloop traffic handling capability.

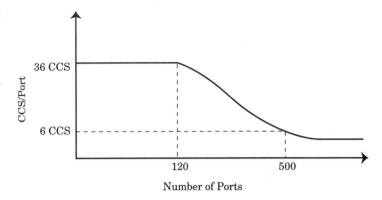

Figure 4-2
Meridian 1 IPC module Superloop segmentation.

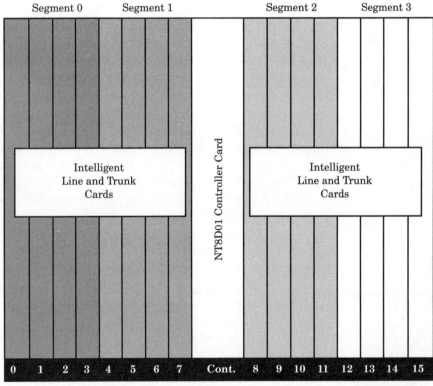

Source: Nortel Networks

Legacy PBX Switch Network Design

Traffic handling capacities for any PBX system local TDM bus are comparable in concept to Meridian 1 Superloop bus ratings:

1. If the number of potential traffic sources is smaller than or equal to the number of available talk slots, then station traffic can be rated at 36 CCS (nonblocking switch network access).
2. If the number of potential traffic sources is larger than the number of available talk slots, then the station traffic rating is less than 36 CCS. The traffic rating will decrease as the number of potential traffic sources increases.

The traffic handling capacity of the local TDM bus declines according to an exponential equation used to calculate probability of blocking levels. Most PBX designers assume a Poisson arrival pattern of calls, which approximates an exponential distribution of call types. The exponential distribution is based on the assumption that a few calls are very short in duration, many calls are a few minutes (1 or 2 minutes) in duration, and calls decrease exponentially as call duration increases, with a very small number of calls longer than 10 minutes. The actual traffic engineering equations (based on queuing models), call distribution arrival characteristics, and station user call attempt characteristics determining the local TDM bus traffic rating at maximum port capacity (if switch network access is not nonblocking) are known only to the PBX manufacturer.

Regardless of the actual traffic engineering equation used by the manufacturer, the calculated traffic rating will be based on three inputs:

1. Potential traffic sources
2. Available talk slots
3. Probability of blocking

A basic assumption used for most traffic analysis studies is a random (even) distribution of call arrivals during the Busy Hour. Traffic analysis studies also must make an assumption about call attempts that are blocked:

1. Station users who encounter an internal busy signal on their first call attempt continue making call attempts until they are successful.
2. Station users who encounter an internal busy signal on their first call attempt will not make other call attempts during a certain period.

In reality, station users who receive a busy signal will immediately redial. The assumption that a station user will not make another call attempt, if the first attempt is unsuccessful, is not realistic. The redial scenario is the assumption used by Poisson queuing model studies. The Poisson queuing model assumes that blocked calls are held in the system and that additional call attempts will be made until the caller is successful. For this reason, Poisson queuing model equations are commonly used by PBX traffic engineers to calculate internal switch network traffic handling capacities.

PBX systems with complex switch network designs (multiple local TDM buses, multitier Highway buses, center stage switch complexes) are far more difficult to analyze and traffic engineer than small PBX systems with a single local TDM bus design. Large, complex PBX switch network designs can provide a traffic engineer with many different switch connection scenarios that must be analyzed. Switch connections across local TDM buses require analysis of at least two switch network elements per traffic analysis calculation. To simplify traffic engineering studies, it is common system configuration design practice to minimize switch connections between different TDMs by analyzing call traffic patterns among stations users and providing station users access to trunk circuits on their local TDM buses. Centralizing trunk circuit connections may facilitate hardware maintenance and service, but it degrades system traffic handling capacity if more talk slots are used per trunk call.

Trunk Traffic Engineering

The number of PBX trunk circuits required to support expected inbound and outbound traffic loads is typically calculated using trunk traffic tables. The most popular trunk traffic table used for telephone system traffic engineering is based on the Erlang B queuing model. The Erlang B model assumes the following:

1. The number of traffic sources is large
2. The probability of blocking is small
3. Call attempts are random
4. Call holding times are exponential
5. Blocked calls are cleared from the system

The last assumption is very important because it says that there is no second call attempt if the first attempt receives a busy signal. The

Legacy PBX Switch Network Design

Poisson queuing model used for PBX station traffic engineering assumes that blocked call attempts are held in the system; that is, subsequent call attempts are made. Another popular telephone system queuing model is Erlang C, based on the assumption that blocked call attempts are held in a delay queue until a trunk is available. The Erlang C model is commonly used in ACD systems to calculate required agent positions used instead of a trunk circuit: inbound calls not immediately connected to an agent position are held in queue until an agent is available.

Another use of the Erlang C model is to calculate the required number of attendant positions to handle incoming trunk calls. Calls not presented to the attendant position are queued by the PBX system. Based on incoming traffic conditions, the average 250-station PBX system may require one, two, or three attendant positions to adequately answer and forward calls with acceptable queue delay times. As the PBX system size increases, the number of attendant positions is likely to increase, but the number of incremental attendant positions does not double when station size doubles because larger attendant position groups are more efficient than smaller groups based on traffic queuing theory conditions.

Erlang B is also a very useful queuing model for analyzing alternate routing on trunk groups within a PBX, where there are usually multiple available trunk circuits across multiple trunk groups. A call that is blocked at one trunk circuit can potentially overflow to another circuit or another trunk group. Erlang B is also used for analyzing traffic conditions across multiswitch networks, where there are many potential call routes per connection.

The Erlang B trunk traffic table consists of three data parameters: probability of blocking, number of trunk circuits, and Erlangs. An Erlang is a unit of measurement for trunk traffic. The maximum traffic load a trunk circuit can handle in 1 hour is equal to 1 Erlang. An Erlang is a dimensionless unit of measure. Knowing any two of the three data parameters allows table look-up of the third data parameter. For customers with existing PBX systems, it is easy to determine the current trunk traffic handling capacities per trunk group because the GoS is a given, as is the number of trunk circuits per trunk group.

The Erlang B trunk traffic table shown in Table 4-1 indicates that five trunk circuits can carry 1.316 Erlangs of traffic with a 1 percent probability of blocking. Ten trunk circuits can carry 4.462 Erlangs, P(0.01), and 15 trunk circuits can carry 8.108 Erlangs, P(0.01). Note that the number of Erlangs is the total traffic handling capacity of 5, 10, or 15 trunk circuits. The Erlang parameter in the table is not the traffic rating per individual trunk circuit. Another important thing to note is that, as the number of trunk circuits per group increases, the traffic per circuit also increases:

5 Trunk circuits = 1.316 Erlangs = 0.2632 Erlangs/trunk circuit
10 Trunk circuits = 4.462 Erlangs = 0.4462 Erlangs/trunk circuit
15 Trunk circuits = 8.108 Erlangs = 0.5405 Erlangs/trunk circuit

TABLE 4-1

Erlang B Traffic

N	P					
	.003	.005	.01	.02	.03	.05
1	.003	.005	.11	.021	.031	.053
2	.081	.106	.153	3.224	.282	.382
3	.289	.349	.456	.603	.716	.9
4	.602	.702	.87	1.093	1.259	1.525
5	.995	1.132	1.361	1.658	1.876	2.219
6	1.447	1.622	1.909	2.276	2.543	2.961
7	1.947	2.158	2.501	2.936	3.25	3.738
8	2.484	2.73	3.128	3.627	3.987	4.543
9	3.053	3.333	3.783	4.345	4.748	5.371
10	3.648	3.961	4.462	5.084	5.53	6.216
11	4.267	4.611	5.16	5.842	6.328	7.077
12	4.904	5.279	5.876	6.615	7.141	7.95
13	5.559	5.964	6.608	7.402	7.967	8.835
14	6.229	6.664	7.352	8.201	8.804	9.73
15	6.913	7.376	8.108	9.01	9.65	10.63

The theoretical maximum traffic handling capacity per trunk circuit is 1 Erlang. This traffic rating will be approached as the number of trunk circuits per trunk group continues to increase. Large trunk groups are more efficient than small trunk groups for this reason; a smaller incremental increase of trunk circuits is required to support increased traffic requirements at the same GoS. Very small trunk circuit groups have minimal traffic handling capabilities and should not be configured in a PBX system unless the number of station users with trunk group access is very small and the traffic load is very light. Very small trunk groups are usually configured for private line connections between two PBX systems. I/O traffic over local telephone company trunk circuits is usually supported by large trunk groups shared by many station users to optimize trunk circuit traffic handling capacity.

Legacy PBX Switch Network Design

Table 4-2 lists the many different types of traffic models that can be used to determine trunk circuit requirements. The model one uses to determine trunk circuit requirements is based on the call scenario and traffic pattern. Figure 4-3 illustrates how a call can be handled if there are no available trunk circuits: cleared, held, or delayed. Random and rough traffic patterns, based on the distribution of calls within 1 hour, are illustrated in Figure 4-4. Note that the random traffic pattern is a tighter distribution of calls than the rough pattern and more closely resembles a symmetrical bell-shaped curve.

TABLE 4-2 Traffic Models

Traffic Model	Sources	Arrival Pattern	Blocked Call Disposition	Holding Times
Poisson	Infinite	Random	Held	Exponential
Erlang B	Infinite	Random	Held	Exponential
Extended Erlang B	Infinite	Random	Retried	Exponential
Erlang C	Infinite	Random	Delayed	Exponential
Engset	Finite	Smooth	Cleared	Exponential
EART/EARC	Infinite	Peaked	Cleared	Exponential
Neal-Wilkerson	Infinite	Peaked	Held	Exponential
Crommelin	Infinite	Random	Delayed	Constant
Binomial	Finite	Random	Held	Exponential
Delay	Finite	Random	Delayed	Exponential

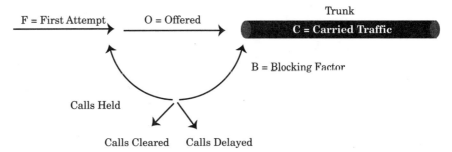

Figure 4-3 How a call is handled when there is no available trunk circuit.

Lost calls cleared (LCC)—Give up on a busy signal.
Lost calls held (LCH)—Redial on a busy signal.
Lost calls delayed (LCD)—Sent somewhere else when busy.

Figure 4-4
Random and rough traffic patterns based on the distribution of calls within one hour.

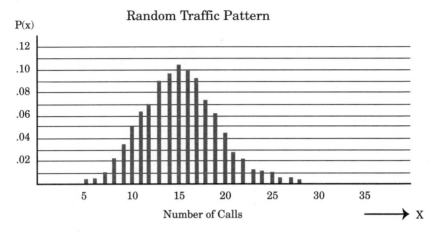

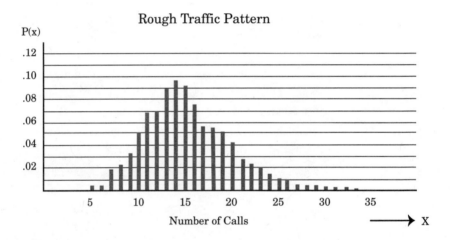

Determining Trunk Circuit Requirements

There are several trunk traffic engineering steps to calculate the required number of trunk circuits needed to satisfy I/O traffic loads at an acceptable GoS level:

1. Collect and analyze existing trunk traffic data
2. Categorize trunk traffic by groups
3. Determine the number of trunk circuits required to meet traffic loads

Legacy PBX Switch Network Design

4. Determine the proper mix of trunk circuit types

Trunk traffic data can be obtained from trunk traffic reports based on CDR data collected by the PBX system. The CDR data is input into a call accounting software program, available from the system manufacturer or third-party software vendor, that generates a variety of billing, internal switch network traffic, and trunk traffic reports. The CDR data does not provide information on calls that were blocked because all trunks were busy. This information is usually available from facility management reports based on optional PBX software programs. Blocked call data is used for determining GoS levels.

Historical trunk traffic data is used to forecast future trunk traffic loads to determine incremental trunk circuit requirements for the following scenarios:

1. Station user growth or contraction
2. Anticipated changing traffic patterns
3. New applications, e.g., centralized VMS

Trunk traffic should be segmented across different types of trunk groups because it is more cost effective to traffic engineer smaller groups of trunk circuits with a common purpose. The first step is to segment trunk traffic into inbound and outbound directions. There are a variety of trunk group types for each traffic flow direction. For example, inbound traffic may be segmented across local telephone carrier CO and DID trunk circuits, dedicated "800" trunk circuits, FX circuits, ISDN PRI trunk circuits, and so on. Outbound trunk circuits are easily segmented into local telephone carrier CO trunk circuits, multiple interexchange carrier trunk circuits used primarily for long distance voice calls, data service trunk circuits, video service trunk circuit circuits, and so on. There is also a variety of private line trunk circuits for PBX networking applications, OPX and other trunk circuits used to support remote station users, and trunk circuits connecting to IVRs and other peripheral systems. Each trunk circuit category can also include several subtrunk groups.

To determine the number of trunk circuits per group trunk type, the traffic load must be calculated. If CDR data reports provide trunk traffic measurements in terms of seconds or minutes, the results must be expressed in terms of hours to determine how many Erlangs of traffic are carried over the trunk circuits to use the trunk traffic tables.

When using the CDR reports to calculate Erlang traffic ratings, it is important to account for call time not tracked by the CDR feature. In addition to the length of a conversation over a trunk circuit, trunk circuit holding time exclusive of talk time includes call set-up (dialing and ringing) time, call termination time, and the time trunk circuits are not available to other callers during busy signal calls and other noncompleted calls (abandons, misdials) that are not recorded and stored by the PBX system's CDR feature. The missing CDR data time is usually calculated by adding 10 to 15 percent to the length of an average call. For example, if the total number of trunk calls is 100 and the total trunk talk time is 300 minutes, the average call length is 3 minutes. With a 10 percent missing holding time factor, an additional 18 seconds per call (3 minutes, or 180 seconds $\times$ 0.1) should be added to the 3-minute average talk time per trunk call. The 10 to 15 percent fudge factor is important and necessary to correctly determine trunk circuit requirements to maintain acceptable GoS levels.

CHAPTER 5

Legacy PBX Traffic Engineering Analysis

The call processing system of a PBX is responsible for all control functions and operations. It is responsible for monitoring and supervising all system design elements, including the switching network, printed circuit boards, and peripheral equipment, including station terminals and trunk circuits. It is also responsible for basic call set-up and teardown, feature/function provisioning, and systems management and maintenance operations. The PBX system call processing design has evolved from a single processing element responsible for all control functions to many processing elements, each with very specific responsibilities and functions. The PBX call processing design may differ from system to system, but there is a similar set of functional elements and responsibilities that is common to all enterprise voice communications systems, regardless of system size or functional complexity. Figure 5-1 illustrates the main PBX processing elements in a typical configuration design.

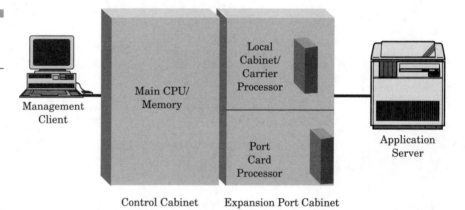

Figure 5-1 Main PBX processing elements.

Common Control Complex

The PBX common control complex can best be described as the brain of the system. Although other systems and subsystems may be responsible for physical operations, such as switch connections and voice signal transport, the common control complex is the command center responsible for issuing orders and supervising operations. There are several main components in a typical PBX common control complex, including:

- Main System Processor
- Main System Memory

Legacy PBX Traffic Engineering Analysis

- System Control Interfaces
- I/O Interfaces

The common control complex can be a single printed circuit board containing all of the listed common control elements, or it can be individual printed circuit boards for processor elements and interfaces and dedicated memory storage elements, such as hard disk or tape drives. The Main System Processor and Main System Memory are the two core elements in the common control. The System Control Interface and I/O Interface provide access to the two main common control elements for other internal system components and external system devices. The System Control Interface provides an intelligent link to the switch network (TDM bus) and processor bus to monitor port circuit activity and pass signals and messages between the main processor and local processors. The System Control Interface also can be used to support external call processing elements or adjunct systems dependent on the PBX common control complex. Examples are a CTI applications server used in call contact centers, and a third-party VMS. The I/O Interface ports typically are used for systems management, maintenance, and reporting functions. Examples are systems management terminals and call accounting systems. Figure 5-2 illustrates the main design elements of the common control complex.

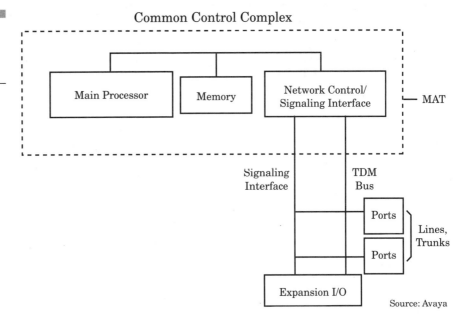

Figure 5-2
PBX common controls and interfaces.

Source: Avaya

Main System Processor

The Main System Processor is responsible for all call processing activities. It may also be responsible for maintenance and diagnostics activities, although some PBX systems are designed with a dedicated processing element for this function. The Main System Processor executes high-level call processing functions based on computer programs stored in system memory, monitors and controls all port-to-port connections, provides status indications to station users, and initiates the operations necessary to implement system features and functions.

The Main System Processor in current PBX systems is typically a 32-bit microprocessor chip from an outside supplier. Intel and Motorola have been the primary suppliers of microprocessor chips used as Main System Processors during the past decade. Several leading PBX systems are currently using Pentium-level microprocessors, although older microprocessors, such as the Intel 386/486 or Motorola 68030/40 chips, are used in many currently marketed systems. Many installed PBXs still operate on older 8-bit or 16-bit microprocessor technology platforms, proof of the long life cycle viability of traditional PBX system common control design.

The main processing element is an important factor for determining PBX call processing power, but it is not the only factor. Software code, call processing system design, and feature/function implementation play important roles in determining the so-called horse-power of a PBX system, known as Busy Hour Call (BHC) capacity (see below). The feature/function capabilities of a PBX system are also relatively independent of the main processor element because the generic software feature/function program in many current PBX systems is processor independent. It must be emphasized that using the latest generation microprocessor chip does not automatically guarantee a PBX system high call processing capacity or advanced feature/function provisioning.

Operating System Platform

The first PBX call processing system designs were based on proprietary operating systems before commonly used platforms such as Windows were developed. The operating system required to support a circuit switched PBX system must meet stringent, real-time, multitasking demands. A PBX system may need to support hundreds, maybe thousands, of simultaneous conversations, with each station user potentially activating a variety of features before or during a call. The real-time nature of PBX-based voice communications requires an operating system designed for its unique operations and features. An operating system such as Windows is not ideally suited for circuit switched, real-time call processing applications. Even today, many years after Windows has become the most popular server operating system for enterprise system applications, most PBX systems use a proprietary operating system. A few small PBX system models designed for customers with fewer than 200 stations are based on a Windows NT or Windows 2000 operating system platform, but there are no announced plans to use a Windows operating system for a large system model based on a circuit switching design.

AT&T was the first PBX manufacturer to use a version of an industry standard operating system, UNIX, when it introduced its System 85 family in 1983. The proprietary operating system developed by AT&T in the early 1980s, known as Oryx/Pecos, is still used by Avaya, the AT&T/Lucent Technologies spin-off, in its current intermediate/large Definity PBX models. In keeping with the company's tradition of innovative operating system platforms, Avaya's small system Definity One and IP600 models were the first circuit-switched PBXs to run on a Windows 2000 operating system. The Avaya PBX system platform will be based on a client/server platform using a version of Linux as its operating system and targeted at customers with significant IP telephony requirements.

Alcatel used a version of the UNIX V operating system, known as Chorus, for its 4200/4400 PBX models during the mid-1990s and continues to use it as the foundation for its new OmniPCX 4400. Nortel Networks began using a UNIX derivative, VX Works, for its Meridian 1 system in the early 1990s, and has successfully migrated its software and operating system to its new Succession CSE 1000 client/server IP-PBX system design.

Mitel was the first traditional PBX system manufacturer to use Windows NT server for a circuit switched system design, but recently changed to a VX Works platform for the Ipera 3000, the latest upgrade of its server-based call processing design that supports the traditional circuit switched SX-2000 peripheral equipment cabinets.

The operating system of a traditional PBX system provides services and system resource allocation to the call processing, feature/function, administration, and maintenance software programs. The operating system coordinates all system processing elements and controls CPU bus activities. An operating system program passes signaling and information between high-level programs running on the Main System Processor and lower-level programs running on localized processors (cabinet carrier and/or port circuit level). Other important operating system programs are used to support maintenance and fault processing programs, mass storage (customer and system database) programs, file management system programs, and I/O programs supporting external devices, such as printers, modems, and alarms.

Basic Call Processing Functions

The primary call processing responsibilities of the Main System Processor are provisioning of dial tone, digit reception and analysis, number analysis, TDM bus talk slot assignments and switch connections for intercom and trunk calls, routing analysis, feature provisioning, and call monitoring. A more detailed discussion of circuit switching functions included in Chapter 3, PBX Switching and Transmission Design. A listing and description of basic PBX features and functions is included in Appendix A, Feature/Function Descriptions.

There are several fundamental main processor management functions used to process calls:

1. Call sequencing control: management of the call sequence logic that takes a call from one state to another
2. Resource management: management of various system resources, such as DTMF receivers; time/talk slots for call connections; tone generators; and internal software records for call processing (including the system dial plan), messaging, measurements, and call detail records
3. Terminal handling: management of different desktop terminal models, including support of line appearances, feature buttons, display fields, adapter modules, and other functional components
4. Routing and termination selection: controls the selection of the terminating endpoint (station, trunk) of the call, including functions such as hunting, bridging, call coverage, and least cost routing.

Call processing is a series of events that result in the completion of a call. Figure 5-3 illustrates the call dialing and connection process. The process begins when a port (station or trunk) changes from an idle to an active state. *Port seizure* occurs when a station goes off-hook, a trunk port circuit receives an incoming call signal from the Central Office or network, or an attendant begins dialing. When a station port has been

Legacy PBX Traffic Engineering Analysis

seized, the main processor seizes a register storage record, instructs a tone sender unit to send dial tone to the caller, and instructs the switch network to establish a connection to the port. When a CO or DID trunk port has been seized, the main processor seizes a register storage record, assigns a tone receiver unit register to the port, and instructs the trunk circuit port to signal that the main processor is ready to receive digits.

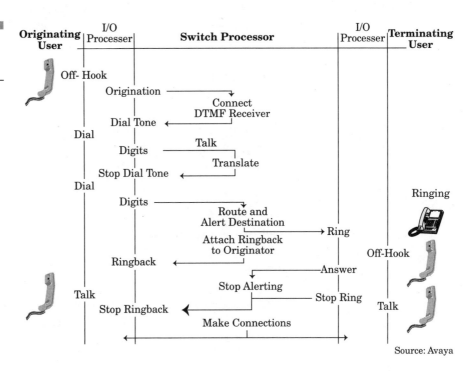

Figure 5-3
Call dialing and connection process.

Source: Avaya

The second step in the call process is digit reception and analysis. Digits are sent by system users to dial another station or activate a specific feature. A call register stores dialed digits or received digits over a trunk circuit. Digits can be received one by one (manual dialing) or in a group if a dialed number has been preprogrammed—speed dial. For station initiated calls, the first digit received by the register suppresses the dial tone. There is typically a time-out period between going off-hook and dialing the first digit or between the first and second dialed digits; if no digits are dialed within the programmed period, an intercept tone is sent to the caller. In addition to seizing a register, a tone sender and receiver are simultaneously seized, a switch connection is made, and the Class of Service of the caller is checked. A no-progress tone is sent to the caller if a tone receiver cannot be seized.

The third step in the call process is number analysis. Number analysis allows the PBX system to identify the number dialed and properly route the call. Internal system number analysis is performed for all calls, both intercom and trunk calls, within the PBX system. Special number analysis processes are performed for DID trunk calls and dialed feature access codes using * and # keys. For incoming DID trunk calls, the main processor analyzes the digits to determine whether the number is an attendant console position. If the number is an attendant console position, the call is processed and routed to the attendant. If the number is not an attendant console position, the internal number analysis program analyzes digits.

The internal number analysis program determines the correct call processing procedure to be implemented. There are many types of dialed or received numbers that can be analyzed for call routing purposes: station number, including hunt group number; individual, group, or emergency attendant number; abbreviated dialing number (speed dial); paging number; automatic route selection access code; Direct Inward System Access (DISA) code; data station number; modem pool group access code; and external destination codes, such as 911. If a number is not defined in the number analysis program, it is treated as a vacant number, and the appropriate intercept treatment is applied.

Another important main processor function is provisioning of call progress tones and indications. Call progress tones and indications are single- and dual-frequency combinations applied in a variety of cadence patterns, such as continuous tone sending, one repeated sequences (tone, pause, tone pause...), two repeated sequence (tone 1, pause 1, tone 2, pause 2...), or three repeated sequences (tone 1, pause 1, tone 2, pause 2, tone 3, pause 3...).

Call progress tone types include:

- Busy tone
- Call diversion indication (call forward indication)
- Call waiting indication
- Conference tone
- Confirmation tone
- Dial tone
- Expensive route warning tone
- Intercept tone
- Intrusion tone
- Message waiting tone
- No-progress tone (congestion tone)

- Off-hook queue tone
- Recall dial tone
- Ringback tone
- Special dial tone (do not disturb tone)
- Special message waiting tone (message waiting)

Main System Memory

The main system memory component of the common control complex consists of several types of memory databases:

1. Generic program
2. Operating memory
3. Customer database

The *generic program* stores the main call processing program consisting of all operating instructions, provides the main processor element with necessary intelligence to perform the tasks required by the system, and executes continuous diagnostics, system measurements, and fault isolation routines. The generic program also includes all feature and function software codes in support of station- or system-initiated call processing features and functions, including the standard feature set and optional software packages.

The *operating memory* is also known as the working memory because it stores all data and information related to the real-time operating conditions of the PBX system, including port circuit status, switch network status (time/talk slot availability and usage), and status of activated features and related data.

The *customer database* memory contains all data and information related to station user profiles, terminal devices, and the system configuration. Customer database information includes customer programmed information such as class of service and restriction assignments; hunt, trunk, and call coverage group assignments; call routes and routing patterns; system dial plan; terminal button assignments; and system access passwords.

There is no standard PBX system memory storage supporting the three basic memory databases. Some PBX systems use a single memory storage element that is partitioned. Some PBX systems dedicate a memory storage element to each memory database or segment the generic program from the operating/customer database memory storage element. PBX systems typically use dynamic random access memory (RAM) for main memory storage. Electronic programmable read only memory (EPROM) might be used by older systems still in operation. Flash ROM is sometimes used in small systems to simplify customer database upgrades and shorten reboot time. Generic programs in small PBX system models typically require at least 24 Mbytes of RAM storage; very large models may require up to 256 Mbytes of RAM. Most PBX systems fall within this memory storage range.

A floppy disk drive unit is typically used to load resident software programs into the mass storage unit. Most current generation PBX systems use hard drives embedded on printed circuit boards; older systems used dedicated hard disk drive units. Other storage options are tape drives, Flash ROM, and magneto-optical drives.

Local Processors

The first generation of stored program control PBX systems was based on centralized call processing system designs. Call processing system designs evolved to include a variety of processing elements outside the common control complex but under the control of the Main System Processor. These processing elements are sometimes referred to as slave controllers or local processors. These processing elements may be used for a variety of functions, such as systems administration and maintenance; function-specific applications, such as messaging, or ACD routing, queuing, and reporting; local switch network access; and diagnostics. Small PBX system models usually centralize all call processing, administration, and maintenance functions within the common control complex, but intermediate/large system models may use dedicated processor elements to offload some call processing operations from the Main System Processor or dedicated processors to handle systems administration and maintenance services.

Port Circuit Card Microcontrollers

The most common local processor element is a microprocessor controller resident on a port circuit card. The primary functions of the on-board microcontroller are to pass control signals originating from the Main System Processor to the individual station/trunk circuits and provide a signaling link between the peripherals and the common control complex. The port circuit board microcontrollers function independently of one another and are responsible only for the port circuit terminations on the printed circuit card. The microcontroller has the primary responsibility for monitoring the status of its colocated port circuit terminations and peripherals. It also provides the processing intelligence for the physical link connections to the local TDM bus under the command of the main control complex and/or cabinet/carrier shelf processors. The very localized processing functions performed by the microcontrollers are generally considered mundane and repetitive but are necessary to support the call processing functions of the main control complex. Localizing some processing operations at the level of the port circuit card reduces the processing load of the main system processor and increases the overall system call processing capacity potential.

The AT&T System 75 integrated the first port circuit card microcontroller into a PBX system call processing design in 1984. Today it is a standard port circuit card design element. The current microcontrollers are not based on the latest processor platforms but on older 8-bit, 16-bit, or 32-bit microprocessors. All port terminations on the printed circuit card depend on the local microprocessor, and processor failure or error will result in loss of service. No PBX system design offers a redundant microcontroller design option because service loss is limited to a small percentage of total system ports. For this reason, it is sometimes prudent for a customers to distribute vital peripheral resources across two or more port circuit cards.

Cabinet/Carrier Shelf Local Controllers

Many, but not all, intermediate/large PBX systems have local control cards that support one or a group of port carrier shelves housed in a port equipment cabinet. The primary functions of the local control cards are passing control signals from the Main System Processor to the port circuit cards and providing a signaling link between the peripherals (via the port circuit card microcontrollers) and the main control complex. It effectively analyzes, controls, and supervises the port circuit cards and peripherals at the cabinet or carrier shelf level. Other common functions of the local controller are local TDM bus talk slot assignments and supervising and monitoring local switch network connections and voice signaling transport. Localizing the switch network access function can greatly reduce the call processing load on the Main System Processor. The reduced processing load can be especially significant for PBX systems with a distributed or dispersed switch network design. Local controllers can also perform a variety of diagnostics and circuit test operations such as passing status updates to the main control complex.

Local controllers, like the port circuit card microcontrollers, do not control call processing functions but merely execute operations under the supervision of the Main System Processor. The type of processing element used as a local controller differs from system to system. Based on the localized and limited role it has in the call processing operation, it is not necessary for the processing element to be a current microprocessor platform; many PBX systems have retained the same local control boards for more than 5 years. The local controllers perform their tasks based on firmware programs resident on the printed circuit board.

Some PBX system call processing designs offer fully duplicated local controller function as a standard or optional system attribute. In an intermediate/large line size system, local controller card problems sometimes can affect hundreds of system ports. Processor failure or error will result in loss of service to all ports supported by the local controller, unless a back-up controller is available.

System Processing Bus

The PBX system processing bus functions as the data transmission path between the common control complex and the dispersed local processors. A dedicated bus interface control card may be used to provide bus access and control monitoring for the Main System Processor and local processors. A few PBX systems have dedicated processing buses that link the individual port circuit card microcontrollers to the higher-level call network, but many systems continue to use several reserved time slots on the circuit switched network TDM buses to transport control signals and messages.

The operating transmission rate of a PBX system processing bus is usually between 1 and 10 Mbps. A few PBX systems, including the Nortel Networks Meridian 1 and the Siemens Hicom 300H, have 10-Mbps system processing buses based on Ethernet communications signaling standards. The Hicom 300H even has a dedicated Ethernet Hub card housed in the common control carrier shelf that is used to link the various system processor elements, including the Main System Processor, Administration Processor, and external CTI applications servers.

PBX Call Processing Design Topologies

There is no PBX design standard that dictates the topology of the call processing network. Every PBX system has a common control complex that includes a Main System Processor, Main System Memory, and System Control and I/O Interfaces, but that may be the only common design element when comparing any two PBX system models. PBX call processing design topologies can be categorized into three general categories:

1. Centralized control
2. Dispersed control
3. Distributed control

These design topologies are shown in Figure 5-4.

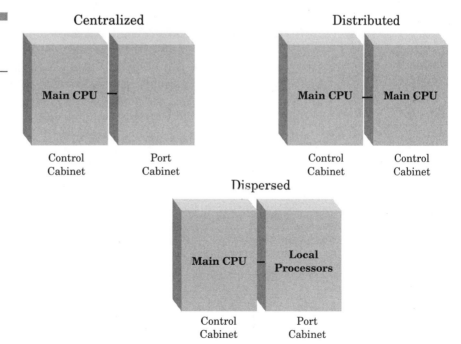

Figure 5-4
PBX processing design topologies.

A fourth design topology can be added to this list—adjunct server control—if PBX systems equipped with an adjunct CTI applications server are taken into consideration, but only for support of features and functions beyond fundamental call set-up and connection functions. A CTI application server is totally dependent on the PBX common control complex to execute any and all communications operations. Adjunct server control design will be discussed separately later in this chapter.

Centralized Control

A PBX system with a centralized control call processing design topology includes the following processor elements:

1. Common control complex, including the Main System Processor
2. Port circuit card microcontrollers

In a centralized control design, the Main System Processor is responsible for all basic call processing functions and the control and execution of all switch network functions. This design specifically excludes local processors at the port equipment and/or port carrier shelf level. Port circuit card microcontrollers interface directly with the Main System Processor via a Service Control Interface.

A PBX based on a centralized control design may have additional processing elements to perform functions and operations not necessary to execute basic call processing functions, such as call set-up and switch connections, although they are necessary to ongoing call processing activities and system availability and survivability. The additional processing elements are typically used for administration, maintenance, diagnostics, and/or measurement operations. The Main System Processor usually handles some, or all, of these functions in small PBX systems with limited processing elements.

Centralized control designs are used most often by small PBX systems targeted primarily at customers with port requirements fewer than 200 stations, although there are a few notable exceptions. For example, all the Avaya Definity models are based on a centralized control design, including the very large G3r model, which can be equipped with more than 20,000 stations. PBXs based on a centralized control design must be equipped with a Main System Processor capable of handling all call processing and switch network functions, without diminished performance when the system is at maximum port capacity and all ports are active.

Legacy PBX Traffic Engineering Analysis

A centralized control design offers advantages and disadvantages. The major advantage of a centralized control design is fewer processor elements that can experience problems and disrupt service. Reduced local processor failures can increase overall system reliability and survivability. The major disadvantage is that the Main System Processor has total responsibility for all call processing and switch network functions, with no local processors to offset the processing load. Unless the Main System Processor is powerful enough to handle current and future call processing requirements, including support of new features and applications, there will be limitations on system performance levels. One solution to offload the call processing burden from the Main System Processor in a centralized control design is to install an adjunct CTI applications server in support of advanced feature and function needs that require significant processing power. There is a more detailed discussion of adjunct server control options later in this chapter.

Most PBX systems are configured at less than 50 percent port capacity, and the number of simultaneously active ports is almost always half that of the equipped ports. A Main System Processor should have little difficulty supporting the call processing requirements of 50 or 100 active ports, even if the processing element is a 16-bit or 32-bit microprocessor. Main System Processor problems can occur in large line size systems, with hundreds or thousands of active ports, if the main CPU is not equipped and designed to handle the potential traffic. When the Definity designers chose a centralized control design for the large and very large system models, a major design change from the older System 85, they also spent much time selecting the right processing element for the Main System Processor. The innovative selection of a Reduced Instruction Set Computing (RISC) microprocessor, the MIPS 3000, came at a time when all other PBX systems were based on a Complex Instruction Set Computing (CISC) microprocessor platform, such as the Intel 386 or Motorola 68030. A few years before Intel made its Pentium microprocessor commercially available, the MIPS 3000 was evaluated as the best available microprocessor in a centralized control design because it could handle the potential processing load at maximum equipped port capacities. The centralized control design resulted in call processing limitations for the Definity G3 models when configured for large, complex ACD-based call center installations. Definity is one of the few PBX systems that does not use an adjunct applications server to handle advanced ACD call analysis and routing functions, and the heavy processing load on the Main System Processor limits the system's optional Expert Agent Selection (EAS) skill assignment programming parameters when compared with PBX systems equipped with an adjunct server to offload the processing burden.

Dispersed Control

A PBX system with a dispersed control call processing design topology includes the following processor elements:

1. Common control complex, including the Main System Processor
2. Local control processors at the port equipment cabinet and/or shelf level
3. Port circuit card microcontrollers

In a dispersed control design the Main System Processor is responsible for all basic call processing functions but may not execute all call processing and switch network functions. Local controllers provide the interface link between the Main System Processor and port circuit card microcontrollers and perform some call processing and switch network functions under the supervision and monitoring of the Main System Processor. The local processor elements function as slave controllers under the Main System Processor, which functions as the master controller. Like a PBX system based on a centralized control design, dispersed control designs can include additional processing elements typically used for administration, maintenance, diagnostics, and/or measurement operations.

A dispersed control design is the most common call processing design for intermediate to very large PBX system models. The primary advantage of a dispersed control processing design is that it offloads processing activities from the Main System Processor to increase overall system call processing performance. Call processing capacity is less dependent on the Main System Processor in a dispersed control design than in a centralized control design. The primary disadvantage is that failure or errors at the local processing level can affect service for all ports under its control. Unless the local controller is available in a redundant or duplicated mode, it is a potential major single point of failure for dozens or hundreds of system ports. For example, the Nortel Networks Meridian 1 Option 81C Controller Card can support one or two port carrier shelves, typically equipped with several hundreds of station ports. If the Controller Card, responsible for some call processing and switch network functions, fails, each port will lose service. The Controller Card is not available in a duplicated mode and is a major point of failure in the

Meridian 1 Option 81C system design. Competing PBX system models from Siemens (Hicom 300H Model 80), NEC (NEAX2400 IMG), and Fujitsu (F9600 XL), for example, generally or optionally duplicate the local controller card at the port equipment cabinet/carrier shelf level to reduce the probability of service loss.

Distributed Control

A PBX system with a distributed control call processing design topology includes the following processor elements:

1. Multiple common control complexes, including multiple Main System Processors
2. Port circuit card microcontrollers

A distributed control design is based on peer-to-peer Main System Processors. It is similar in operation to a centralized control design because each Main System Processor is responsible for all basic call processing functions and the control and execution of all switch network functions. There is a major difference, however, in that each Main System Processor has control over a limited number of system ports. Each Main System Processor has responsibility for one or more port equipment cabinets but not all installed port equipment cabinets in the PBX configuration. This design excludes local processors at the port equipment and/or port carrier shelf level because the Main System Processor functions as a local processor to the one or two port cabinets it controls. In a distributed design port circuit card, microcontrollers interface directly with the Main System Processor via a Service Control Interface, if no local controllers are included in the design.

There are several important distributed control design advantages:

1. Multiple common control complexes increase system reliability and survivability; each Main System Processor is responsible for a limited number of system ports.
2. System call processing capacity is a function of the number of installed Main System Processors; adding an additional Main System Processor will increase the total system call processing capacity.
3. Multiple common control complexes are ideally suited for system configurations with multiple equipment rooms (campus, multilocation) because locations remote from the main equipment room are

not dependent on a centralized Main System Processor for call processing operations.

A distributed control design requires synchronization and coordination among the multiple common control complexes for call processing, switching, and administrative functions. A true distributed control design supports transparent features and function operations across the system, with the option of using a single administration and maintenance interface for the entire system. The design is a difficult one to develop and operate successfully. One of the first distributed control designs was attempted by Rolm in the early 1980s, and the technical problems in synchronizing the VLCBX's multiple Main System Processors and memory databases significantly delayed commercial availability of the product after its announcement. Rolm eventually solved the problems and was successful in marketing and selling its multinode CBX II 9000 system, a successor to the VLCBX, later in the decade.

Another early distributed control design that has been very successful and continues to be marketed and installed today is the Ericsson MD-110. First introduced in the early 1980s, the MD-110 was originally based on the Ericsson AXE 5 central office switching system design. Each LIM port equipment cabinet has a dedicated control complex; LIM cabinets communicate with each other over PCM-based FeatureLinks. In theory, a single MD-110 can be installed with more than 200 LIM cabinets and Main System Processors. LIM cabinets can be centralized or dispersed across multiple customer locations, with each cabinet dependent only on its local Main System Processor for all call processing functions. The MD-110 is currently the only circuit-switched PBX system based on a fully distributed Main System Processor design.

Call Processing Redundancy Issues

One of the most important call processing design attributes for a large number of customers is redundancy. The term *redundancy* can be an ambiguous term if not properly defined. For example, a PBX with dispersed local controllers can be characterized as a redundant call processing design, simply because failure of one local controller usually has no effect on the other local controllers. A localized failure that does not affect systemwide operations can be considered a form of redundant design because the loss of 100 ports due to local controller failure is not as catastrophic as

100 percent system loss should the Main System Processor fail. Loosely defined, all dispersed control call processing designs are redundant call processing designs. This definition, however, may not satisfy the reliability and survivability requirements of many PBX customers.

If call processing redundancy is more clearly defined by a customer as having a readily available back-up processor element should an active processor element fail, then redundancy requires a duplicate local controller for each local processor element in the dispersed control design example. Duplication of processing elements is a high degree of redundancy.

Regardless of the call processing design category, a PBX system with a fully duplicated call processing design may include any or all of the following:

1. Fully duplicated common control complex, including the Main System Processor, the Main System Memory (software program, customer database), and the Mass Storage Device
2. Fully duplicated local controllers at the port cabinet and/or carrier shelf level
3. Fully duplicated processor bus, including intercabinet communications links

Although the reliability level of the typical PBX system common control complex is very high, usually 99.999 percent (about 5 minutes average annual downtime), hardware and software failures and problems can occur. If there is a problem with the Main System Processor, then all system operations and all system ports can be affected. For this reason, a duplicated Main System Processor is usually required by customers who wish to avoid even minimal service disruptions. In addition to the Main System Processor, customers may request duplicated Main System Memory elements, especially the generic program. A duplicate Mass Storage Device may also be requested because loss of customer database records will affect call processing operations to the same extant as Main System Processor failure.

In a PBX system with a duplicated common control complex, if the active Main System Processor or Main Memory experiences problems, then the back-up (passive) call processing element should instantaneously take control of system operations without interruption of service; active calls remain connected and all activated features continue operating. This is commonly known as a *hot-standby duplicated common control system*. The only call processing event that is disrupted when the passive element assumes control is a call in the process of being set-up, before call connection to the called party; otherwise, all system functions and operations continue as if nothing happened. If the passive Main System Processor assumes call processing control but all existing switch connections are lost, then it is said to be a *cold-standby duplicated control system*. A cold-standby system also may require a few seconds or minutes before it is available to begin new call processing operations.

The passive common control elements in a hot-standby design are said to be *shadowing* the activities of the active elements. For example, the active and passive Main System Processors monitor port status and switch connections, but only the active Main System Processor issues control commands for call processing operations. The passive Main System Processor is merely an observer. Downloads to the active main customer database are simultaneously downloaded to the passive database. Some duplicated common control system designs support operations between the back-up passive Main System Processor and the active Main System Memory and between the active Main System Processor and the back-up passive Main System Memory. This form of shadowing is known as *crossover arbitration*. Common control complexes with basic shadowing capability transfer all call processing and system operations to the passive processor and memory elements when any of the active common control elements fails; the more advanced shadowing design allows system operations between active processor or memory elements that do not experience problems and the back-up passive processor or memory element—active processor and passive memory or passive processor and active memory. The crossover operation allows for four modes of full function system operation:

1. Active elements (only one)
2. Passive elements (only one)
3. Mix of active and passive (two)

A duplicate common control complex is usually available only in intermediate/large PBX system models. Almost all intermediate/large PBX system models offer duplicated common control as a standard or optional capability. Small PBX system models have traditionally been designed without a duplicate common control complex, even as an option, because manufacturers originally decided that the added system cost to the customer would result in limited sales potential. Likewise, limited sales would not justify the research and development dollars expended for the design. Small system customers who require a fully duplicated common control complex are forced to buy a larger system model to satisfy that need. These customers pay a price penalty for installing a PBX system model with a greater than needed port capacity, because duplicate common control is not available in small system models better suited (and less costly) for their port capacity requirements. For example, the Avaya, Nortel Networks, Siemens, and NEC small system PBX models targeted primarily at customers with fewer than 200 stations are not available with duplicate common control as a standard or an option. Customers must step up to the larger models, sometimes two models above the entry model, if duplicate common control is a requirement. Avaya, Siemens, and NEC offer duplicate common control only as an option on their intermediate/large system models. The duplicate common control option may add as much as 25 percent to the basic system price for small system customers, in addition to the higher cost for the larger system model. Of the four manufacturers, only Nortel Networks offers duplicate common control standard on its intermediate/large system models (Meridian 1 Options 61C and 81C). Figure 5-5 shows the core module complex of the large Meridian 1 systems.

Figure 5-5
Meridian 1 option 61/81C core complex.

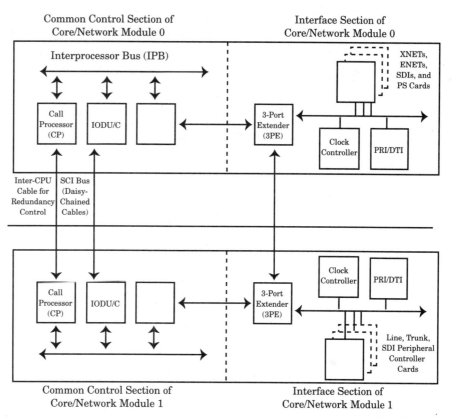

Source: Nortel Networks

Dispersed control designs with duplicated local controllers are available in many intermediate/large system models. Most, but not all, PBX systems using a dispersed control design with a duplicated common control complex capability also offer duplicated local controllers. For example, the Siemens Hicom 300H Model 80 can be equipped with duplicated common control and duplicated local control function as an option, but the Nortel Meridian 1 Options 61C and 81C with duplicated common control modules are not available with a duplicated Controller Card. It is possible to have duplicated common control but nonduplicated local controllers. PBX systems equipped with duplicated local controllers are usually available with duplicated common control.

The duplication of the processor bus and intercabinet links is another important redundant call processing system design capability. Processor bus problems can affect call processing operations the same way as Main System Processor or local controller problems. Duplicated processor bus design is inherent to particular PBX system models and is usually available with systems offering a duplicate common control complex. Intercabinet links are less likely to be fully duplicated as a standard design capability, and duplication of the links usually depend on installation of a duplicate common control complex and/or duplicated local controllers. Intercabinet links are part of the system design, whether the call processing design is centralized, dispersed, or distributed, because signaling and communications among the processing elements dispersed among multiple common equipment cabinets may depend on the links for a variety of call processing operations. A single link failure to the Main System Processor may affect hundreds and possibly thousands of system ports housed in the isolated port equipment cabinet. Intercabinet links in a PBX system with a distributed control design may not affect call processing operations in the isolated cabinet, but all intercabinet communications will be affected.

PBX Call Processing Power: BHC Rating

PBX system call processing capability is rated with the BHC benchmark. BHC is simply defined as the maximum of number of calls processed in 1 hour by the PBX system. There are two types of BHC measurements, Busy Hour Call Attempts (BHCAs) and Busy Hour Call Completions (BHCCs). BHCAs indicate the total number of placed calls that can be processed by the PBX system. BHCA calls include successful and unsuccessful calls. Unsuccessful call attempts include the following call types: no answer, busy, misdial, and abandoned. BHCCs indicate the only total number of successful placed calls. Measured call attempts and completions include station-originated calls and incoming trunk calls.

The BHCA rating of a PBX system will always be greater than its BHCC rating because unsuccessful calls included in the BHCA measurement have a lesser call processing burden than successful calls. The larger the number of unsuccessful call attempts, the greater the BHCA rating. BHCCs are successful calls that require a switched call connection, which is continually supervised and monitored by the call processing system, and require a teardown process when the call is terminated. PBX manufacturers may provide data on either BHC measurement but do not describe the benchmark tests used to determine the rating.

A manufacturer can test its PBX system to determine its BHCA or BHCC rating by using the system test parameter of provisioning a dial tone within a target period. Station users expect to receive dial tone immediately upon picking up their handsets, but some manufacturers perform BHC rating tests with acceptable dial tone delays of 1.5 seconds. In addition to dial tone delay parameters, BHC ratings are heavily dependent on the following three parameters:

1. System design configuration
2. Call connection type
3. Feature/function activity

The system design configuration defines the number and type of station terminals used for call testing purposes, the types of trunk circuits used for incoming or outgoing calls, and adjunct equipment that may be used to answer or support calls, such as VMSs and recorded announcers. Terminal type and complexity may have a major effect on the call processing rating. For example, a digital multiline telephone model with a softkey display field and add-on module options typically will require more processing power to place and receive calls than a basic analog telephone with no display or options. There are different call processing load factors associated with two-way analog trunks as compared with digital trunk circuits used for ISDN PRI services. An intercom call between two analog telephones is more likely to use less processing resources than an incoming ISDN PRI trunk circuit call, with ANI received by a digital displayphone.

The type of call connections determining the BHC rating will also affect the results. For example, trunk calls are usually more processing intensive than intercom calls. Calls between stations connected to the same local TDM bus may require less processing than calls between stations connected to different local TDM buses that require a center stage switched connection. Availability and physical location of incoming registers, outgoing registers, and sender tone receivers also will affect the BHC rating.

The call processing burden imposed by some PBX features and functions may be the most significant factor affecting BHC ratings. ACD and networking features require significantly greater processing resources than simpler features such as hold or transfer. ACD operations typically will require call screening, routing, queuing, and treatment operations before a connection is completed to an agent position. Intensive implementation of complex ACD features can reduce BHC ratings by factors greater than 50 percent. Figure 5-6 shows feature/function load factor effects on BHC rating.

Figure 5-6
Busy hour call processing.

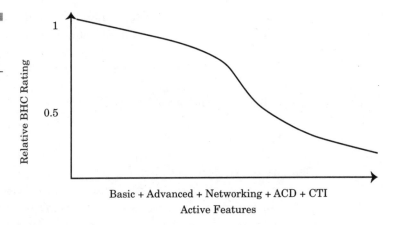

Networking operations typically require routing table look-up and analysis before trunk circuits can be selected and calls are routed. If a manufacturer's benchmark testing procedure does not include some type of feature/function activation and implementation during a reasonable percentage of placed calls, the resulting BHC rating will not mirror the reality of a customer's actual system installation.

The published PBX system BHC ratings, BHCA or BHCC, based on testing procedures may not adequately reflect the call processing capability of a customer-installed system configuration. The actual rating of an installed PBX system likely will be less than the published number. Fortunately for customers today, even if the true installed system rating is half of the published BHC ratings, the call processing capacity will be far greater than required by the customer. For example, small PBX system models from the leading manufacturers typically have call processing ratings in the range of 10,000 to 50,000 BHCC. Assuming a system configuration of 100 stations with associated trunking, the call processing rating of the system will far exceed the realistic call handling requirements. It is highly unlikely that, during the BHC period, all 100 stations will place 100 or 500 calls per hour. It is more likely that the total number of BHCs for a typical PBX system with 100 stations will be a few hundred.

Many of today's intermediate/large PBX systems have call processing ratings greater 100,000 BHCC, and some are greater than 500,000 BHCC. Only very large systems approaching maximum port capacity will come close to approaching the maximum call processing capacity of the system. If the PBX is used for ACD applications, customers may need to install a larger system model than necessary for its greater BHCC rating, but it is highly unlikely that call processing limits will be reached, except in extreme circumstances.

CHAPTER **6**

Legacy PBX Common Equipment

PBX common equipment can be divided into two main hardware categories: cabinets and printed circuit cards. There are many cabinet and card category types, and each PBX system model is designed and configured with the use of unique and proprietary hardware equipment. Neither cabinets nor cards are interchangeable between different PBX systems from different manufacturers. Proprietary common equipment hardware is currently the major reason the open design platform of new IP-PBX client/server systems is attracting the attention of many customers. Although the industry trend is toward reduced dependence on proprietary common equipment, the very large installed base of traditional PBX systems will guarantee the continued requirement for new equipment cabinets and port circuit interface cards for many years to come.

The PBX starter package is usually called the *main system assembly* and includes the necessary hardware elements and software for basic system operation. Small port size customers may require only a single cabinet equipped with all of the necessary equipment to support their station and trunk interface needs, but very large port size customers may require a dozen or more colocated or distributed cabinets. For any specific customer port size requirement, common equipment costs are relatively fixed regardless of feature or function requirements because the same core hardware components that support plain old telphone system (POTS) applications are required for advanced call center or networking applications. The advanced application configurations may include adjunct application servers, but the common equipment hardware supporting port station and trunk requirements is usually application independent. For example, desktop telephone instrument prices can differ by hundreds of dollars across a family portfolio, but cabinet and port interface circuit card costs supporting an inexpensive analog telephone are about the same as compared with an expensive ACD-based displayphone model with a headset interface. Figure 6-1 shows the basic PBX common equipment components and their relation to one another.

PBX common equipment installations can range from a wall-mounted small cabinet design supporting fewer than 100 stations to floor-based multiple cabinet designs supporting 20,000 station users, but each configuration provides the same fundamental call processing, switching, port interface, and applications support capabilities. Single-carrier cabinet designs are usually based on a common control and switching complex consisting of a limited number of multifunction printed circuit boards to conserve cabinet real estate and a few port interface card slots. The large system models may have one or more carriers dedicated to common control functions, dedicated switch network carriers, and

Legacy PBX Common Equipment

Figure 6-1
Basic PBX architecture.

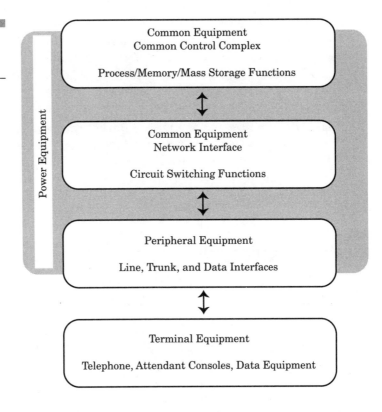

numerous expansion cabinets primarily supporting port interface circuit cards. The basic PBX operational elements are the same in the small and large models, although the common equipment design is significantly different.

System Cabinets

There are several types of PBX system cabinets, categorized by carrier size and function. A system cabinet may be a single-carrier or multicarrier cabinet. Most very small PBX models are based on a single-carrier cabinet common equipment design. The wall-mountable single-carrier cabinets are sometimes referred to as modules. Single-carrier cabinet designs are typically targeted at customers with port size requirements of fewer than 100 stations. The Avaya Definity One model is representative of this design type. Customers with port size requirements between 100 and 400 stations are usually supported by stackable carrier designs including a control carrier and a few expansion port carriers, like the Nortel Networks Meridian 1 Option 11C model. Some PBX manufacturers also use a stackable cabinet design for intermediate and large system configurations, like the NEC NEAX2400 IPX that can support port capacity requirements approaching 20,000 stations. Multi-carrier cabinet designs, which are not cost effective for small and intermediate line size customers, are fairly prevalent in the large and very large line size market. For example, the small/intermediate line size Hicom 300H Model 30 from Siemens is based on a stackable carrier design, but the larger Hicom 300H Model 80 is based on a six-carrier shelf cabinet design (Figure 6-2).

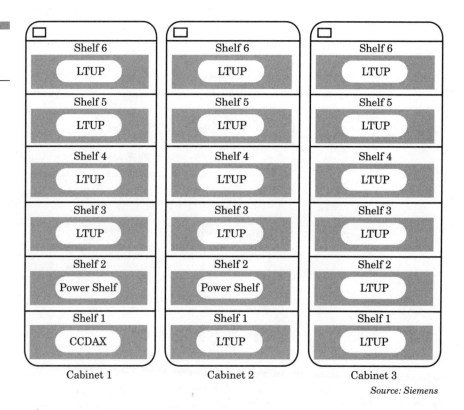

Figure 6-2
Multicabinet Hicom 300H Model 80.

Source: Siemens

Major cost benefits of the new client/server IP-PBX system designs are due to minimal telephony system cabinet requirements. A 100 percent IP desktop configuration may consist of a single server cabinet with integrated gateways for PSTN trunk access because port cabinets and port circuit interface cards are not required. The switching functions, however, are dependent on Ethernet switch carrier shelves or stackable carrier cabinets, but in almost all situations these common equipment components have been installed to support data communications applications. The migration from a traditional circuit-switched PBX design to the emerging IP-PBX designs will slowly reduce common equipment requirements dedicated solely to telephony applications.

The first generation of digital PBX systems was based on multicarrier cabinets. Multicarrier cabinet designs are still used for many large and very large PBX system models. A multicarrier cabinet typically has four to six carrier shelves. The primary cabinet, often referred to as the *control cabinet*, houses the common control complex, the center stage switch complex (if incorporated into the switch network design), and may also house port circuit cards. Expansion cabinets, often referred to as *port cabinets*, house the bulk of the port circuit interface cards. The typical number of printed circuit board slots is between 16 and 20 per carrier shelf. Although the number of carrier shelves and card slots has remained relatively constant during the past 15 years, port cabinet density has increased significantly because port circuit interface card density has increased. Several currently available expansion port cabinets can house between 60 and 100 port circuit interface cards, with each card typically supporting 24 port interfaces. Simple mathematics indicates that a single port cabinet can support between 1,000 and 2,000 stations with associated trunking. Port cabinets during the early 1980s were usually restricted to a few hundred ports.

The first stackable PBX system cabinet design was introduced by NEC in 1983. The original NEC NEAX2400 was an evolutionary design because it offered an cabinet design alternative to traditional multicarrier cabinets. A stackable cabinet design offers customers several benefits:

- **Optimizes common equipment requirements**—Before the stackable cabinet design, a customer might have installed a full-size multicarrier cabinet but use a fraction of the available carriers and card slots. Unless a customer had near-term growth requirements, too much hardware was installed to satisfy port requirements at system installation.
- **Reduces common equipment costs**—Stackable cabinets more cost effectively satisfy customer port size requirements.
- **Simplifies system upgrade and enhancements**—Customer port growth may be accommodated easily by adding another port carrier cabinet. Manufacturer design changes can focus on individual carriers,

so that customers can perform system upgrades by changing out one or two carriers, instead of replacing multicarrier cabinets.

The single-carrier cabinet design is most popular for PBX systems targeted at customers with very small to intermediate line size requirements, although a few manufacturers use the stackable cabinet design for their large and very large system models. NEC, the innovator of the stackable carrier cabinet design, uses stackable single carrier cabinets for its small/intermediate NEAX2000 IVS^2 system and its intermediate to very large NEAX2400 IPX models.

Carrier Types

A PBX cabinet or a single carrier may support one or more of the following functions.

Control

The control carrier contains printed circuit board slots that house specific control circuit cards or modules supporting processing and memory functions. A single circuit pack may include processor and memory chips. Additional circuit cards may include tone clocks, maintenance and diagnostic circuit packs, auxiliary equipment circuit packs, and a variety of expansion interface boards for system designs with multiple cabinets and/or stackable carriers. An expansion interface board may be used to link to a duplicated control carrier shelf in a fully redundant common control complex design or to other cabinet carriers, such as switching network, port, and application carriers. Tone clock and maintenance and diagnostic functions may be embedded in the processor/memory circuit pack in very small system designs. The control carrier shelf is equipped with one or two local power supply modules, if a centralized power supply carrier shelf is not included in the design.

Switching

The switching carrier contains printed circuit board slots that house switch network circuit cards or modules. The switch network circuit

Legacy PBX Common Equipment

cards or modules may support center stage switching or local TDM bus functions. Switch network function may be embedded on the processor/memory circuit pack in small system designs. For a dedicated switching carrier shelf expansion, interface boards are also required to provide control and signaling links to the control carrier and TDM bus transmission links to expansion port carriers/cabinets. A dedicated switching carrier shelf is equipped with one or two local power supply modules, if a centralized power supply carrier is not included in the design. The Definity G3r ECS Switch Node Interface Carrier is an example of a dedicated switch carrier (Figure 6-3).

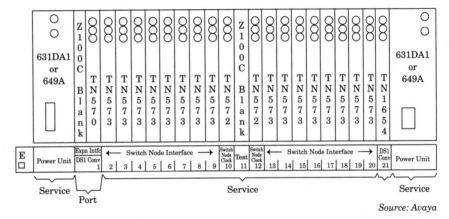

Figure 6-3
Definity G3r ECS switch node interface carrier.

Source: Avaya

Port

The port carrier contains printed circuit board slots that house port interface circuit cards. There are a variety of port interface cards supporting the many types of station and trunk peripherals. Some port carrier shelves may also support application circuit packs/modules. A maintenance/diagnostics circuit pack may also be supported on the carrier shelf. The carrier shelf is also equipped with one or two local power supply modules, if a centralized power supply carrier is not included in the design. Port carriers typically support between 16 and 20 port circuit card slots, although some PBX system model port carriers may support fewer or more slots. The number of port card slots has remained relatively constant during the past 15 years, although port circuit card density (port terminations per card) has increased significantly. The Definity G3si/r ECS Port Carrier is an example of a dedicated port carrier (Figure 6-4).

Figure 6-4
Definity G3si/r ECS port carrier.

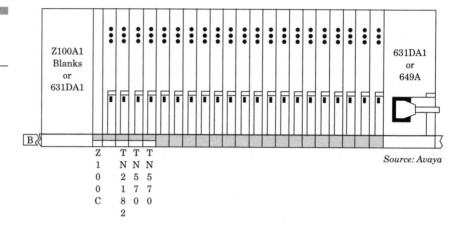

Source: Avaya

Auxiliary

A dedicated auxiliary carrier/cabinet is available as a common equipment option and typically available only with large PBX system designs. Auxiliary equipment housed in the carrier/cabinet may include call accounting systems, recorders, announcers, paging systems (loudspeaker, code calling), and music-on-hold. Small and intermediate PBX system designs may support some of these functions with the use of circuit packs housed in the control carrier shelf. Local power supply modules power the carrier/cabinet.

Application

Application carriers contain printed circuit board slots that house circuit card packs/modules dedicated to specific system applications, such as system administration, messaging (voice, integrated, unified), or call center (processing, reporting). The carrier shelf or cabinet is also equipped with one or two local power supply modules, if a centralized power supply carrier is not included in the design. The industry design trend has been to use an adjunct server cabinet, instead of an integrated application carrier/cabinet, to support advanced optional application requirements. The early adjunct server options were linked to the PBX common equipment via a proprietary cable, but the more common current method is a TCP/IP-based Ethernet LAN interface. The first adjunct server cabinet options were usually limited to a single applica-

Legacy PBX Common Equipment

tion. Current adjunct servers now support multiple application software programs. For example, the Siemens HiPath AllServe 150 applications server supports system administration, ACD, messaging, and call accounting.

Power

The power carrier distributes power from the main power distribution to the PBX common equipment components. Dedicated power carriers are usually available with large systems only. Based on the port size configuration and number of port cabinets, more than one power carrier may be required.

A very small or small PBX system may have a cabinet design based on a single multifunction carrier that supports the functions mentioned above. In this type of PBX cabinet design, all the basic control, switching, and port interface boards are housed in a single carrier sharing a common backplane, including power converters distributing current to the provisioned circuit cards. Intermediate line size systems are more typically based on stackable carrier or multicarrier cabinet designs that include a control/switching carrier and one or more expansion carriers. Each carrier shelf in an intermediate/large system design would include power supply modules. The control/switching carrier may also support a limited number of port interface card slots for customers with limited station/trunk requirements at system installation to minimize upfront common equipment requirements. Large/very large PBX system models are commonly designed with dedicated carriers for control and/or switching functions. These dedicated control/switch carriers do not support port port interface cards, which are housed in dedicated port carriers/cabinets.

Cabinet Power System

Alternating (AC) or direct (DC) current power sources may power PBX systems. Of the two, the current popular power option is AC. AC-powered systems plug directly into commercial AC power outlets. DC-powered systems require external rectifiers to convert the incoming AC current provided by the electrical utility carriers to DC voltage. The external current flows into a centralized power distribution unit that distributes power to the common equipment carriers/cabinets. If reserve power is required, customers can install an Uninterruptible Power Supply (UPS), with its associated batteries, in series with the utility AC power feed. Figure 6-5 shows a Meridian 1 DC-powered cabinet.

Figure 6-5
Meridian 1 DC-powered cabinet.

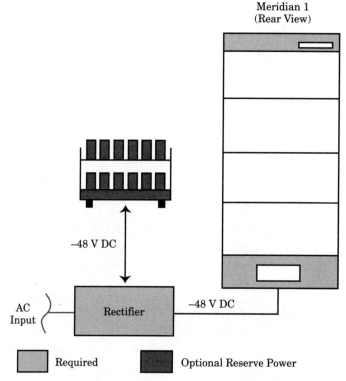

Source: Nortel Networks

AC-powered systems always require internal rectifiers to convert commercial AC power into the distributed DC power used to support internal common equipment hardware via carrier shelf power units. Power units convert AC input to DC output power for distribution to carrier backplanes to support the housed printed circuit boards. AC-powered system power units typically convert AC commercial power into –48 V DC for internal system distribution. External AC/DC rectifiers distribute –48 V DC voltage directly to the centralized power distribution unit, thus eliminating the need for internal rectifier equipment. Carriers and cabinets are equipped with power filter units to control power distribution. Circuit breakers or electronic fuses are located in the centralized unit to prevent short circuits in one carrier from affect-

Legacy PBX Common Equipment

ing power distribution to other carriers. Figure 6-6 shows a Meridian 1 AC-powered cabinet.

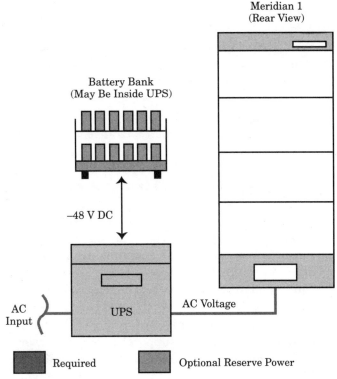

Figure 6-6
Meridian 1 AC-powered cabinet with reserve power.

Source: Nortel Networks

At the carrier shelf level, the power unit converters produce the DC operating voltages required by the various circuit boards. Most printed circuit boards use +5 V DC power. If AC commercial power fails, batteries in the power units provide the –48 V DC necessary to support system operation. The batteries are connected in parallel to provide required voltage. When relay circuits detect a commercial power failure, the battery backup system can continue to power the entire the system for a very limited period, typically 5 to 15 seconds, to avoid interruptions in service. A standard battery backup system powers the control complex for a longer period, from 5 to 30 minutes for most systems, to avoid loss of stored database memory. A UPS option can be used to support full system operation when commercial AC power is lost for extended periods. Relay circuits and circuit breakers are used to detect commercial power loss and distribute UPS power.

Cabinet Backplane

The carrier shelves in a multicarrier cabinet connect to a common backplane at the rear of the cabinet. At the rear of each printed circuit board card slot is a connector that attaches to the backplane. The backplane wiring, which uses wire-wrapped connections or printed circuit board technology, contains the system bus structure for intra- and intercarrier processor control and communications signaling transmission. The processor control signals are multiplexed onto the backplane bus structure via time slot assignments. The time slot assignments to the port circuit card slots are usually arranged to allow maximum flexibility in the placement of different device boards in the carrier, although some systems are designed with a fixed number of time slot access connections per port card slot. Physical backplane connectors, such as 25-pair connectors, are used to interface the port circuit card packs. There are also connectors for intercabinet or intercarrier connections. The shelf backplane of a control carrier may have multiple RS-232C connectors to support management and maintenance terminals or call detail recording equipment. Another connector is used for connecting a tip-and-ring pair and maintenance leads to a crossconnect field.

PBX cabinets and/or distributed carriers are usually connected with optical fiber cabling links. Optical fiber links provide very high transmission bandwidth in support of switch network traffic requirements and call processor signaling. Optical fiber distance limitations between cabinets vary from system to system, but the typical loop length is about 1,000 feet without repeaters or special interface cards.

Cabinet/Carrier Expansion Requirements

PBX system cabinet and carrier costs, without housed printed circuit boards, can account for 10 to 15 percent of the total system price. To keep system costs down, customers should seek PBXs based on cabinet/carrier designs that limit the necessary common equipment hardware. Carriers with many port card slots may be more cost effective than carriers with fewer card slots. Multicarrier cabinets with more carrier shelves may be more cost effective than cabinets that house fewer carriers. Regardless of the number of installed cabinets or available card slots at system cutover, customers will likely be required to purchase additional common equipment over the life of the system.

The following are the most common reasons for installing additional cabinets and/or carriers.

Port Growth

The most common reason for cabinet/carrier expansion is port capacity growth requirements for stations and/or trunks. Current carriers can typically support a few hundred ports, and cabinets can usually support more than 1,000 ports.

Increased Traffic Requirements

Some PBX systems require additional port cabinets/carrier common equipment to satisfy customer traffic engineering requirements even if port capacity does not increase. This is a common requirement for PBX systems based on blocking switch network designs, such as the Avaya Definity ECS and Nortel Networks Meridian 1 Option 61C/81C models.

Increased Call Processing Requirements

Although the call processing capacity of most PBX systems is based on the main common control complex, a few systems with distributed call processing designs require additional control cabinets to satisfy increased customer call processing capacity needs. The Ericsson MD-110 is a good example of a distributed call processing PBX system design with limited call processing capacity per LIM cabinet but significant collective call processing capacity across multiple configured LIMs.

New Application Requirements

An important PBX design trend during the 1990s was offloading of call processing–intensive applications, such as ACD, from the common control complex to adjunct applications servers. With rare exceptions, most current PBX systems require an adjunct server to support advanced ACD features. All PBXs require an adjunct server to support advanced ACD MIS reporting capabilities. Other applications that may require dedicated carriers or adjunct servers include messaging, wireless communications, and systems administration.

Printed Circuit Boards

PBX circuit cards are based on solid-state integrated circuitry mounted on printed wiring boards. A label, usually color coded to simplify installation and maintenance operations, identifies each circuit card. The circuit cards usually have fault, test, and busy multicolor LED indicators. A metal latch for electrostatic discharge protection is typically included. Circuit cards can be categorized into three basic types: control, service, and port. Control circuit cards support PBX call processing and switching functions. Service circuit cards enable some call processing features and functions. Port circuit cards serve as connection gateways between peripheral equipment and the PBX call processing, switching network, and power distribution systems.

Each port circuit card supports a unique type of peripheral endpoint, but all share several common design elements. Each port circuit card has a TDM bus buffer, control channel interfaces, an onboard microprocessor controller with limited memory storage, and additional processing elements to perform functions such as voice quality control.

Port circuit card bus buffers are the digital interface between the backplane TDM transmission line wires and the onboard integrated circuitry. The buffers have wire leads that transmit or receive electrical signals from the transmission line. The control channel interfaces access signaling information on the TDM bus. Control channel interface wire leads interface to the TDM bus, pickup common channel information intended for the port circuit pack, and transmit it to the onboard microprocessor controller. When the microprocessor controller transmits information to the TDM bus control channel, it is transported by the control channel interface. In addition, the control channel interface initiates power-on start-up procedures, performs diagnostics tests on the onboard microprocessor controller, and follows Main System Processor instructions to disable the port circuit pack in case of onboard problems. It also can perform several localized processing functions at the station user desktop, such as control of voice terminal LED/LCD status indicators.

Legacy PBX Common Equipment

The main responsibilities of the onboard microprocessor controller are relatively low-level processing and monitoring functions, such as scanning for changes and relay operations. It generally carries out the instructions of the Main System Processor and reports back status changes. Additional onboard processing elements are responsible for voice quality control, such as conference and gain-adjust functions, and port circuit termination access to the TDM bus time slots based on instructions from the onboard microprocessor controller. Figure 6-7 shows a Siemens Digital Line Card with the various design elements of a port interface card.

Figure 6-7
Digital line card block diagram.

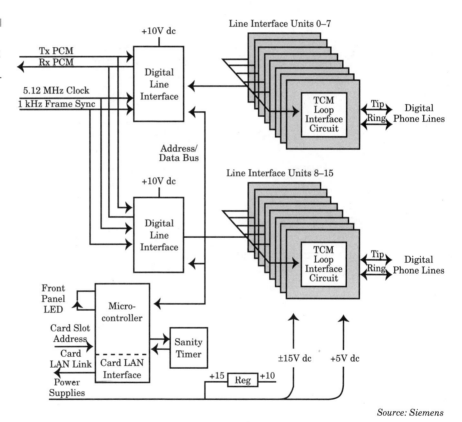

Source: Siemens

The following section contains descriptions of circuit interface cards typically used in PBX systems. Some PBX systems may combine the functions for several circuit cards into single multifunction card.

Control Cards

Processor. The processor card has the main Central Processing Unit (CPU) microprocessor chip that controls the entire system and executes the stored program control commands to perform call processing operations and administrative control functions. Integrated memory chips store the generic system program. In a duplicated control system design, the card allows the memory of the passive standby processor to be continuously updated to reflect the memory in the active processor. Additional card functions usually include provisioning of alarm LEDs for system status; monitoring and control of all port circuit pack conditions; and an interface to systems management and maintenance systems.

The processor card may also perform the following functions: transmission of control channel messages to the port circuit cards over the TDM bus; control signaling to customer-provided equipment, such as CDR and accounting devices; time-of-day clock; battery backup; and monitoring of system clocks. Some PBX systems may use an additional system control card for these functions, if not performed by the processor circuit card.

Power system monitor. A power system monitor card monitors the status of power-related hardware elements, such as power supplies, fan operation, power failure transfer, circuit breakers, and LEDs. Power error messages may be routed to the processor card for subsystem switchover operations or to the administration and maintenance system to generate alarms.

System memory. The system memory card has one or more memory chips for storage of the generic system program. The system memory may be partitioned to also support the customer database. Small and intermediate line size PBX systems may include the memory chips on the processor card, instead of having a dedicated memory card. Some systems require multiple system memory cards.

Memory storage drives. The PBX memory storage system usually requires a tape drive circuit pack or a floppy disk unit. The memory drives load the generic system program into system memory.

Switch network. The switch network cards may include the center stage switch network and/or local TDM bus switch network interfaces to the center stage switch network.

Local Loop Interfaces: Maintenance/Diagnostic

The maintenance/diagnostic card performs basic maintenance and diagnostic operations, such as monitoring of port circuit packs, TDM bus interfaces, and power distribution.

Service Cards

Tone clock. The tone clock supplies a variety of tones, including call progress tones, touch tones, answer-back tones, and trunk transmission tones. The card provides multiple clock signals for transmission over the TDM buses. Tone clock cards may also provide an interface to the external Stratum 3 synchronization clock.

Serial data interface. The serial data interface card can provide interfaces to the main processing element for one or more I/O devices.

Tone receiver. The touch-tone receiver card supports multiple channels of DTMF or multi frequency (MF) detection for analysis and interpretation of incoming tone signals from 2500-type terminals and certain DID and tie trunks. General-purpose tone receivers detect call progress tones and answer tones, modem answer-back tones, transmission test tones, and noise. Tone detection capability is usually required for ARS and OPX dialing applications. Tone receivers also detect answer on outgoing analog trunk (CO, FX, WATS) calls so that CDRs can be generated. More than one tone receiver card may be required per system, based on port size requirements and customer applications.

Conference. The conference card supports the multiparty conference feature. Only a select number of PBX system models require a dedicated card to implement a conference call among three or more system ports.

Speech synthesizer. The speech synthesizer card stores customer-programmed audio messages that are required for several PBX features, such as automatic wake-up, voice message retrieval, and call intercept treatments. The speech synthesizer card supports a limited number of communications channels, usually between two and eight.

Call classifier. The call classifier card supports tone detectors that are used for call prompting applications. The call prompt feature prompts callers to input touch-tone digits for call screening and analysis purposes. It is generally used in ACD system configurations instead of a stand-alone automated attendant, VMS, or IVR system for digit collection, analysis, and call redirection applications. The card can also detect special intercept tones for outbound call management applications.

Expansion interface. The expansion interface card extends internal system buses (call processing, switch network) between stackable carriers and/or cabinets.

Control buffer. The control buffer card is housed in a port carrier card slot and provides an interface between the local TDM bus and the center stage switch network complex. The card may also include a microprocessor that supports local processing control and maintenance functions.

Announcer board. An announcer board stores customer-programmed recorded announcements. PBX/ACD features requiring an announcer board may include call intercept treatments, estimated wait time, and the number of callers in queue. The number of distinct announcements per announcer board varies by manufacturer, but most boards can support between 32 and 128 announcements. The total recording time for announcements is usually limited to between 4 and 10 minutes per board. The number of communications channels per board varies by manufacturer, but is typically between 16 and 32 channels. Several callers can usually listen to the same channel announcement.

CTI link. The CTI link card supports a signaling interface link between the PBX system and a CTI applications server or host. First-generation CTI signaling links were based on proprietary PBX interface protocols. Current PBXs commonly use a TCP/IP signaling link and an Ethernet LAN interface connection supporting CTI applications. The TCP/IP interface cards used for CTI application requirements can be used for a variety of other PBX applications, such as VMS signaling interface, adjunct ACD MIS reporting system server signaling interface, and IP call control signaling to IP station/trunk peripherals. The first-generation proprietary CTI link cards were usually limited to a signaling interface link for a single server; the current TCP/IP LAN

Legacy PBX Common Equipment

interface cards can support signaling interfaces to multiple LAN-connected servers.

Port Cards

Analog line. The analog line card provides the interface to the local TDM bus for analog communications terminals and devices. These terminals and devices include 500-type rotary dial telephones, 2500-type DTMF dial telephones, facsimile terminals, modems, external alerting devices, paging systems, dictation machines, and recorded announcements. Analog line cards may also provide PBX system communications links to external VMSs, IVR systems, and wireless communications system controllers. One or more analog line cards may be required to support any or all of the listed peripherals. The port density of current PBX analog line cards is typically 16 or 24 single bearer communications channel circuit terminations.

OPX line. The OPX line card provides the interface to off-premises wiring with rotary or DTMF dialing in support of an OPX. The port density of a current PBX OPX line card is typically 16 single bearer communications channel circuit terminations.

Digital line. The digital line card provides the interface to the local TDM bus for proprietary digital telephones or data modules. For many PBX systems, the digital line card also can be used to support proprietary attendant consoles. Each digital circuit port interface supports a 2B+D transmission protocol to the desktop device. The port density of a current PBX digital line card is typically 16 or 24 dual bearer communications channel circuit terminations.

Digital attendant console. The attendant console card provides the interface to the local TDM bus for proprietary digital attendant consoles. (A digital line card, instead of a dedicated digital attendant console card, currently supports most PBX attendant consoles.) Each digital circuit port interface supports a 2B+D transmission protocol to the attendant console. The port density of current PBX digital attendant consoles is usually limited to one or two dual bearer communications channel circuit terminations.

Data line. The data line card provides the interface to the local TDM bus for proprietary data modules. Data modules support a modemless communications link between customer-provided Data Terminal Equipment (DTE) or Data Communications Equipment (DCE) and the PBX system. There are a variety of data line cards; each data line card is designed to support asynchronous and/or synchronous data endpoints at different transmission rates. The port density of a current PBX data line card is typically 16 single bearer communications channel circuit terminations.

Power failure transfer. The power failure transfer card provides direct physical connections between central office trunk line circuits and analog station equipment in the event of a failure of the PBX power system. The power failure transfer lines function as normal CO trunks in the normal PBX state. Power failure transfer cards typically support a limited number of central office trunk connections, such as two or four lines.

ISDN BRI line. The ISDN BRI line card provides the interface to the local TDM bus for DTE conforming to National ISDN standards. There are two types of ISDN BRI line cards: U-interface and S/T-interface. A U-interface line card supports a 1B+D transmission protocol to the desktop device; an S/T-interface supports a 2B+D transmission protocol to the desktop device. The port density of a current PBX ISDN BRI line card typically supports 16 single or dual bearer communications channel circuit terminations.

IP line. The IP line card provides the interface to the local TDM bus for IP telephones and IP PC client softphones. The IP line card functions as a media gateway to convert IP communications signaling format to TDM/PCM communications signaling format and may be used to convert proprietary PBX signaling protocol to an IP call control protocol, such as H.323 supported by the IP telephone. A dedicated port circuit card may also support call control signaling transmission to and from the IP telephone. The IP line card is embedded with numerous DSP resources that function as the media gateways. IP line cards may support more physical IP telephones than available media gateway channel connections to the local TDM bus. The number of local TDM channel connections depends on available DSP resources and the type of audio coder used by the IP telephone. Current IP line cards support between 24 and 500 IP telephones and between 24 and 64 channel connections;

the actual numbers depend on each manufacturer's PBX system platform. Some IP line cards may function as universal IP port cards, and support IP trunks (see discussion below).

CO trunk. CO trunk cards provide interfaces to the local TDM bus for the following types of local exchange carrier trunk circuits: loop-start or ground start CO, FX, or WATS. The CO trunk card may also be used for paging systems or other external communications systems, such as IVR systems. The port density of a current PBX CO trunk card is typically 8 or 16 communications channel circuit terminations.

DID trunk. DID trunk cards provide interfaces to the local TDM bus for immediate-start or wink-start DID trunks. The port density of a current PBX DID trunk card is typically 16 communications channel circuit terminations.

E&M tie trunk. E&M tie trunk cards provide interfaces to the local TDM bus for two-wire or four-wire E&M trunk circuits. There are several types of E&M tie trunk circuit categories, but the category currently used by most PBXs for networking applications is Type 1. Some PBXs may also use the E&M tie trunk card to support a Centralized Attendant Service (CAS) configuration requiring a Release Link Trunks (RLT). The port density of a current generation PBX E&M Tie Trunk card is typically eight communications channel circuit terminations.

DS1 digital trunk. DS1 digital trunk cards provide interfaces to the local TDM bus for T1 trunk circuits operating at 1.544 Mbps (North America standard) or E1 trunk circuits operating at 2.048 Mbps (Europe standard) digital trunk circuits. Digital T1 trunk circuit cards support 24- to 64-Kbps channels; digital E1 trunk circuit cards support 32- to 64-Kbps channels. The DS1 digital trunk card can be programmed to support voice-grade digital trunks using inband bit-oriented signaling on a per-channel basis or Alternate Voice Data (AVD) digital trunks using common channel signaling. Some DS1 digital trunk cards can also be programmed to support ISDN PRI services over T1/E1 trunk circuits. A DS1 digital trunk card supports one T1/E1 trunk circuit connection and 24 to 32 communications channel circuit terminations.

ISDN PRI trunk. ISDN PRI trunk cards provide interfaces to the local TDM for T1/E1 trunk circuits programmed to support ISDN PRI services. A T1-based ISDN PRI trunk card supports 23 bearer communications channels and one signaling channel (23B+D); an E1-based ISDN PRI trunk card supports 30 bearer communications channels and one signaling channel. Some PBX systems require a dedicated circuit card to support D-channel signaling protocol. The D-channel handler card, as it is sometimes called, supports a manufacturer-defined, limited number of ISDN PRI trunk cards. An ISDN PRI trunk card supports one ISDN PRI service T1/E1 trunk circuit.

ISDN BRI trunk. ISDN BRI trunk cards provide interfaces to the local TDM for digital subscriber lines conforming to ISDN BRI signaling standards. An ISDN BRI trunk card supports two bearer communications channels and one signaling channel (2B+D). An ISDN BRI trunk card supports one ISDN BRI digital subscriber line. ISDN BRI trunk cards are not commonly available for intermediate/large PBX systems.

IP trunk. The IP trunk card provides the interface to the local TDM bus for IP trunking services via WAN IP routers. The IP trunk card functions as a media gateway to convert IP communications signaling format to TDM/PCM communications signaling format and may also be used to convert proprietary PBX signaling protocol to an IP call control protocol, such as H.323, supported by IP routers. A dedicated port circuit card may also support call control signaling transmission to and from the IP router. The IP trunk card is embedded with numerous DSP resources that function as the media gateways. The number of local TDM channel connections depends on available DSP resources and the type of audio coder used by the IP telephone. Current IP trunk cards typically support 24 channel connections using the H.323 call control protocol.

ATM trunk. The ATM trunk card provides the interface to the local TDM bus for ATM trunking services via customer premises ATM switching systems. The card performs a cell assembly/disassembly operation to convert between TDM/PCM communications signals and packeted ATM cells. ATM trunk cards can also be used for T1/E1 Circuit Emulation Service (CES); a single ATM trunk card can support multiple T1/E1 trunk circuit connections.

Auxiliary trunk. Auxiliary trunk cards provide the interface to the local TDM bus for on-premises trunk connections to customer-provided equipment that supports communications applications, such as loudspeaker paging, code calling, recorded dictation access, recorded announcements, and music-on-hold. The port density of current auxiliary trunk cards typically supports 4, 8, or 16 communications channel circuit terminations.

Pooled modem. The pooled modem card converts resources for switched connections between data modules and modems for off-premises trunking applications. A pooled modem card may support one or two modem connections.

Circuit Card Provisioning Issues

Control and service circuit cards are usually housed in designated card slots in the control carrier, although some service circuit cards may be housed in certain port carrier card slots as designated by the manufacturer. Most port circuit cards are housed in any available port circuit card slot, but there are exceptions to this generalization for certain PBX system models. A PBX system that can support any port circuit card in any open card slot is said to have a *universal port circuit card slot design*. If certain port circuit cards have designated port card slot assignments, the system fails to satisfy the universal port circuit card design standard.

There are several reasons port circuit card housing may have to conform to card slot mapping rules.

TDM Bus Time/Talk Slot Restrictions

This is the most prevalent card slot mapping factor for designating a PBX system a nonuniversal port circuit card slot design. PBXs that have nonblocking, segmented TDM bus designs restrict the number of active port circuits housed across contiguous card slots. For example, the NEAX2400 IPX PIM has a local TDM bus with 384 talk slots. The system software is based strictly on nonblocking access to talk slots for all potentially configured port circuit terminations. The 18-card slot PIM can house a mix of 8-port and 16-port digital line cards; each port circuit is equipped with two bearer communications channels. Although a PIM can house an 8-port digital line card in each card slot and remain within the talk slot limitations of the local TDM bus, it cannot support a 16-port digital line card (32 bearer channels, total) in each card slot. The PIM is therefore limited to twelve 8-port (12 × 8 ports × 2 channels/port = 182 channels) and six

16-port (6 × 16 ports × 2 channels/port = 192 channels) digital line cards. The maximum number of communications channels per PIM is the limiting factor for housing the higher density digital line cards. PIM card slot mapping guidelines require 16-port digital line cards to be housed in very specific card slots (ports 7 to 9 and 16 to 18). The remaining card slots are available for the 8-port cards.

Another type of talk slot limitation affecting port circuit card housing is one that requires nonblocking access to the TDM bus for a certain type of port circuit card. The Nortel Networks Meridian 1 Option 61C/81C consigns DS1 digital trunk cards to control/network carrier card slots instead of the port carrier card slot. The Meridian 1 Digital Trunk Interface (DTI) card must be housed in a network card slot to have guaranteed access to a 32 talk slot network loop. This provides nonblocking switch network access to digital trunk circuit terminations, thereby reducing the need for port carrier shelf traffic engineering.

Call Processing Limitations

Some PBX systems limit the number of port circuit card types because of system call processing limitations. Many of the early digital PBX systems supported a limited number of electronic telephones because of limited call processing capacity. When digital telephones became available, a similar limitation was made for select PBX system models. The same call processing limitation factor is in effect today with IP telephones. The processing requirements to support complex port circuit cards can limit the number of these port circuits.

Power Distribution Limitations

The internal PBX power distribution system may limit the number and type of port circuits per carrier or cabinet, based on power design limitations.

Card Slot Requirements

Some port circuit cards may require more than one card slot because of the size of the printed circuit board or the required number of talk slot connections. If two card slots have access to a limited number of talk slots and one port circuit card requires more than an average number, the contiguous card slot may have to remain open, or only a low-density port card can be housed. For example, the Fujitsu F9600 intermediate and large system models are based on segmented local TDM buses. Every two card slots per carrier has access to 32 talk slots. Although there are no hard and fast rules for port circuit card assignments, the housing of a high-density digital trunk card requiring 24 talk slots in one of two card slot pairs requires the second port circuit card to have eight or fewer port circuit terminations.

CHAPTER 7

Introduction to IP-PBX Systems

The evolution from a circuit switched to packet switched communications system platform has been one of the most widely discussed and analyzed events within the enterprise communications system market for the past few years. Globally, customers have invested more than one quarter trillion dollars in currently installed circuit switched premises communications system hardware, and migrating to the new packet switched technology platform will be a slow and gradual process because investment protection is a very important buying decision criteria. Customers are beginning to realize that enterprise communications systems based on an IP control signaling and communications transmission platform are not a new product type but merely another category of PBX systems. A customer's enterprise communications system will continue to be referred to as a PBX system, regardless of the underlying technology used to support real-time communications between station users. The emergence of IP packet switched communications technology is not the first time a major technology disruption in the PBX status quo has occurred.

Analog PBXs transformed into digital PBXs during the late 1970s and early 1980s, just as circuit switched digital PBXs are currently transforming into packet switched PBXs, but remnants of the older technology will not immediately disappear. When Rolm, a pioneering interconnect equipment supplier since acquired by Siemens, introduced the first PBX system based on a digital circuit switching platform in 1975, it was not the beginning of the end of the sale and use of analog telephone instruments. Rolm's digital circuit switched Computerized Branch Exchange (CBX) system did not support digital telephone instruments at time of initial availability. Several other digital PBXs were introduced in the latter part of the decade, but digital telephones were not available. Electronic telephones using analog transmission technology were the most advanced desktop instruments until Intecom introduced its IBX S/40 system in 1980. The IBX S/40 was the first PBX system to support digital telephones with integrated codecs for digitization of voice communications at the desktop. By the late 1980s every leading PBX system supported digital telephones, but many analog instruments were still being sold and configured on new system installations. Digital telephone shipments did not exceed analog station ports on PBXs until about 1990.

More than 25 years after the introduction of PBX digital transmission and switching technology, a very sizable number of analog telephones, fax terminals, and modems remain in place. Likewise, analog-based trunks are still tariffed and countless millions are leased. Voice communications system customers have traditionally adapted new technology, but do not quickly abandon the tried and true when something new comes along. A distinguishing characteristic of the voice communications market space is a slow and gradual evolution between technology platforms.

Customers who accept and implement a new technology and/or design platform shortly after first availability are known as *early adopters*, but mainstream acceptance traditionally takes several years for a variety of reasons:

- Elimination of new technology/design bugs
- Definition and acceptance of standards
- Price curve economics
- Depreciation of existing assets

Legacy voice communications systems have achieved a very high level of reliability and support and an extremely robust set of features and functions. Unless a customer can balance the risk of new technology against potential cost savings and/or productivity improvement, the natural tendency is to minimize risk and proceed slowly into the future. For most customers, incremental evolution is preferable to wholesale revolution when it comes to their real-time communications requirements. It is a very rare technology paradigm that succeeds in the blink of an eye.

Although the basic PBX functions—call processing, switching, transmission, and memory storage—have remained constant over time, the method and technology used to perform the functions have changed. It has often taken several years for a newly introduced PBX design element or attribute to become an industry standard. For example, most new PBX system sales were still based on analog switching system designs until the early 1980s, although digital PBX shipments started in 1976. Digital telephones did not dominate the landscape until the early 1990s. Shipments of CTI and wireless system options are limited, although these PBX options were introduced in the early 1990s. LAN-based PBXs incorporating an IP client/server design currently account for a very small percentage of the market based on total system/station shipments despite widespread publicity and hype in the trade press. It is the collective wisdom of the equipment suppliers and industry analysts that IP-based packet switching will eventually supplant circuit switched technology as the primary means of call connection and transport, but the change will not happen overnight.

ToIP and IP-PBX Systems

The IP-based packet switching network is the underlying principle for the technology known as Telephony over IP (ToIP). ToIP is simply described as an IP-based LAN/WAN infrastructure that supports telephony system communications and applications. A customer's packet switching LAN/WAN infrastructure can carry control and/or voice communications signals in IP packet format between a call processing complex and peripheral endpoints or between two peripheral endpoints. A traditional PBX common control complex can logically manage a switch connection between two endpoints, but the physical switching function between the IP peripherals may be handled external to the PBX common equipment by Ethernet switches. The internal PBX circuit switched network may have no role in the voice communications call.

A PBX system using ToIP technology in support of any or all control or voice communications signaling can be categorized as an IP-PBX system. In theory, an IP-PBX system can have any or all of the following attributes:

- LAN-connected common control complex, such as a telephony server that provides call signaling control to port cabinets and/or peripherals and stores the generic software feature program for feature/function provisioning operations.
- LAN-based control signaling of IP peripherals (stations and/or trunk circuits) with or without an integrated gateway/gatekeeper function port circuit card(s). This form of ToIP would be available with a PBX system based on a traditional common control complex or a LAN-connected telephony server.
- Ethernet switched support of IP peripheral to IP peripheral calls or of non-IP peripheral to IP peripheral calls. The former call type typically would involve no circuit switched connections. An example of the latter call type is a call connection between a traditional analog or digital telephone and an IP telephone. Voice communications signals are carried across the LAN infrastructure between IP peripherals or to gateways interfacing to non-IP communications equipment.

A circuit switched PBX system based on a traditional common control complex that supports IP peripherals using an external gateway and/or gatekeeper network element would not qualify as an IP-PBX. A PBX based on a CTI design that supports analog telephones logically integrated with a desktop computer equipped with PC telephony software would

Introduction to IP-PBX Systems

not be an IP-PBX, unless the system also supported IP telephones or had a fully integrated gateway function and interface for IP WAN trunk calls.

An IP-PBX is not necessarily based on a LAN/WAN data communications client/server design. A traditional PBX system, with an integrated gateway module board used for intersystem networking across a customer WAN, can also be categorized as an IP-PBX system based on the above listed criteria. There are some PBXs categorized as an IP-PBX that support IP telephones but lack integrated trunk support. A PBX system that supports neither IP stations nor trunk ports but does support a distributed cabinet design using a LAN/WAN infrastructure for control signaling and intercabinet communications can also be categorized as an IP-PBX even if there are no IP telephones or IP private line trunk networking capabilities. Just as the design architecture and feature capabilities of traditional circuit switched PBXs are not uniform across product models from the same supplier or competing suppliers, IP-PBXs are based on a variety of architecture designs and hardware configurations that support different feature/function capabilities.

Recognizing that there are some significant differences in PBX design and function across the spectrum of product offerings and that there is a natural tendency to label a product, a simple classification system must be defined. As a consequence, PBX systems currently can be classified into three basic categories based on call control and switching platforms:

1. **TDM/PCM circuit switched**—TDM/PCM circuit switched PBX system with proprietary common control complex and internal circuit switch network.
2. **IP packet switched**—IP packet switched PBX system based on a client/server design consisting of a telephony server that uses a LAN/WAN infrastructure for call control and communications signaling operations. A circuit switched communications network is not included as part of the standard system design.
3. **Converged**—Integrated circuit/packet switched PBX system with a proprietary common control complex, an internal circuit switched network, and integrated gateway interfaces to support port cabinets and/or IP peripherals connected directly to an external LAN/WAN switch network. Traditional PCM-based peripherals and new IP peripherals can be supported.

The second and third PBX categories are classified as IP-PBX systems because each uses ToIP technology for some or all system processing and/or switching functions. Just as there are many design variations of a traditional circuit switched PBX system, there are design variations of IP packet switched and converged IP-PBXs. For example, the Cisco System AVVID CallManager solution has a radically different design than the 3Com NBX 100. Both systems are clearly categorized as IP packet switched IP-PBXs, but the telephony server and peripheral support options differ greatly. The Cisco solution requires a variety of gateway options to support analog stations and PSTN trunk circuit interfaces, many of which are housed in Ethernet switches exclusive to Cisco and require a dedicated messaging server in addition to the primary telephony server. The 3Com solution is based on a telephony server cabinet with an integrated digital T1 trunk gateway and fully integrated unified messaging support and supports analog stations using desktop adapter modules. Even within the 3Com IP-PBX family of switches there are distinct differences between model designs. The NBX 100 is based on a server cabinet design, and the SuperStack 3 (SS3) model uses a modified SuperStack Ethernet switch chassis. The functions and applications of the two 3Com models are essentially identical, although the common control hardware equipment is different.

A client/server IP-PBX design does not necessarily preclude support of traditional circuit switched PBX equipment hardware. The Mitel Networks 3300 (MN 3300) is based on a closed server cabinet that can support a traditional Mitel SX-2000 Light port equipment cabinet with a dedicated optical fiber cable or digital T1 trunk connection. The MN 3300 supports direct call control signaling to LAN-connected IP telephones and depends on the Ethernet switch network for voice communications connections and transmission, but also provides the call control signaling for analog and digital telephones supported by traditional port circuit interface cards. Switch connections for calls between analog and digital telephones are circuit switched over the SX-2000 TDM backplane. The MN 3300 T1 gateway interface is used for calls between LAN-connected IP telephones and non-IP peripherals housed in the port equipment cabinet.

Support of TDM/PCM port cabinet equipment may not be unusual for IP-PBXs based primarily on a client/server design architecture. As customers continue to migrate their PBX systems from a circuit to packet switched platform, a telephony server directly supporting LAN-connected IP telephones will become more popular, but traditional desktop instruments will need to be supported until the migration is complete. If not making use of modified existing port cabinet or carrier equipment with a TDM backplane, the telephony server will support newly designed LAN-connected port carriers with integrated gateways for customers who wish to continue to using analog or proprietary digital telephones originally supported on traditional TDM/PCM port interface equipment. For example, several manufacturers of traditional circuit switched PBX systems plan to make available a telephony server to optionally replace the traditional common control cabinet but continue support of existing TDM/PCM port equipment cabinets. Direct signaling support over the LAN of IP telephones eventually will be supported, as will the legacy port equipment cabinets. The port equipment cabinets will have integrated gateway interfaces for call control signaling and intercabinet communications. Avaya and Alcatel have indicated that they plan to use a LAN-connected telephony server to support traditional port equipment cabinets in near-term future releases.

The converged IP-PBX system design is based on a traditional circuit switched design platform with proprietary common control, internal TDM buses for local switch network access, and a center stage switch complex, if required. Station and/or trunk interface circuit cards with integrated gateways are used for IP peripheral support. The gateway function provides channel connections to the internal local TDM for calls between IP peripherals and non-IP peripherals. The interface card may also support gatekeeper call control signaling over the LAN to the IP peripherals. The gateway and gatekeeper functions sometimes can be divided between two port circuit cards. For example, the Nortel Networks Meridian 1 ITG card supports control signaling and TDM bus channel connections for IP peripherals; the Avaya Definity requires an IP Media Processing Board to support gateway channel connections and a CLAN card for call control signaling.

A converged IP-PBX system may also support multicarrier TDM/PCM multicarrier or single-carrier port cabinets over a LAN/WAN infrastructure similar to that described above with a telephony server, except the converged system is based on a traditional common control complex. This design arrangement would require a gateway/gatekeeper interface card in the centralized common equipment cabinet and an integrated gateway interface function in the distributed LAN-connected port cabinet/carrier equipment. All call processing functions would originate in the common control complex and be transported over a LAN/WAN infrastructure in support of a remotely located port equipment cabinet. A distributed or dispersed switch network design lends itself better to a converged IP-PBX system as opposed to a centralized switch network design because there is less dependency on the LAN/WAN for call switching requirements. Only intercabinet calls would require use of the gateway and LAN switches in this converged design.

IP-PBX SYSTEM: Benefits and Advantages

There are several important reasons why customers may decide to implement an IP-PBX system for enterprise communications.

Leverage Existing Investment in LAN/WAN Infrastructure

During the past decade customer investment in LAN/WAN infrastructure has greatly exceeded investment in traditional circuit switched voice communications systems. The current installation life of a typical customer's LAN/WAN equipment is much shorter than their PBX system. Financially, it makes sense to leverage the more up to date LAN/WAN equipment, rather than the older PBX system, to satisfy current and future growth and application requirements. The residual value of circuit switched PBX system components is rapidly declining because the future of voice communications increasingly will depend on packet switched LAN/WAN infrastructures.

Using a single cabling and network infrastructure for voice and data communications applications instead of the traditional two network approach offers customers several advantages: reduced upfront capital expenditures and ongoing maintenance and service expenses, simplified installation and management/maintenance operations, and use of LAN/WAN by an increased number of station users.

Reduce Capital, Network, and Operating Expenses

Using a single network infrastructure for all communications media requirement reduces upfront capital expenditures, monthly network service costs, and operating expenses. Although the current price for an IP-PBX system may not always be less than that of a circuit switched PBX system, the projected cost curve favors the former solution in the future. IP telephone instrument prices will continue to decline as product levels increase. Eventually, many of the equipment cost components of a circuit switched PBX system will be eliminated in a ToIP environment, including port carriers and printed circuit boards, in support of desktop peripherals and trunk circuit interfaces. The cost of a telephony server will be less than that of existing common control complexes.

Using the existing data communications network to support voice networking requirements across multiple customer locations may significantly reduce telecommunications service expenses. WAN-based trunk connections require less bandwidth because of voice compression techniques and packet routing network configurations. Fewer trunk circuits translate to reduced expenses. The growth of broadband optical fiber networks will make bandwidth a non-issue for voice communications. Voice over IP (VoIP) trunking on IP-PBXs is currently limited to private network applications, but eventually will expand across different customer enterprise networks.

Operations, administration, and maintenance (OA&M) expenses are always greater than equipment and telecommunications service costs because of personnel expenses and outlays (salary, benefits, training, and tools). Price curves for equipment and telecommunications services typically decline, but personnel costs always rise. Reducing OA&M expenses is possible only by reducing personnel requirements. One communications network will save on maintenance personnel costs, and the greater centralization of management administration staff will further reduce costs. The days of separate voice and data communications management staffs are slowly coming to an end.

Simplify System and Network Configuration Upgrades and Expansion

The most beneficial ToIP technical advantage is the ease of adding station users and supporting dispersed geographic locations to an existing system configuration. Using the ubiquitous LAN/WAN infrastructure to interface individual station users or groups of station users simplifies network configuration upgrades and expansion. Remote individual IP station users do not require local PBX common equipment (carriers, port circuit cards) to interface to the voice communications system. Using the LAN/WAN to support remote carrier cabinet requirements for station users within a campus setting or across multiple premises eliminates the need for dedicated circuit switched trunk links for signaling and communications. The distributed capabilities of an IP-PBX can also reduce the need for multiple systems in a network configuration. A single system configuration offers several significant benefits over a multisystem network solution, including full feature/function transparency across locations, simplified systems management and administration, and reduced system upgrade costs over the life of the system—one system upgrade, when needed, instead of system by system upgrades.

The underlying technology of an IP-PBX system offers greater configuration scalability than traditional circuit switched systems because it may be simpler and more cost effective to add stations and/or port carriers using the LAN/WAN infrastructure as the cabling and transmission system. For example, IP telephones can use existing Ethernet data ports to connect into the voice communications network, or remote station users using a PC softphone can dial into the LAN/WAN for system access and connectivity. An IP-PBX system reduces the need for interface outlets and wiring dedicated to voice-only communications.

Conforms to Standards

Circuit switched PBXs were traditionally designed to use proprietary signaling protocols in support of call processing and switching functions. Except for ISDN BRI telephones, all digital PCM-based telephones are proprietary to a unique PBX system. The operating system of each PBX system was once proprietary and closed to third-party software programmers. Although CTI has slightly opened up PBX systems, the common control complex remains based on proprietary cabinet and printed circuit board hardware elements. Most IP-PBX systems, however, have been designed to conform to published call control signaling protocol standards, such as H.323 or SIP, and can support third-party telephone instruments for basic call operations and feature operation. The emergence of client/server design IP-PBXs allows customers to use less proprietary hardware equipment and more off-the-shelf system components, such as third-party servers running telephony call processing and management software. LAN switches from a variety of suppliers provide the switching function for call connections.

Support of Applications across the Enterprise

A centralized telephony call server theoretically can support dozens of geographically dispersed locations, as can a LAN-connected applications server for systems management, messaging, or contact center applications. The same features, functions, and applications available at one location can be provided across the entire customer enterprise network without replicating expensive processing equipment. Centralizing systems management, messaging, and contact center processing functions saves money and provides consistent performance potential across the enterprise. The most efficient and cost-effective method to provide access to the centralized processing system(s) is over a LAN/WAN infrastructure using IP signaling and control.

Availability of New and Improved Station User Features and Applications

An IP-PBX system can support an array of new desktop and system features and applications not available with traditional circuit switched PBX systems. The second generation of IP telephones will support many features and options not available on traditional digital telephones. Among these new capabilities are integrated Web browser, unified messaging access, integrated Ethernet switch ports, and multiple directory access. IP softphones will be equipped with integrated computer telephony features and applications without a dedicated desktop voice instrument. An IP softphone also simplifies provisioning of mixed media contact center agents capable of handling voice calls, e-mails, and interactive Web calls.

An IP-PBX supports increased station user mobility and flexibility in accessing the communications network, because there are more methods of system connectivity for communications and/or information access through a greater number of communications devices: desktop IP telephone, IP softphone, wireless telephone handset, and wireless PDA. Each communications device automatically identifies itself to the communications system when it attempts to establish a connection, and regardless of physical location or connection method, the IP-PBX can confirm the identity of the station user with Dynamic Host Control Protocol (DHCP) and allow station user access to system features and functions. An IP-PBX system that shares the same network infrastructure as data communications systems and terminals can support a single directory number for each network subscriber regardless of how many desktop devices are installed, if they all share a common telecommunications outlet and Ethernet switch port.

An IP-PBX system lends itself better to the general PBX design trend of increased use of peripheral applications processors to support optional and/or advanced features and functions, particularly those related to messaging and contact centers. As multiple LAN-based servers support an increased number of communications applications, it makes sense from a design perspective to move toward LAN-based common control and desktop terminal equipment. The ubiquitous presence of the Internet, connecting station users regardless of terminal device, makes IP control and communications signaling the logical choice for a PBX system using a LAN/WAN infrastructure.

The Case for Converged IP-PBX Systems

There are several advantages for implementing a converged IP-PBX system or a client/server IP-PBX. First let's review the case for the converged IP-PBX system.

A converged IP-PBX system occupies the gray zone between the traditional PBX system and the evolving client/server design because it can offer customers the best of both worlds: the reliability, redundancy, and feature/function performance benefits of legacy circuit switched PBXs and the unique capabilities of ToIP technology to leverage LAN/WAN infrastructure in support of dispersed common carrier equipment and desktop terminals. A converged PBX system is best viewed as a bridge between the existing circuit switched world of voice communications and the emerging packet switched world. Instead of attempting to leap from the old world to the new world in one jump, it makes more sense to travel a longer, but less risky, road.

A converged IP-PBX system and an IP packet switched client/server platform can provide each of these customer benefits. The latter solution is usually optimized in a green field situation—a new customer location without a previously installed system, although a customer may still elect to install a converged system for incremental migration from a circuit to a packet switched solution. Many customers with an existing circuit switched system installation are likely to upgrade to a converged system with a minimal investment in new hardware and/or software. In contrast, a client/server design represents a major design break from previously installed circuit switched PBX systems and entails replacement of most existing common equipment and desktop terminal instruments.

The following are probable reasons a customer requiring ToIP capabilities might choose a converged IP-PBX system solution instead of an IP packet switched client/server design:

Introduction to IP-PBX Systems

1. ToIP requirements
2. Investment protection
3. Critical reliability
4. Private network compatibility
5. Feature requirements
6. Pricing

ToIP Requirements

A large number of PBX system customers may not currently require ToIP capabilities for their enterprise communications system, although future plans are very likely to require support of ToIP options. A converged IP-PBX system is fully capable of being configured without any IP-based elements at initial installation but can support such a requirement when needed. The group of customers who are risk-averse may also choose to delay installation of ToIP options at the present time because they are taking a wait-and-see attitude toward the emerging technology. Early versions of products incorporating new technology may not satisfy accepted benchmark reliability standards for telephony applications. In addition, ToIP standards are still in a state of flux, and customers may wish to wait until control signaling protocol and voice codec standards have stabilized. Some first-wave customers who installed IP-PBX systems shortly after product introduction have watched as evolving standards and system upgrades obsoleted their installed terminal equipment, voice codecs, and/or telephony servers. A converged IP-PBX reduces this risk by allowing customers to upgrade and enhance their system for active ToIP capabilities when the time is acceptable.

A converged IP-PBX system is also ideal for customers who wish only to test ToIP technology without installing a 100 percent IP peripheral solution. For trial purposes it is less costly to use a converged IP-PBX system that is also functioning as the full-time circuit switched communications system than to purchase and install a dedicated IP packet switched client/server design. For trial purposes, a select number of IP peripherals can be installed on a converged IP-PBX system. If the trial is successful, additional IP ports can be equipped and activated when needed. A very important factor in the successful implementation of an IP-PBX system is the LAN/WAN infrastructure, because the network must be properly designed and programmed to provide an acceptable Quality of Service (QoS) level for real-time voice communications applications. A small-scale IP-PBX trial is usually necessary to discover in advance whether or not a major LAN/WAN overhaul is needed. What works on paper does not always work in real life. Using a converged IP-PBX system is the most efficient solution for trial testing ToIP QoS levels for station and/or trunk traffic communications.

Investment Protection

All customers make substantial financial and organizational investments in their installed premises communications system and usually seek to maximize their return on investment (ROI). Replacing an installed circuit switched PBX system with a new IP packet switched client/server design solution at the present time may be counter to this customer objective. For example, a customer may not have fully depreciated the installed system's common equipment and desktop terminals. Replacing cabinet and port interface hardware with a telephony call server and gateways and digital telephones with new IP telephones is an expensive proposition, as are major upgrades to the LAN/WAN infrastructure to support real-time voice-grade QoS levels. Training expenses for station users, system managers, and maintenance personnel also add to upfront cost outlays. Additional costs are associated with a disruption of communications operations and procedures when migrating from an installed system to an entirely new platform. Although the data communications world has grown accustomed to constant change, one of the most fundamental characterizations of the voice communications world is reluctance to change.

Upgrading an installed circuit switched PBX system to a converged IP-PBX system is less costly than overhauling the entire communications system and network and is typically a far less disruptive experience because most station users are likely to retain their legacy desktop terminal equipment over the short term. The few station users who migrate to a ToIP desktop are more easily trained and supported in this environment. All of the features and functions available to station users before the system upgrade will remain available afterward, which is not usually the case if the legacy common control design is replaced with a new telephony call server loaded with a different software package.

PBX life cycles have fluctuated during the past several decades. Before the introduction of the computer-controlled digital PBX in the 1970s, the typical installed life of a PBX system was greater than 10 years. PBX systems did not change significantly over short periods. During the 1980s, the typical PBX life expectancy shrunk to between 6 and

8 years because of continual major design changes and dramatic software program upgrades. Since the early 1990s, the life cycle of an installed PBX system has slowly been increasing because system designers have become more cognizant of product migration strategy and its effect on market positioning and sales. Another reason for increased system life is that an increasing number of recent system performance enhancements and upgrades had minimal effect on the core processing and switching system design, because they were focused on peripheral server applications. The concept of upgrading an older PBX system to the current platform instead of the earlier system forklift approach makes possible a cost effective and operationally efficient migration from a circuit switched design platform to a converged circuit/packet switched system.

Upgrading a circuit switched PBX system to a converged IP-PBX system may be as simple as adding a few new port interface cards that provide ToIP gateway and gatekeeper functions for IP peripherals (stations, trunk). A *gateway* used for ToIP applications is defined as a device that provides a translation, or conversion, between IP and TDM/PCM signaling protocol and communications signals. There must be a physical and logical interface between circuit and packet switched communications networks. A gatekeeper performs or facilitates several basic call control functions within an IP communications network: peripheral equipment registration with the network; address translation of LAN aliases (i.e., telephone directory numbers into IP addresses); and bandwidth management of LAN/WAN network resources. The gatekeeper function in a pure IP communications design is similar to the call control processor in a traditional PBX system. Additional interface cards may be required for support of remote cabinets and port carriers when using the LAN/WAN to transport intercabinet signaling for call control and voice communications traffic. In some cases there may be a requirement for a call processing and/or system memory upgrade. It is most likely that a generic software release upgrade would be required unless a very recent software release is already installed.

A sizable percentage of the current installed base of circuit switched PBX systems can be upgraded to a converged IP-PBX system platform with the hardware/software additions just described. It is estimated that more than 65 percent of the current installed PBX base can be upgraded to a converged IP-PBX system without a major system overhaul. As the oldest installed PBXs are replaced or upgraded, the percentage of upgradeable systems will continue to increase. More than two-thirds of currently installed circuit switched PBXs can be upgraded in place to a converged IP-

PBX system platform while protecting up to 90 percent of a customer's original hardware/software investment. Examples of circuit switched PBXs installed during the past decade that are easily upgraded to a converged system include Avaya Definity ProLogix and ECS models, Nortel Networks Meridian models (Option 11C/51C/61C/81C), Siemens Hicom 300 models, and NEC NEAX 1000/2000/2400 models. These systems alone represent almost two-thirds of the North American installed PBX system base.

Critical Reliability

Traditional circuit switched PBX systems are known for their high reliability standards. The frequently quoted 99.999 percent system reliability benchmark is not a marketing gimmick but the reality. Five "9s" does not apply to every system PBX system component, such as a port circuit card, but to overall system availability. Individual port circuit cards or telephone instruments occasionally may fail, and software glitches may cause a feature failure or call disconnect, but total system failure is a rarity in the world of telephony. The reason for high system reliability, as a result of very low PBX system component failure rates, commonly expressed as Mean Time Between Failures (MTBF) or Mean Time Between Outages (MTBO), is that many system operations are supported by redundant hardware and/or software elements and great care in the electronic design and manufacture of hardware components. During the past decade, most PBX customers have installed and operated their communications systems without ever experiencing a catastrophic failure. Traditional circuit switched PBX technology has reached the bottom plateau of the failure rate curve. In case of failure, redundant system elements, such as main processor boards, system memory storage, switch network and transmission paths, and power supplies, are usually duplicated in a hot standby mode for intermediate and large PBX system models from many, but not all, system suppliers. For example, the Siemens HiPath 4000 and NEC NEAX 2400 IPX systems offer optional duplication of the listed critical system elements.

Although a converged IP-PBX is designed and equipped with several new hardware components and its software generic program includes new features and functions, the system, for the most part, is based on tried and true technology. The most important architecture element of any PBX system design is its common control complex, whether it is based on a proprietary cabinet carrier equipped with several printed circuit board modules or a third-party processing server. All PBX functions begin and end under the control and supervision of the common control complex. Traditional circuit switched PBXs and the upgraded converged IP-PBX version are based on the same common control complex, with perhaps a few modifications, such as an upgraded processor board. The hardware and software components and the core internal diagnostics, maintenance, and management functions remain basically unchanged when the system is upgraded to support ToIP capabilities and applications. The catastrophic failure rate should remain at (or be very close to) the expected 99.999 percent level. Unless an expensive fault tolerant server is used in the client/server IP-PBX system design or a dedicated back-up server is available, the common control reliability level will be less than that offered by a converged IP-PBX system. It should be noted that no client/server IP-PBX system currently uses a fault tolerant server, and very few are currently available with a backup server option.

The reliability of the PBX center stage switch complex is also a very important system factor for minimizing occurrences of catastrophic failure and supporting communications connections. In a converged IP-PBX, the center stage switch function is necessary for all calls among circuit switched peripherals and calls between circuit switched and packet switched peripherals. A client/server IP-PBX design makes use of LAN/WAN switches and routers for all call connections, even calls originating or terminating at non-IP ports. Although some converged IP-PBXs are based on a redundant internal switch network design with numerous duplicated hardware elements, a redundant LAN/WAN design requires multiple switches and routers at the Layer 2 and Layer 3 network levels for redundant connections and transmission paths for calls. More equipment is needed, more communications links must be supported, and more overhead costs are incurred. A configuration with a sizable number of non-IP peripherals in the IP-PBX configuration favors a design with traditional circuit switching capabilities in addition to ToIP capabilities. Except in a few isolated situations, there are very few intermediate/large customer installations that have mandatory requirements for 100 percent IP station users. Until the percentage of IP desktops outnumbers traditional analog and digital desktops, the converged IP-PBX system solution may be the preferable solution for customers with ToIP requirements.

Private Network Compatibility

There has been significant growth in the number of PBX private networks during the past 20 years. Private networks initially were limited to Fortune 500–type customers who had sufficient traffic volume to justify expensive private line facilities. The boom in private networks can be traced to the introduction of virtual private networking (VPN) services in the 1980s, such as the AT&T's Software-Defined Network (SDN), and the continuing decline in private line lease rates that coincided with the increased availability of wideband and broadband digital carrier facilities. Many present-day private networks are based on a mix of private lines and virtual network facilities and provide a high degree of feature-transparent operation across PBX system locations.

Intelligent feature transparent networking requires a common PBX system platform for optimal transparency of feature/function operation. The evolving Qsig.931 standard for interoperability between dissimilar PBX systems currently provides a limited level of transparency among most of the leading PBX systems. Although a Qsig-based private network implementation may be adequate for some customers, it is not an acceptable solution for most customers. A multisystem network, including PBXs from a variety of suppliers, does not lend itself to a unified systems management solution. The option to centralize network and systems management functions at one location is not currently available in a mixed system platform network. There are other network issues to consider, such as feature/function and desktop terminal uniformity. Unique features on one system cannot be supported transparently across the network to be enjoyed by all stations users. If station user interfaces and telephone instruments are not standardized across the network, training costs increase and station user productivity is affected when an individual moves between network locations.

Taking into account these private networking issues, a customer wishing to migrate from a circuit switched PBX to an IP-PBX is best served by upgrading the installed system to a converged system platform. The upgraded and converged IP-PBX system will retain existing networking capabilities, and station users will have continued access to their accustomed feature sets. A few converged IP-PBX offerings, such as the Siemens HiPath 3000/4000 families and NEC NEAX 1000/2000/2400 families, provide IP station users the option of simply adding an IP adapter to the installed digital telephone or installing a replacement IP telephone with the same look and feel as the older digital telephone model. Replacing a circuit switched PBX system with a client/server IP-PBX system from another supplier will reduce networking performance and force station users to learn how to use a new telephone to access and implement a different feature set.

Feature Requirements

Each customer has unique feature requirements, although there is a common core of features on most everyone's shopping list that is available with most every PBX system, regardless of switching technology or architecture design. A survey of the leading circuit switched PBX systems indicates that there are at least 500 software features that support a wide range of customer applications, ranging from basic station user desktop features, such as Call Forward, to advanced contact center ACD features, such as ACD agent skill profiling. Most station users, when surveyed, will come up with a list of fewer than a dozen PBX features that they commonly use in their everyday workplace. This does not mean that PBX systems should have a much smaller set of features because the typical station user uses a small percentage of currently available features.

Different station users use different features, and it is likely that in a large system environment a very high percentage of the available PBX features are used by at least one of the system's station users or administrators. Unique user populations within the PBX system make use of different features groups, such as attendant features, message center features, or ACD features. There are also many PBX features that station users are not aware of that support high-level system operations that are activated concurrently with many station users operations, such as CDR for off-premises calls, or Automatic Route Selection (ARS) when placing long distance calls.

Most of the currently marketed circuit switched PBX systems have software feature packages based on more than 20 years of additions, upgrades, and enhancements. A converged IP-PBX based on its antecedent circuit switched system platform would share the same highly developed and refined feature package, with no sacrifice in performance potential. Very few client/server IP-PBXs are based on previously available circuit switched PBX software feature packages. Most of the first-generation client/server designs are from system suppliers relatively new to the communications system market, with less than 5 years of software feature development. In addition, a few experienced PBX system manufacturers currently offer client/server IP-PBXs with far fewer features and functions than are available with their traditional circuit switched systems.

An analysis of the currently available client/server IP-PBX systems reveals that there are several functional areas where these new system designs are likely to be feature deficient when compared with circuit switched or converged system designs. Feature/function gaps are most common in attendant position, ACD, and private networking. There is also the occasional missing desktop station user feature that has been a longtime standard offering on circuit switched that customers cannot do without.

It has been the standard practice for customers to always ask for more features and functions in their new communications systems in addition to the features they have in their currently installed PBXs. Upgrading to a converged IP-PBX would satisfy this requirement. Replacing a circuit switched PBX system with a client/server IP-PBX that lacks more than a few traditional features and functions should not be acceptable because a smart customer does not trade a few new ToIP features for longtime available features.

Pricing

Price is a concern for all customers across all market segments. Very few customers have unlimited funds for their next communications system purchase. Among all the customer purchase criteria, pricing is the most prevalent. Every PBX system configuration has its own price point, and few customer configurations are identical. One cannot say that a converged IP-PBX is always priced higher or lower than a client/server IP-PBX because it is configuration dependent. Based on current pricing schedules, however, comparisons between the two IP-PBX design types can be made for specific defined configurations, and there are several pricing model assumptions that are usually valid regardless of system model:

1. An IP telephone is priced higher than a digital telephone with comparable capabilities, such as line appearances, programmable feature buttons, display field, speakerphone option.
2. Analog station (telephones, fax terminals) and PSTN trunk circuit connections are more expensive with a client/server design because gateways are more expensive than the traditional port circuit cards used in converged IP-PBXs.
3. Emergency power costs are greater for a client/server design because there is more distributed hardware equipment, such as telephony call servers, database servers, Ethernet switches, routers, and desktop terminals.
4. The cost to add an incremental IP port to a converged IP-PBX is greater than a client/server design because the converged system

Introduction to IP-PBX Systems

requires a port circuit card housed in a port carrier to support the added port.

5. Cabling costs for a green field installation are less for a client/server design than for a converged design.

Adding a few IP ports to a circuit switched PBX upgraded to a converged system will naturally be far more cost effective than replacing the entire system with a client/server design. If a significant number of IP ports are required, however, the cost of a client/server design is likely to be less expensive because the only significant variable cost is the price of a telephone. Converged systems require gateway/gatekeeper function port circuit cards to support IP telephones and/or IP trunk circuit connections for non-IP stations.

Figure 7-1 and 7-2 shows the two types of IP-PBX designs at a fixed port capacity. The customer configuration assumes a mix of single-line analog telephones, multiple-line telephones (digital or IP), PSTN local trunk circuits, and private network trunk circuits (PSTN or IP WAN). Design requirements such as redundancy (call processing, memory storage, power, switching) can significantly influence the shape of the two curves, as can advanced application requirements, such as contact centers, but the general trend lines would remain the same.

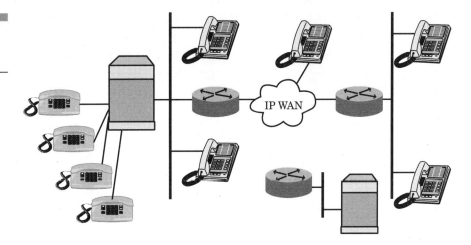

Figure 7-1
IP-enabled circuit switched PBX.

Major differences in the design types are best exemplified at the two extremes of the price plots, 0 percent and 100 percent IP ports. At minimal mandatory IP port requirements, a converged system is priced significantly less than a client/server design that is ill-suited to support non-IP ports, despite a lower cost for common control and reuse of LAN

Figure 7-2
TDM/LAN IP-PBX.

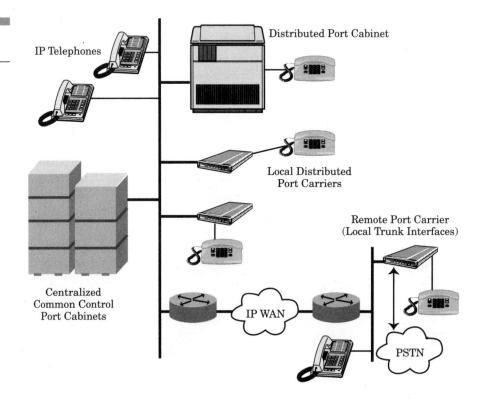

switches and IP routers, because IP telephones are more expensive and support of analog devices is very expensive. At maximum IP port requirements, a converged system is priced significantly more than a client/server design because of the requirement for gateway/gatekeeper port circuit cards. As mandatory IP port requirements increase, a client/server design is the favored solution because higher priced IP telephones are offset by reduced common equipment and wiring costs as compared with a converged solution.

A few closing comments regarding IP-PBX system pricing:

1. Client/server designs are priced lower than converged solutions as mandatory IP port requirements increase and/or traditional circuit switched port requirements decrease.
2. Green field environments are optimal for client/server designs because the newly installed LAN/WAN infrastructure can be initially designed and voice-grade QoS, and a single cabling system can be installed for all media communications needs.

Introduction to IP-PBX Systems

3. Upgrading a circuit switched PBX to a converged IP-PBX system is more cost effective than replacement with a client/server system, regardless of IP port requirements, because the common control complex and cabinet equipment is already in place and paid for.

The Case for the Client/Server IP-PBX

A client/server IP-PBX is likely to be the standard design platform of the future for enterprise communications systems. Although there are several drawbacks to current client/server IP-PBXs, there may be several benefits to customers who select it as their enterprise communications solution. Recalling the pricing discussion in the preceding paragraphs, the upfront capital investment for a client/server design is favorable for customer solutions with the following characteristics:

- Green field location
- Large percentage of IP peripherals, such as telephone instruments
- Remote location requirements with small port capacities
- Small port capacity requirements with PBX performance requirements

Green field locations without existing PBX equipment and with the newest generation of LAN switches and IP routers that can be programmed to support voice-grade QoS levels are best suited to cost effectively support a client/server design. A client/server design that supports a significant number of IP stations and minimal analog communications equipment will be more optimally priced, and support of remote IP stations likely will be less expensive than circuit switched alternatives. Customers with KTS/Hybrid port capacity requirements, but who require the performance capabilities of a larger, more complex PBX system, are likely to discover that the price of a client/server design is a more attractive alternative to a full-featured circuit switched or converged system solution.

The most common reasons given by system suppliers to promote the performance value of a client/server IP-PBX are:

1. Using a converged network for a variety of communications media: voice, data, video, text, graphics
2. Universality of IP transport

Figure 7-3
Basic client/server IP/PBX design.

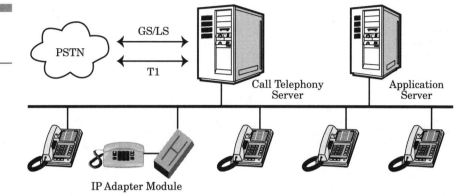

3. Lowered communications bandwidth requirements and more efficient use of existing bandwidth
4. Simplified centralized management and administration
5. Rapid deployment of new technology and applications
6. Fully distributed network design
7. Scalability

Converged Network

The concept of the integrated voice/data PBX system in the early 1980s was driven by the increasing amount of data traffic generated by geographically dispersed desktop CRT terminals linked to a centralized host processing system. It was thought that the circuit switched telephony network could be used to carry voice and data enterprise communications traffic. The advent of Ethernet LAN hubs and switches in the mid-1980s effectively ended the dream of the PBX system as an all-media office systems controller. Today the pendulum has swung the other way; packet switched LANs carry telephone-generated voice communications in addition to PC client data traffic. To data communications network designers and managers, the telephone is viewed as just another client, and voice features and functions are just other applications supported by a LAN-based server. If a customer is seriously thinking about migrating to an IP-PBX system, then a client/server design using the existing LAN/WAN infrastructure is the optimal solution. As LAN bandwidth capacity continues to increase, more video communications traffic will be carried between desktops with decreasing dependence on larger, more expensive, room-based videoconferencing systems.

Introduction to IP-PBX Systems

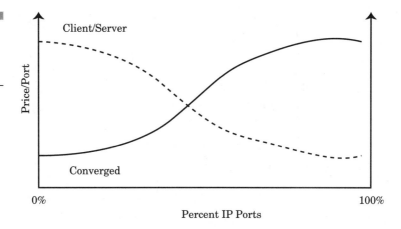

Figure 7-4
IP-PBX pricing: converged versus client/server.

Universality of IP Transport

Ten years ago the Internet was used almost exclusively by government and higher education institutions. Today the Internet is everywhere, and IP control and transmission signaling have become the standard for data communications networks. There is universal access to LANs and WANS in all medium and large enterprises across all industry sectors. The client/server enterprise data communications network design has replaced the host mainframe design that was dominant in the 1970s and most of the 1980s, and IP networking has replaced IBM's Systems Network Architecture (SNA) as the standard. If a customer is looking at an IP-PBX system solution, the current data networking infrastructure is favorable to a client/server design.

Network Bandwidth

Using the same communications network for voice and data traffic reduces overall bandwidth requirements because the two traffic streams can be interleaved and QoS levels can be engineered and programmed to satisfy real-time voice communications requirements. There will be cost savings and increased network efficiency due to economies of scale as a customer migrates an increasing percentage of traditional circuit switched PBX traffic to the packet switched LAN/WAN. The major bandwidth cost savings and optimization will be attributed to off-premises communications because circuit switched PSTN trunk carrier facility requirements can be reduced.

Simplified Centralized Management and Administration

A single communications management system is less costly to operate and more easily administered than separate systems for voice and data communications. The primary elements of an IP-PBX client/server design—desktop IP telephones and telephony call server(s)—are indistinguishable to a data network management system when compared to with PC clients and a data processing server or database manager. All voice system moves, adds, and changes are performed from the data network management workstation, as are maintenance and service operations.

Rapid Deployment of New Technology and Applications

A client/server design lends itself more to rapid deployment of new technology and applications because there are fewer hardware elements in the system architecture than in a converged IP-PBX that retains the traditional proprietary common equipment components of a circuit switched PBX. It is far easier to implement a technology upgrade for a client/server IP-PBX because there are fewer if any, proprietary switching network elements, and new advanced applications can be implemented through a software upgrade or an optional applications server. Client/server designs based on third-party servers are easily upgraded and enhanced by replacing the existing server with a newer, more powerful server. Migration between client/server IP-PBX generations will be smoother and less disruptive than between converged system generations that are based on more proprietary and costly common equipment hardware.

Fully Distributed Network Design

A client/server IP-PBX is by definition a distributed network design. A single telephony call server can support premises and off-premises IP stations. Premises stations can be distributed in a single building or across a campus. Multiple server designs can be programmed to support redundant emergency call processing in case of individual server failure. The servers can be colocated or distributed for disaster recovery situations. Customers have multiple layers of LAN/WAN switching and routing that preclude single points of failure.

Scalability

A client/server design has the potential to be highly scalable because IP telephones are easily added to the system without the need for specialized port interface circuit cards, and port capacity can be expanded through the addition of another server. Converged IP-PBXs require port interface cards, if only for control signaling, and, except for the rare system model based on a distributed processing design, there are call processing and port capacity limitations when using a single common control complex. The switching and transport limits of a client/server design are virtually boundless because a customer can continually install switches and routers to the LAN/WAN infrastructure to support increased traffic or more stringent QoS levels.

There are many customers for whom a converged or client/server IP-PBX system design will satisfy financial, feature performance, and applications requirements. Many enterprise communications system customers may also decide that their circuit switched PBX system will continue to be a satisfactory solution for current and near-term needs. A customer may not be ready to immediately replace his legacy PBX system with an IP-PBX system, but should seriously consider testing ToIP technology very soon. Chapter 8 discusses the technical design differences of the two competing IP-PBX system solutions in greater depth and reviews the basic ToIP standards currently used by currently available IP-PBXs.

CHAPTER 8

VoIP Standards and Specifications

Internet telephony is the transport of telephone calls over the Internet. The process of supporting voice calls over the Internet using the Internet communications protocol IP, is known as Voice over IP (VoIP). The first business customer implementations of VoIP were long distance calls over an IP WAN as an alternative to traditional PSTN trunk facilities. Network gateway servers had converted PBX TDM/PCM signals into IP format for transport through a router network. VoIP offered no new features or functions, merely an alternative means of transport. VoIP was not used for station-to-station on-premises calls, and IP control signaling was not used for call set-up or feature activation. As the new technology evolved to include PBX system desktop call control signaling and communications over an IP network infrastructure, ToIP gradually started to replace VoIP. VoIP is still the most commonly used term, although ToIP is being used more to describe the workings of an IP-PBX system.

There are several sets of VoIP communications protocols that can be used by an IP-PBX system for call control signaling and communications transmission. Circuit switched PBXs were based on proprietary call control signaling to the digital desktops and used TDM/PCM standards for transmission across internal switching networks and to network with other communications systems over PSTN trunk carrier facilities. It was the intention of the early IP-PBX system developers to use open standard communications protocols over packet switched networks as a counter to the closed, proprietary nature of circuit switched PBXs. The communications protocol of choice used by most first-generation IP-PBX system designers was the ITU-T H.323 series of protocols. A competing standards body, the IEFT, developed Session Initiation Protocol (SIP). Many IP-PBX manufacturers are planning to offer SIP versions of their current systems based on H.323, but the first product offerings will not be commercially available until later this year. SIP may offer several advantages over the ITU-T's recommendation, but the momentum of H.323 will delay saturation of the IEFT's solution to VoIP communications systems for several years according to the major IP-PBX manufacturers.

ITU-T H.323

ITU-T H.323 is a set of protocols for voice, video, and data conferencing over packet switched networks such as Ethernet LANs and the Internet that do not provide a guaranteed QoS. The H.323 protocol stack is designed to operate above the transport layer of the underlying network. H.323 uses the IP for internetwork conferencing.

H.323 was originally developed as one of several videoconferencing recommendations issued by the ITU-T. The H.323 standard is designed to allow clients on H.323 networks to communicate with clients on other videoconferencing networks. The first version of H.323 was issued in 1996 and designed for use with Ethernet LANs. H.323 Version 1 borrowed much of its multimedia conferencing aspects from other H.32x series recommendations. H.323 is part of a larger series of communications standards that enable videoconferencing across a range of networks. This series also includes H.320 and H.324, which address ISDN and PSTN communications, respectively.

H.323 is known as a broad and flexible recommendation. Although H.323 specifies protocols for real-time point-to-point voice communication between two terminals on a packet switched network, it also includes support of internetwork multipoint conferencing among terminals that support not only audio (voice) but also video and data communications.

H.323 recommendations can be summarized as followed.

Point-to-Point and Multipoint Conferencing Support

H.323 conferences may be set up between two or more clients without any specialized multipoint control software or hardware. If an MCU is used, H.323 supports a flexible topology for multipoint conferences. A multipoint conference can be centralized so that new participants can join all the others in the conference. A multipoint conference may also be decentralized so that new participants can elect to join one or more participants, but not all participants, in the conference. The centralized approach is a star topology; the decentralized one is a mesh topology.

Internetwork Interoperability

H.323 clients are interoperable with clients on circuit switched networks such as those based on recommendations H.320 (ISDN), H.321 (ATM), and H.324 (PSTN/Wireless). For example, it is possible to call from an H.323 client to a regular telephone on a PSTN. At the corporate level, this internetworking capability allows enterprises to migrate voice and video from existing networks to their own data networks.

Heterogeneous Client Capabilities

An H.323 client must support audio communication. Support of video and data communications is optional. During call set-up, capabilities are exchanged and communication established based on the lowest common denominator.

Audio and Video Codecs

H.323 specifies a required audio and video codec, but there is no restriction on the use of other codecs. Clients are allowed to decide which codec to use.

Management and Accounting Support

H.323 calls can be restricted on a network based on the number of calls already in progress, bandwidth limitations, or time restrictions. Policy management guidelines are used for H.323 traffic control. H.323 also provides accounting facilities that can be used for billing purposes.

Security

H.323 provides authentication, integrity, privacy, and nonrepudiation support.

Supplementary Services

Recommendation H.323 provides a basic framework for the development of application services, similar to the call processing features typically supported by a PBX system. This effort began with H.323 Version 2, which standardized a few services with recommendation H.450, including call transfer and call forwarding.

H.323 Benefits

IP-PBX system designers gain several important benefits by using the H.323 recommendations for support of IP telephony. One of the most important is an open system design because H.323 is not proprietary to any hardware equipment or operating system. There is a multitude of products that support H.323 standards and provide IP-PBX system designers with a flexible choice of system components and solutions. Major system hardware elements, such as servers and clients (telephones, PCs), are available from dozens of sources. Proprietary hardware is associated with circuit switched PBXs and is something IP-PBX system designers hope to avoid. Standards for voice and video codecs and for the compression and decompression of digital bit streams are also established under H.323, ensuring communications and networking between systems from different manufacturers. H.323 also establishes methods for receiving clients to communicate their capabilities to sending clients to establish common call set-up and control protocols, because it is important for clients in different corporate networks to communicate with each other.

The fact that H.323 is designed to operate on top of common network architectures is another important factor in the IP-PBX system open design concept. Network technology is constantly evolving, as are bandwidth management and applications services over the network. Systems based on H.323 standards can evolve as networks evolve. Dependency on one type of network or the currently installed network would limit the functional growth of an IP-PBX system. Many users may want to conference LAN clients to a station user at a remote site. H.323 establishes a means of linking LAN-based desktop systems with ISDN-based group systems. H.323 uses common codec technology from different audio- and videoconferencing standards to minimize transcoding delays and provide optimum performance.

An important PBX feature is multiparty conferencing. H.323 supports conferences of three or more endpoints without requiring a specialized MCU, but the H.323 MCU element can support a more powerful and flexible architecture for hosting multipoint conferences. The MCU element can be embedded in a variety of H.323 system components. H.323 also can support multicast transport in multipoint conferences. Multicast sends a single packet to a subset of destinations on the network without replication. Multicast transmission uses bandwidth more efficiently than unicast or broadcast methods of transmission because all stations in the multicast group read a single data stream. An H.323 conference also can include endpoints with different communication media capabilities. For example, a telephone with audio-only capabilities can participate in a conference with workstations that have audio, video, and data capabilities. Further, an H.323 multimedia terminal can share the data portion of a video conference with a T.120 data-only terminal while sharing voice, video, and data with other H.323 terminals

H.323 also addresses bandwidth management in support of bandwidth-intensive audio and video traffic. The network manager can control the number of simultaneous H.323 connections within a network or the amount of bandwidth available to H.323 applications. Limiting connections (conversations) across the network ensures that critical traffic will not be degraded. The QoS issue is one of the most important facing IP-PBX growth because many customers have problems addressing increased traffic load on their networks. Voice communications is very sensitive to delays, packet loss, and other network problems that do not affect data traffic to the same degree. Assuming that a significant amount of voice traffic will be transmitted on LAN networks based on a mix of old and new equipment originally configured for non–real-time communications, achieving acceptable QoS levels is one of the greatest barriers facing IP-PBX market penetration. Security of voice communications over LANs and WANs is another customer concern addressed by H.323. The H.323 recommendation provides authentication, integrity, privacy, and nonrepudiation support.

H.323 Architecture Components

H.323 specifies components, protocols, and procedures for real-time point-to-point and multipoint multimedia communications over packet-based networks. It also sets interoperability guidelines for communication between H.323-enabled networks and the H.32X-based family of conferencing standards.

An H.323 implementation requires four logical entities or components. These are terminals, gateways, gatekeepers, and MCUs. A fifth component element, known as *border elements*, is optional. Terminals, gateways, and MCUs are collectively known as *endpoints*. Although an H.323 network can be configured with only terminals, the other components are essential to provide greater practical usefulness of the services.

Terminal

A terminal, or a client, is an endpoint where H.323 data streams and signaling originate and terminate. In an IP-PBX system, it may be an IP telephone; a non-IP communications device, such as an analog telephone with an IP adapter module; or a PC client softphone with an H.323 compliant stack. A terminal must support audio communication; video and data communication supports are optional.

Gateway

A gateway is an optional component in an H.323-enabled network. A gateway will be required if at least one of the terminals does not conform to H.323 standards but is designed for a different type of network. The gateway is usually located at the interface between the two networks. Through the provision of gateways in H.323, it is possible for H.323 terminals to interoperate with other H.32X-compliant conferencing terminals. For example, a PC client softphone station user can talk to a station user on an analog telephone. A gateway provides data format translation, control signaling translation, audio and video codec translations, and call setup and termination functionality on both sides of the network. Gateway equipment is composed of a Media Gateway Controller (MGC) and a Media Gateway (MG), which may co-exist or exist separately. The MGC handles call signaling and other nonmedia-related functions, and the MG handles the media. There are different types of gateways required to support H.310, H.320, H.321, H.322, or H.324 endpoints.

Gatekeeper

A gatekeeper is an optional component of an H.323-enabled network. Gatekeepers ensure reliable, commercially feasible communications and are used for admission control and address resolution. The gatekeeper may allow calls to be placed directly between endpoints or it may route the call signaling through itself to perform functions, such as follow-me/find-me, forward on busy, etc. A gatekeeper is often referred to as the *controller* of the H.323 enabled network, because it provides central management and control services. When a gatekeeper exists, all endpoints (terminals, gateways, and MCUs) must be registered with it. Registered endpoints' control messages are routed through the gatekeeper.

The gatekeeper and the endpoints it administers form a management zone. A gatekeeper provides several services to all endpoints in its zone.

Address translation. A gatekeeper maintains a database for translation between aliases, such as international phone numbers, and network addresses.

Admission and access control of endpoints. This control can be based on bandwidth availability, limitations on the number of simultaneous H.323 calls, or the registration privileges of endpoints.

Bandwidth management. Network administrators can manage bandwidth by specifying limitations on the number of simultaneous calls and by limiting authorization of specific terminals to place calls at specified times.

Routing capability. A gatekeeper can route all calls originating or terminating in its zone. This capability provides numerous advantages: accounting information of calls can be maintained for billing and security purposes; a gatekeeper can reroute a call to an appropriate gateway based on bandwidth availability; and rerouting can be used to develop advanced services such as mobile addressing, call forwarding, and voice mail diversion.

Multipoint Control Unit

The MCU is an optional component of an H.323-enabled network. It is responsible for managing multipoint conferences (two or more endpoints engaged in a conference). It consists of a mandatory Multipoint Controller (MC) that manages the call signaling and optional Multipoint Processors (MPs) to handle media mixing, switching, or other media processing. Although the MCU is a separate logical unit, it may be combined into a terminal, gateway, or gatekeeper.

The MC provides a centralized location for multipoint call set-up. Call and control signaling are routed through the MC so that endpoint capabilities can be determined and communication parameters negotiated.

An MC may also be used in a point-to-point call, which can later be extended into a multipoint conference. The MC may also determine whether to unicast or multicast the audio and video streams depending on the capability of the underlying network and the topology of the multipoint conference.

The MCU is required in a centralized multipoint conference, where each terminal establishes a point-to-point connection with the MCU. The MCU determines the capabilities of each terminal and sends each a mixed media stream. In the decentralized model of multipoint conferencing, an MC ensures communication compatibility, but the media streams are multicast and the mixing is performed at each terminal.

Border Elements

Border elements are often colocated with a gatekeeper. They exchange addressing information and participate in call authorization between administrative domains. Border elements may aggregate address information to reduce the volume of routing information passed through the network and assist directly in call authorization/authentication between two administrative domains or via a clearinghouse.

H.323 Architecture Protocols and Procedures

H.323 is an umbrella recommendation that depends on several other standards and recommendations to enable real-time multimedia communications. The following are the key H.323 reference recommendations.

Call Signaling and Control

- H.323: Packet-based multimedia communications systems
- H.225: Call control protocol
- H.235: Security
- H.245: Media control protocol

- Q.931: Digital subscriber signaling
- H.450.1: Generic functional protocol for the support of supplementary services in H.323
- H.450.2-11: Supplemental features: blind call transfer and call diversion, consulting (450.2); call forward, activation/deactivation, interrogation (450.3); call hold (450.4); call park, call pickup (450.5); call waiting (450.6); message waiting (450.7); name interrogation (450.3), call hold (450.4); call park/call pickup (450.5); call waiting (450.6); message waiting (450.7); name identification service (450.8); call completion (450.9); call offer (450.10); call intrusion (450.11)

H323 Annexes

- Annex D: Real time fax over H.323
- Annex E: Multiplexed call signaling
- Annex F: Simple Endpoint Terminal (SET)
- Annex G: Text SET
- Annex H: Mobility
- Annex I: Operation over low QoS networks
- Annex J: Secure SET
- Annex K: HTTP service control transport
- Annex L: Stimulus signaling
- Annex M: Qsig tunneling
- Annex N: QoS

Audio Codecs

- G.711: PCM audio codec 56/64 kbps
- G.722: Audio codec for 7 Khz at 48/56/64 kbps
- G.723: Speech codec for 5.3 and 6.4 kbps
- G.728: Speech codec for 16 kbps
- G.729: Speech codec for 8/13 kbps

Video Codecs

- H.261: Video codec for $\geq$ 64 kbps
- H.263: Video codec for $\leq$ 64 kbps

VoIP Standards and Specifications

H.323 Version 1 was approved in 1996. Version 1 centered on multimedia communications, such as voice and video over IP data networks. Version 2 was approved in January 1998, and addressed deficiencies in Version 1 and introduced new functionality within existing protocols, such as Q.931, H.245, and H.225, and entirely new protocols. The most significant advances were in security, fast call set-up, supplementary services, and T.120/H.323 integration. There was more efficiency about getting media streams to transfer at a faster rate. The major functions introduced were Fast Connect (also known as Fast Start) and H.245 tunneling (transferring minimal call signals to more quickly establish connection). Version 3, approved in May 1999, had several new annexes or sections focused on H.323-compliant devices for large-scale production networks. It covered bandwidth management and QoS issues and focused on "smart" networks and "dumb" endpoints (master/slave relationship). Some specific areas addressed were:

- Connection over UDP
- Simple endpoint type
- Interdomain communications
- H.263 and packetization
- H.GCP decomposition architecture

H.323 Version 4 was approved in November 2000 and contained many enhancements in a number of important areas, including reliability, scalability, and flexibility. Many of the new features facilitate more scalable gateway and MCU solutions to meet the market requirements for IP telephony. Version 4 addressed the following issues:

- Gateway decomposition
- Multiplexed stream transmission
- Supplementary services
- Additive registrations
- Alternate gatekeepers
- Usage information reporting
- Endpoint capacity
- Tones and announcements
- Mapping aliases
- Indicating desired protocols
- Bandwidth management
- Reporting call status

- Real-time fax
- Call linkage
- Tunneling
- QoS
- H.245 in parallel with Fast Connect
- Generic extensibility framework
- H.323 URL
- Call credit-related capabilities
- DTMF relay via RTP

H.323 Audio Codecs

H.323 specifies a series of audio codecs ranging in bit rates from 5.3 to 64 kb/s. The mandatory codec is G.711, which uses traditional PCM to produce bit rates of 56 and 64 kb/s. G.711 is fine for premises networks, where bandwidth is not an important issue, but it is less appropriate for communication over WANs, where network bandwidth is a more expensive and limited resource. Sample delay for G.711 is minimal. Other codecs include G.723.1, which is much more bandwidth efficient, offers sufficient quality audio at 5.3 kb/s and 6.3 kb/s, but has the largest sampling delay factor. The G.728 and G.729 codecs use advanced linear prediction quantization of digital audio to produce high-quality audio at 16 and 8 kb/s, respectively. G.729 codecs have emerged as the most popular for voice calls across the WAN, because it is almost as bandwidth efficient as G.723.1, but its sampling delay is far more acceptable. Table 8-1 summarizes the audio codec specifications, including audio bandwidth, sampling time, and total IP bandwidth.

H.323 Control and Signaling Mechanisms

The flow of information in an H.323-enabled network consists of a mix of audio, video, data, and control packets. Control information is essential for call set-up and tear-down, capability exchange and negotiation, and administrative purposes. H.323 uses three control protocols: H.245

VoIP Standards and Specifications

TABLE 8-1
Audio Codec Specifications

Coding Algorithm		Bandwidth	Sample	IP Bandwidth
G.711	PCM	64 kbps	0.125 ms	80 kbps
G.723.1	ACELP	5.6 kbps 6.4 kbps	30 ms	16.27 kbps 17.07 kbps
G.726	ADPCM	32 kbps	0.125 ms	48 kbps
G.728	LD-CELP	16 kbps	0.625 ms	32 kbps
G.729 (A)	CS-ACELP	8 kbps	10 ms	24 kbps

media control, H.225/Q.931 call signaling, and H.225.0 registration, admission, and status (RAS). The Q.931 protocol was originally developed for ISDN control signaling, and is currently used for inter-PBX networks implementing Qsig standards (see Networking chapter).

Figure 8-1 shows the H.323 protocol stacks for control and signaling processes.

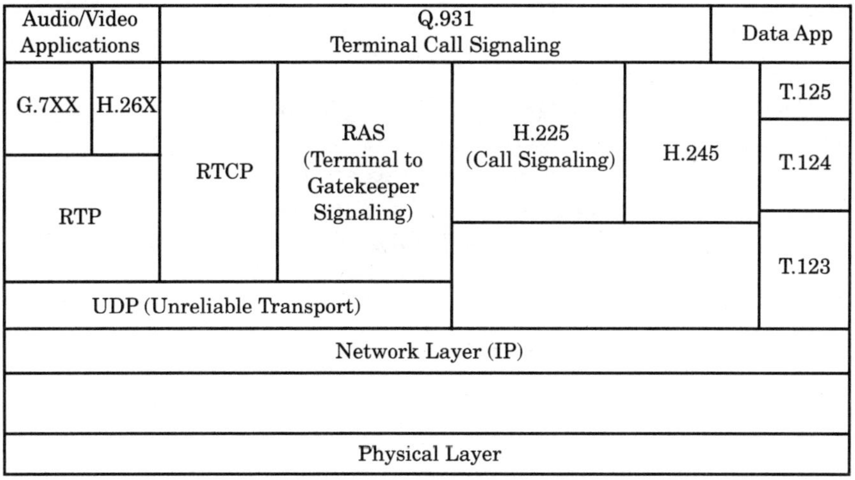

Figure 8-1
H.323 system: protocol architecture.

Audio/Video Applications		Q.931 Terminal Call Signaling			Data App	
G.7XX	H.26X				T.125	
RTP	RTCP	RAS (Terminal to Gatekeeper Signaling)	H.225 (Call Signaling)	H.245	T.124	
					T.123	
UDP (Unreliable Transport)						
Network Layer (IP)						
Physical Layer						

The diagram follows the ISO OSI seven-layer model. H.323-specific protocols are above the transport layer (Layer 4). Real-Time Transport Protocol (RTP)/RTCP, RAS, H.225, and H.245 span across the fifth and sixth layers. Q.931, sometimes included as part of the H.225 protocol set, is at the terminal applications layer (Layer 7). Data communications are supported by multiple T.120 protocols and use TCP/IP transmission protocols, which are different from the H.323 protocols that support real-time audio and video communications requirements.

H.225.0 RAS

H.225.0 RAS messages define communications between endpoints and a gatekeeper. H.225.0 RAS is only required when a gatekeeper exists. Unlike H.225.0 call signaling and H.245, H.225.0 RAS uses unreliable transport for delivery. RTP is used to guarantee delivery transport.

Gatekeeper Discovery

Gatekeeper discovery is used by endpoints to find their gatekeeper. An endpoint needing to find the transport address of its gatekeeper(s) will multicast a gatekeeper request (GRQ) message. One or more gatekeepers may reply with a GCF message containing the gatekeeper transport address.

Endpoint Registration

Once a gatekeeper exists, all endpoints must be registered with it. This is necessary because gatekeepers need to know the aliases and transport addresses of all endpoints in its zone to route calls.

Endpoint Location

Gatekeepers use this message to locate endpoints with a specific transport address. This process is required, for example, when the gatekeeper updates its alias transport address database.

Other Communications

A gatekeeper performs many other management and control duties such as admission control, status determination, and bandwidth management, which are all handled through H.225.0 RAS messages.

VoIP Standards and Specifications

Figure 8-2 shows RAS. The terminal sends a request to the gatekeeper for registration and admission. The endpoints acknowledge and confirm the requests. When the call is completed, the terminal notifies the gatekeeper of the call status and receives confirmation that the request to disconnect has been received.

Figure 8-2
RAS.

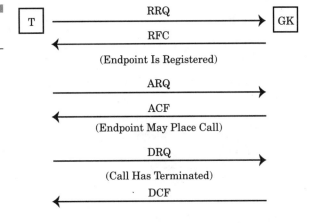

H.225.0 Call Signaling

Call signaling is a basic requirement for setting up and tearing down a call between two endpoints. H.225.0 uses a subset of the Q.931 signaling protocol for this purpose. H.225.0 adopts Q.931 signaling by incorporating it in its message format. H.225.0 call signaling is sent directly between the endpoints when no gatekeeper exists. When a gatekeeper exists, then the call may be routed through the gatekeeper. The exchange of messages needed during a basic H.323 call is described in detail in the next section.

H.245 Media Control

H.323 requires that endpoints negotiate compatible settings before audio, video, and/or data communication links can be established. H.245 uses control messages and commands that are exchanged during the call to inform and instruct. The implementation of H.245 control is mandatory in all endpoints. H.245 provides the following media control functionalities:

Capability Exchange

H.323 allows endpoints to have different receive and send capabilities. Each endpoint records its receiving and sending capabilities (media types, codecs, bit rates, etc.) in a message and sends it to the other endpoint(s).

Opening and Closing of Logical Channels

H.323 audio and video logical channels are unidirectional end-to-end links (or multipoint links in the case of multipoint conferencing). Data channels are bidirectional. A separate channel is needed for audio, video, and data communications. H.245 messages control the opening and closing of such channels. H.245 control messages use logical channel 0, which is always open.

Flow Control Messages

These messages provide feedback to the endpoints when communication problems are encountered.

Other Commands and Messages

Several other commands and messages may be used during a call, such as a command to set the codec at the receiving endpoint when the sending endpoint switches its codec. H.245 control messages may also be routed through a gatekeeper if one exists.

Figure 8-3 shows H.245 signaling. After establishing a control signaling link between two gateways, media bandwidth is negotiated. After the two terminals agree, with acknowledgments, an open link is established for the call.

The Need for RTP/RTCP

The IP is a relatively low-level protocol. It was originally developed for delivery of packets (or datagrams) between host computers across the

VoIP Standards and Specifications

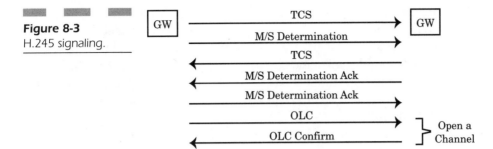

Figure 8-3
H.245 signaling.

ARPAnet (Internet) packet network. For an IP telephony application, datagrams are transmitted between desktop voice terminals. IP is a connectionless protocol that does not establish a virtual connection through a network before commencing transmission. Establishing a communications path between endpoints is the responsibility of higher-level protocols.

IP makes no guarantees concerning reliability, flow control, error detection, or error correction. As a result, datagrams could arrive at the destination computer out of sequence, with errors, or not even arrive at all. This is known as *jitter*. IP does succeed in making the network transparent to the upper layers involved in voice transmission through an IP-based network.

VoIP transmission, by definition, uses IP, although it is not well suited for voice transmission. Real-time applications such as voice and video require guaranteed connection with consistent delay characteristics. Higher-layer protocols address these issues. There are two available protocols at the transport layer when transmitting information through an IP network. These are Transmission Control Protocol (TCP) and User Datagram Protocol (UDP). Both protocols enable the transmission of information between the correct processes (or applications) on network endpoints. These processes are associated with unique port numbers.

TCP is a connection-oriented protocol. It establishes a communications path before transmitting data. It handles sequencing and error detection, thus ensuring that a reliable stream of data is received by the destination application. TCP can address real-time voice applications to a certain extent but would require higher-layer functions. Voice applications require that information is received in the correct sequence, reliably, and with predictable delay characteristics. With this in mind, the ITU-T decided that the alternative protocol, UDP, should be used. UDP is also a connectionless protocol. UDP routes data to its correct destination port but does not attempt to perform any sequencing or ensure data reliability.

To provide feedback on the quality of the transmission link, the RTP/RFTP protocols, developed by the IETF, are used. Real-Time Transport Protocol (RTP) transports the digitized samples of real-time information, and Real-Time Control Protocol (RTCP) provides the mechanism for quality feedback. RTP and RTCP do not reduce the overall delay of the real-time information. Nor do they make any guarantees concerning QoS.

When an IP voice terminal transmits datagrams across the LAN/WAN, the IP, UDP, and RTP headers are followed by the data payload of the RTP header. The data payload is comprised of digitized voice samples. The length of these samples can vary, but for voice samples representing 20 ms are considered the maximum duration for the payload. The number of transmitted datagrams varies indirectly with the sampling rate—the longer the sampling period, the fewer the number of packets transmitted per second. The selection of the payload duration is a compromise between bandwidth requirements and quality. Smaller payloads demand higher bandwidth per channel band because the header length remains at 40 octets. However, if payloads are increased, the overall delay of the system will increase, and the system will be more susceptible to the loss of individual packets by the network.

Real-Time Transport Protocol

The RTP provides end-to-end network transport functions suitable for applications transmitting real-time audio or video packets over multicast or unicast network services. It was developed by the IETF and is used with the H.323's recommended H.225 protocols to provide reliable communications. RTP by itself does not address resource reservation and does not guarantee QoS for real-time services. The packet transport is supplemented by a control protocol (RTCP) that monitors data delivery in a manner scalable to large multicast networks and provides minimal control and identification functionality. RTP and RTCP are designed to be independent of the underlying transport and network layers. The protocol supports the use of RTP-level translators and mixers. The following are the elements of RTP.

RTP Payload

The media payload is transported by RTP in a packet, such as audio samples or compressed video data. The payload format and interpretation are beyond the scope of this document.

RTP Packet

A packet consists of the fixed RTP header, a possibly empty list of contributing sources (see below), and the payload data. Some underlying protocols may require an encapsulation of the RTP packet to be defined. Typically, one packet of the underlying protocol contains a single RTP packet, but several RTP packets may be contained, if permitted by the encapsulation method.

RTCP Packet

A control packet consists of a fixed header part similar to that of RTP packets, followed by structured elements that depend on the RTCP packet type. Typically, multiple RTCP packets are sent together as a compound RTCP packet in a single packet of the underlying protocol; this is enabled by the length field in the fixed header of each RTCP packet. Figure 8-4 shows the RTP header.

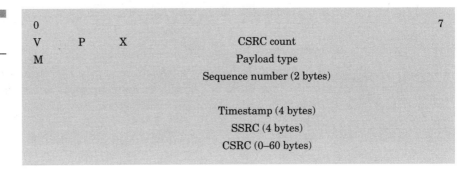

Figure 8-4 The RTP header.

The RTP fixed header fields have certain functions. V (version): Identifies the RTP version. P (padding): When set, the packet contains one or more additional padding octets at the end that are not part of the payload. X (extension bit): When set, the fixed header is followed by exactly one header extension with a defined format. CSRC count: Contains the number of CSRC identifiers that follow the fixed header. M (marker): The interpretation of the marker is defined by a profile. It is intended to allow significant events such as frame boundaries to be marked in the packet stream. Payload type: Identifies the format of the RTP payload and determines its interpretation by the application. A profile specifies a default static mapping of payload type codes to payload formats. Additional payload type codes may be defined dynamically through non-RTP means. Sequence number: Increments by one for each RTP data packet sent and may be used by the receiver to detect packet loss and restore packet sequence. Timestamp: Reflects the sampling instant of the first octet in the RTP data packet. The sampling instant must be derived from a clock that increments monotonically and linearly in time to allow synchronization and jitter calculations. The resolution of the clock must be sufficient for the desired synchronization accuracy and for measuring packet arrival jitter (one tick per video frame is typically not sufficient). SSRC (synchronization source): Identifies the synchronization source. This identifier is chosen randomly, with the intent that no two synchronization sources within the same RTP session will have the same SSRC identifier. CSRC (contributing source): Contributing source identifiers list. Identifies the contributing sources for the payload contained in this packet.

Real-Time Transport Control Protocol

The RTCP is based on the periodic transmission of control packets to all participants in the session, with the same distribution mechanism as that for the data packets. The underlying protocol must provide multiplexing of the data and control packets, for example, separate port numbers with UDP.

The format of the header is shown in Figure 8-5.

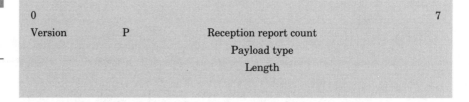

Figure 8-5 Format of the header.

Version: Identifies the RTP version, which is the same in RTCP packets and RTP data packets. Version 2 is defined by this specification. P (padding): When set, this RTCP packet contains some additional padding octets at the end, which are not part of the control information. The last octet of the padding is a count of how many padding octets

VoIP Standards and Specifications

should be ignored. Padding may be needed by some encryption algorithms with fixed block sizes. In a compound RTCP packet, padding should be required only on the last individual packet because the compound packet is encrypted as a whole. Reception report count: The number of reception report blocks contained in this packet. A value of zero is valid. Packet type: Contains the constant 200 to identify this as an RTCP SR packet. Length: The length of this RTCP packet in 32-bit words minus one, including the header and any padding. (The offset of one makes zero a valid length and avoids a possible infinite loop in scanning a compound RTCP packet, and counting 32-bit words avoids a validity check for a multiple of four.)

Figure 8-6 shows the complete packet header for IP, UDP, and RTP. The headers of the three payload-carrying protocols are sent sequentially before the digitized voice samples, which are actually the payload of the RTP header. The result is a 40-octet overhead for every information data packet.

Figure 8-6
Packet header for IP, UDP, and RTP.

Bits	0-3	4-7	8-15	16-23	24-31
	Octet 1, 5, 9...	Octet 2, 6, 10...		Octet 3, 7, 11...	Octet 4, 8, 12...
1–4	Version	IHL	Type of Service	Total Length	
5–8	Identification			Flags	Fragment Offset
9–12	Time to Live		Protocol	Header Checksum	
13–16	Source Address				
17–20	Destination Address				
21–24	Source Port			Destination Port	
25–28	Length			Checksum	
29–32	V=2 P X CC	M	PT	Sequence Number	
33–36	Timestamp				
37–40	Synchronization Source (SSRC) Number				
	The headers are followed by a payload of digitized voice/video samples.				

Figure 8-7 shows an H.323 call setup between two H.323 terminals. The gatekeeper server in the diagram could represent an IP-PBX call telephony server if it were an IP-PBX system, and the H.323 terminals could just as well be IP telephones. The gatekeeper and H.323 terminals reside on a LAN. The first steps in the call set-up process are terminal registration and admission with the gatekeeper. The calling terminal establishes a TCP signaling connection with the called terminal and receives a connection acknowledgment. Bandwidth requirements and management are controlled by TCP-based H.245 signaling. UDP voice packets are transmitted across the LAN between the terminals under the control of RTP and RTCP protocols.

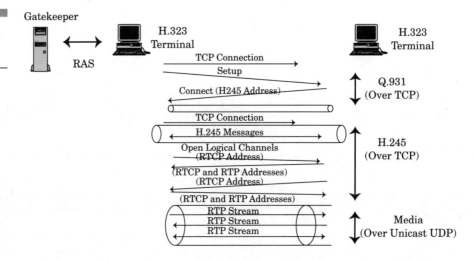

Figure 8-7
H.323 protocol and call setup.

Table 8-2 shows the precise control messages that are exchanged between terminals from call set-up to call termination. The originating terminal (1) initiates a call to the destination terminal (2) directly, without any intermediate gateway or gatekeeper. H.225 and H.235 messages are indicated. Some messages overlap each other (Messages 4/5 and 9/10). H.225 messages are Messages 1, 2, 3, and 12. 12; the remaining messages are H.245.

Terminal 1 initiates the call by sending the Setup message to Terminal 2. Terminal 2 replies with Alerting and Connect messages to Terminal 1, indicating that it is ready for the call. H.225 call signaling is followed by H.245 capability exchange messages (beginning at Message 4). Each terminal sends a termCapSet message to communicate its media settings to the other terminal. The terminals then acknowledge each other's settings by the termCapSetAck messages. Master and slave ter-

VoIP Standards and Specifications

TABLE 8-2
Terminal Message Exchange During a Call

Message	Terminal 1	Terminal 2
1	Setup	
2		Alerting
3		Connect
4	termCapSet	
5		termCapAck
4		termCapSet
5	termCapAck	
6	masterSlvDet	
7		masterSlvDetAck
8	masterSlvDetConfirm	
9	openReq	
10		openAck
9		openReq
10	openAck	
11	endSession	
11		endSession
12	Release	

minals are determined by using the masterSlvDet and masterSlvDetAck messages. The master (Terminal 1) leads the opening of a logical channel using the openReq message. Terminal 2 follows by opening a logical channel in the other direction. The terminals can open as many channels as is practically possible. The call is terminated after the exchange of H.245 endSession messages (Message 11) with the H.225.0 Release message (Message 12).

SIP

The IETF developed SIP in reaction to the ITU-T H.323 recommendations. The IETF believed H.323 was inadequate for evolving IP telephony requirements because its command structure was too complex and its architecture was too centralized and monolithic. SIP is an application layer control protocol that can establish, modify, and terminate multimedia sessions or calls. SIP transparently supports name mapping and redirection services, allowing the implementation of ISDN and Intelligent Network telephony subscriber services. The early implementations of SIP have been in network carrier IP-Centrex trials. IP-PBX manufacturers are in the process of developing SIP-based versions of their current product offerings.

SIP was designed as part of the overall IETF multimedia data and control architecture that supports protocols such as Resource Reservation Protocol (RSVP), RTP, Real-Time Streaming Protocol (RTSP), Session Announcement Protocol (SAP), and Session Description Protocol (SDP). Figure 8-8 shows SIP and its associated protocols.

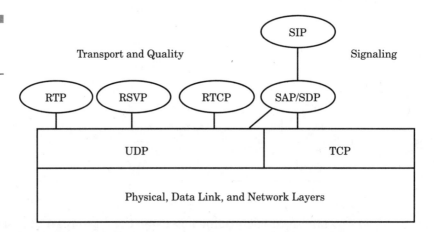

Figure 8-8
SIP signaling protocols.

SIP provides the necessary protocol mechanisms to support the following basic functions:

- **Name translation and user location**—Determination of the end system to be used for communication
- **Feature negotiation**—Allows station users involved in a call to agree on the features supported, recognizing that not all features are available to all station users
- **Call participation management**—During a call, a station user can conference other station users into the call or cancel connections to conferenced parties; station users can also be transferred or placed on hold
- **Call feature changes**—A station user should be able to change the call characteristics during the course of the call; new features may be enabled based on call requirements or new conferenced station users

VoIP Standards and Specifications

The two major components in a SIP network are User Agents and Network Servers. A User Agent Client (UAC) initiates SIP requests, and a User Agent Server (UAS) receives SIP requests and return responses on user behalf. A Registration Server receives updates regarding the current user location, and a Proxy Server receives and forwards requests to a next-hop server, which has more information regarding called party location. A Redirect Server receives requests, determines next-hop server, and returns an address to client.

SIP request messages consist of three elements: Request Line, Header, and Message Body. SIP response messages consist of three elements: Status Line, Header, and Message Body.

Figure 8-9 shows the basic steps for a SIP call set-up.

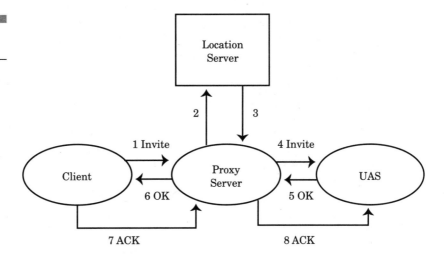

Figure 8-9
SIP call setup.

SIP Call Process

SIP Addressing

The "objects" addressed by SIP are users at hosts, identified by a SIP URL. The SIP URL takes a form similar to that of an e-mail address: user@host. The user part is a user name or a telephone number. The host part is a domain name or a numeric network address.

Locating a SIP Server

When a client wishes to send a request, the client sends it to a locally configured SIP proxy server (as in HTTP), independent of the request-URL, or to the IP address and port corresponding to the Request-URL. For the latter case, the client must determine the protocol, port, and IP address of the server receiving the request.

SIP Transaction

Once the host part has been resolved to a SIP server, the client sends one or more SIP requests to that server and receives one or more responses from the server. A request (and its retransmissions) and the responses triggered by that request make up a SIP transaction.

SIP Invitation

A successful SIP invitation consists of two requests: INVITE, followed by ACK. The INVITE request asks the callee to join a particular conference or establish a two-party conversation. After the callee has agreed to participate in the call, the caller confirms that it has received that response by sending an ACK request. If the caller no longer wants to participate in the call, it sends a BYE request instead of an ACK.

SIP itself only defines the initiation of a session. All other parts of the session are covered by the other parts of the aforementioned protocol, some of which come from other applications or functions not necessarily designed for real-time multimedia over IP. Compared with H.323, SIP is less defined and more open, which can result in interworking difficulties because of different implementations of the standard. Every SIP developer can implement a unique version with different extensions that aren't included in the basic standard. Although H.323 and SIP handle call set-up, call control, and media in different ways, H.323 defines all of these processes, whereas SIP defines call set-up only, and uses other protocols for call control and media. Using SIP, call control and set up are handled separately from media. This becomes an important issue when interworking with the PSTN, which uses SS7 for signaling, although SS7 can be translated to SIP through a gateway or softswitch. Otherwise, intelligent networking services such as caller ID and call forwarding will not work with SIP. Media and signaling are handled separately in SIP, requiring separate media gateway and signaling gateways for interoperability with the PSTN. This can create a major problem in the case of common PSTN services like DTMF or touch tones, in which signaling is carried in the media. There is also no SIP equivalent of ISUP message transport from SS7.

H.323 or SIP?

Despite SIP's limitations, there are several often-cited reasons demonstrating why SIP is superior to H.323 in supporting IP telephony processes and applications.

Emerging Dominance of IP

H.323 is designed to be interoperable with other ITU-T standards in support of ISDN and ATM networks and includes all of the necessary mechanisms to support interoperability across multiple networks. SIP is designed primarily for IP networks, and the continuing growth of the Internet and its associated protocol makes interoperability a less important issue. Although the Internet has become a ubiquitous presence, the dominance of PSTN services for voice communications will continue for some time, making interoperability an issue for years to come. PSTN signaling interoperability and the more mature nature of H.323 are advantages over SIP at present. H.323 products currently far outnumber SIP offerings, although this is expected to change over time.

Signaling Reliability Mechanism

SIP provides its own reliability mechanism, is independent of the packet layer, and only requires an unreliable datagram service. H.323 requires RTP/RTCP for reliability but provides a better QoS level (see below).

Client/Server Design

SIP messages are exchanged between a client and a server in the same way as HTTP messages. H.323 has a more complex call control protocol.

It takes more time to establish an H.323 call than a SIP call. SIP's client/server operation mode allows security and management features to be implemented easily in SIP, when compared to H.323 calls. SIP uses distributed multicasting signaling support, a more flexible method than the H.323 centralized, unicasting signal method.

Addresses

SIP addresses are like e-mail addresses. Each user is identified through a hierarchical URL that is built around the elements such as a station user's telephone number or host names. A SIP URL can be easily associated with a user's e-mail address, greatly simplifying dialing plans. SIP's use of any URL address for station user identification is less cumbersome than H.323's requirement for host addressing. Using a single identifier for voice and text communications has its benefits, but only if everyone is using SIP. Private network calls may use the SIP address, but dialing off-network stations will still require the use of the traditional multiple-digit dialing codes.

Complexity and Cost

SIP is a much smaller and less complicated standard that is based on the architecture of existing popular protocols such HTTP and FTP, whereas H.323 is large and complicated. As a result, H.323 products and services are more expensive to develop, and license fees will also be more expensive. Although SIP cost savings are not apparent in the first generation of products, particularly telephones, this appears to be a longer-term advantage for SIP versus H.323.

Command/Message Format

Compared with H.323, SIP uses a simple command format, and the text strings are easier to decode and debug. The entire set of messages is also much smaller than in H.323. This gives SIP an advantage for future SIP software development efforts, particularly development of new features and application services.

QoS Management

SIP supports loop detection, unlike H.323. SIP's algorithm using "via header" is somewhat better than H.323's PathValue approach. In other areas, H.323 has the upper hand. Regarding fault tolerance, H.323 supports redundant gatekeepers and endpoints, unlike SIP. H.323 support of Call Admission Control is better than SIP's reliance on other protocols for bandwidth management, call management, and bandwidth control. H.323 also has limited support of Differentiated Services (Diffserv), but SIP does not. Overall, H.323 wins in this arena.

Firewall/Proxy Design and Configuration

SIP commands can easily be proxied and firewalls can be designed to allow/disallow SIP communications. Getting H.323 through firewalls and proxies is much more complicated.

Extendible and Scalable

SIP is more scalable than H.323 because it is based on a distributed client/server architecture. H.323 often requires peer-to-peer communication, making it more difficult to expand networks. Extending SIP is also easier because of its simpler message format and greater experience with similar protocols such as HTTP. SIP's use of hierarchical feature names and error codes, which can be IANA registered, is more flexible than H.323's vendor-specific single extension field.

H.323 offers the benefits of superior network interoperability, better QoS management, and redundancy. SIP is a less complex protocol that is more easily adapted to expansive and growing networks, supports a faster call set-up time, and uses an addressing scheme that leverages the existing DNS system instead of recreating a separate hierarchy of telephony name servers. The two protocols will likely exist side-by-side for many years, until they either merge or are supplanted by something newer and better.

Other Protocols: MGCP and MEGACO (H.248)

The MGC Protocol, or MGCP, was designed to address the requirements of VoIP networks that are built using "decomposed" VoIP gateways. MGCP is used by external call control elements called MGCs for controlling MGs. Decomposed MGCP-compliant VoIP gateways appear to the outside as a single VoIP gateway. Examples of these gateways are trunking gateways that interface the PSTN and VoIP network, desktop gateways that provide traditional analog 2500-type interfaces to VoIP networks, and access gateways that provide traditional analog or digital PBX interfaces to VoIP networks. MGCP-based VoIP solutions separate call control (signaling) intelligence and media handling. MGCP functions as an internal protocol between the separate components of a decomposed MGCP-compliant VoIP gateway.

MEGACO (H.248) is a standard protocol for interfacing between external call agents (MGCs) and MGs. The standard is the result of a unique collaborative effort between the IETF and ITU standards organizations. H.248 was derived from MGCP, which in turn was derived from the combination of Skinny Gateway Control Protocol (SGCP) and IPDC. SGCP is sometimes known as Skinny Protocol and is currently used by Cisco systems to support their proprietary IP telephone instruments. H.248 is based heavily on MGCP but includes several enhancements. MEGACO offers the following key enhancements over MGCP:

- Supports multimedia and multipoint conferencing-enhanced services
- Improved syntax for more efficient semantic message processing
- TCP and UDP transport options
- Allows text or binary encoding, formalized extension process for enhanced functionality
- Formalized extension process for enhanced functionality

H.248 has the same architecture as MGCP. The commands are similar, but the main difference is that H.248 commands apply to terminations relative to a context rather than to individual connections, as is the case with MGCP. Connections are achieved by placing two or more terminations into a common context. It is the concept of a context that facilitates support of multimedia and conferencing calls. The context can be viewed as a mixing bridge that supports multiple media streams for enhanced multimedia services.

H.248 packages include more details than MGCP packages. H.248 packages define additional properties, statistics, and event and signal information that may occur on terminations. With H.248, the primary mechanism for extension is by means of packages. To accommodate expanded functionality, MEGACO specifies rules for defining new packages.

Even though MGCP was deployed first, H.248 is gaining momentum and is expected to achieve wider industry acceptance as the official standard for decomposed gateway architectures sanctioned by the IETF and ITU. MGCP is currently maintained under the auspices of the Packet-Cable and the Softswitch Consortium. There are no plans for the MGCP specification to be enhanced by any international standards body. MGCP is like an orphan without a parent, looking for approval from a standards body.

At the time of this writing, only one IP-PBX supplier (Sphere Communications) has announced support of MGCP for signaling to its proprietary IP telephones, and no IP-PBX supplier plans to use H.248 for proprietary voice terminal support. MGCP and H.248 are sometimes supported by external gateway equipment used as system options to support analog communications devices. MGCP and H.248 are supported for networking applications by a limited number of IP-PBX suppliers, although the more dominant interworking protocol for multiple system network configurations is, and is expected to remain, H.323.

CHAPTER 9

Converged IP-PBX System Design

The two IP-PBX design categories, converged and client/server, are based on very different architecture platforms. Individual product models within both categories also exhibit different design architectures. Just as no two circuit switched PBX systems are based on identical call processing, switching, and common equipment topologies and constructs, no two IP-PBX system models have the same design attributes. The IP telephony hardware options of a converged IP-PBX system are influenced by the basic design of the circuit switched platform on which it is based. Beyond total dependency on a LAN/WAN infrastructure for all switching and transport, currently available client/server IP-PBXs have sufficiently unique design attributes to easily distinguish themselves from competitive offerings. Support of IP control signaling and packeted voice communications is sometimes the only common thread between many of the IP-PBX systems currently being marketed to the public.

A converged IP-PBX system design is based on a traditional circuit switched design platform that supports analog and digital ports with a TDM/PCM transmission and voice coding format. What distinguishes a converged IP-PBX system from a traditional PBX system is the integration of ToIP technology to support call control or voice communications signaling. Signaling is packeted by using IP format, and voice coding may or may not involve compression of the 64-Kbps PCM sample traditionally used to digitize analog voice waveforms.

Converged IP-PBXs can include any or all of the following system design attributes:

1. Integrated support of IP station ports
2. Integrated support of VoIP trunking
3. Integrated support of TDM/PCM common equipment port carriers/cabinets over an IP-based LAN/WAN infrastructure

IP Station Ports

The key word in the above design attributes is *integrated*. A circuit switched PBX that supports IP peripherals using external gateway equipment is not considered an IP-PBX. A PBX system must be configured with integrated signaling or port interface circuitry to support ToIP communications if it is classified as an IP-PBX; otherwise any older PBX system with a T1/E1 interface and a third-party VoIP gateway could be considered an IP-PBX.

Converged IP-PBX System Design

Most first-generation converged IP-PBX systems were based on a traditional proprietary common control complex. The common control complex, specifically the main system processor, functions as the gatekeeper for IP clients. A gatekeeper was originally defined as a component in an H.323 communications system used for admission control and address resolution. Gatekeepers allow calls to be placed directly between two IP endpoints with the use of peer-to-peer LAN switched connections; two IP telephones can communicate over a LAN/WAN infrastructure without using the PBX's internal circuit switching network. A converged IP-PBX system also supports calls between an IP endpoint and a non-IP endpoint using gateway resources. The primary functions of a traditional gatekeeper are address translation, admissions control, bandwidth management, and zone control. The IP-PBX main processing unit assumes this responsibility for all IP communications. Most converged IP-PBXs support H.323 standards for LAN-based real-time multipoint communications networks.

A traditional circuit switched Meridian 1 Option 81C model, with its traditional core module cabinet can be upgraded to a converged IP-PBX system platform through the simple addition of an optional ITG line card and a software upgrade capable of supporting IP telephony. The optional interface card supports IP telephones by providing physical connectivity to the LAN, which links IP-PBX common equipment to the desktop IP voice terminal for control and voice communications signaling. The ITG line interface card converts on-board gateway resources for conversion between PCM and IP packet signals connectivity to an Ethernet LAN through a 10BaseT or 10/100BaseT interface, and passes gatekeeper control signals between the PBX call processor and IP endpoints.

In some system designs the IP port card with gateway resources also functions as a proxy server for the system gatekeeper to transmit control signals to and from IP endpoints. Nortel's ITG line card is used for TCP/IP control signaling connections between the PBX system and the LAN. IP line interface cards from Siemens and Alcatel also support the server as proxy gatekeeper servers. Gatekeeper signaling also may be transmitted with a dedicated Ethernet TCP/IP interface circuit card, such as Avaya's Definity/IP 600 CLAN interface card, or through an Ethernet connector located within the common control complex, such as a daughterboard on the CPU board (NEC's NEAX2000 IPS). The terms *gatekeeper* and *gateway* were originally defined for H.323 LAN-based telephony systems but are now also used to describe the role of the PBX common control complex and new IP port interface cards used by circuit switched PBXs to support IP endpoints and connections. Although IP port interface cards might not support gatekeeper signaling, they all support gateway functions.

A gateway is composed of two elements: an MGC and an MG. The MGC handles call signaling and other non–media-related functions. The MG handles the communications media transmission. H.323 gateway functions typically include protocol conversion, connectivity, compression, decompression, fax demodulation, and fax remodulation. The latter two gateway functions are not commonly used with converged IP-PBX IP port interface cards because fax terminals are supported with traditional analog line interface circuit cards. The PBX gateway function is extremely important because TDM/PCM ports are likely to remain a significant percentage of the total number of converged IP-PBX peripherals endpoints for several years to come. Communications between IP ports and non-IP ports require protocol conversion between the different voice coding formats and connectivity to the local TDM bus via the gateway bearer channels.

The IP port card gateway resources can also be used to establish voice communications links between IP endpoints using different voice codecs. For example, an IP endpoint using G.729/A format for voice packeting cannot communicate with an IP endpoint using only a G.711 format unless there is a conversion process between the endpoints. IP port card gateway resources can perform this conversion process, which is known as *transcoding*. Transcoded IP calls may require circuit switched connections on a local TDM bus, unless the IP port card can redirect the reformatted communications signals back across the LAN to the IP endpoints.

IP line card capabilities for a converged IP-PBX differ across manufacturers. There are several parameters that can differentiate IP port interface capabilities:

- Maximum number of IP stations supported (based on gatekeeper signaling)
- Gateway channel resources
- Voice codecs supported

The number of gateway channel connections per card is based on the number of on-board DSP resources, referred to by some system manufacturers as *digital compressors*. The term *compressor* signifies the compression algorithm function that converts the 64-Kbps PCM transmission format to the lower transmission rates supported by different IP voice codecs, such as those conforming to G.729 formatting standards (8 Kbps). PCM transmission may or may not be compressed to conserve LAN/WAN bandwidth resources. IP station transmission across the gateway to the local TDM bus needs to be decompressed (if not using the uncompressed, 64-Kbps G.711 format standard) to communicate with non-IP stations or access PSTN trunk circuits. The most common IP voice codecs supported by IP port interface cards are G.711, G.723.1, and G.729/A. The specific voice codec used for each IP call can be predefined by the system administrator, based on the call type (on-premises, off-premises) or specific number dialed.

There are different approaches used by converged IP-PBX system manufacturers in their IP station interface card designs. One approach is to limit the gateway channel capacity to support physical IP stations and provide nonblocking access to the local TDM bus in support of IP to non-IP calls. Siemens employs this strategy by limiting IP port capacity, based on gatekeeper signaling capacity, to 30 stations per interface card, the same as the maximum number of gateway channel links to the local TDM bus. The 1:1 ratio between IP station and gateway connections to the local TDM bus eliminates the possibility of blocked communications signaling to and from the TDM bus and LAN. Ericsson also implements this design strategy on its MD-110 for its IP station card.

Other leading converged IP-PBX system suppliers take a different approach. The Nortel Networks ITG line card supports 96 IP telephones, but only 24 gateway channels to the local TDM bus. The Alcatel OmniPCX 4400 INT-IP line interface card can support up to 500 IP clients, but the maximum number of gateway connections to the local TDM bus is limited to 60 gateway channels. The actual number of gateway channels per INT-IP card can be configured based on the number and type of daughterboards (8 or 30 compressors). An Alcatel system is unlikely to be configured with 500 IP Reflexes telephones supported by a single INT-IP board because of the extreme ratio of potential IP telephones to local TDM channel connections, unless there are minimal requirements for IP port to non-IP port connections. The reason Alcatel designed their interface card with a limited number of channels for a large number of physically connected telephones, and other manufacturers such as Nortel Networks designed IP station interface cards with potential contention for limited channels, is that gateway channels serve as pooled resources for any IP endpoint and can be configured based on traffic engineering requirements. For all the systems under discussion, IP interface cards are not dedicated to specific endpoints for gatekeeper or gateway operations. IP telephone communications services can be performed by any of the IP port interface cards in the system, a major difference from the support of traditional analog and digital telephones by dedicated port interface cards.

NEC and Avaya have designed IP port interface cards that handle only gateway functions because gatekeeper signaling is transmitted to IP endpoints through other means. The Avaya IP Media Processor Interface, sometimes referred to as a *prowler board*, supports between 32 and 64 channel connections, based on the IP voice codec implemented per call, G.711 or G.729/A. The Avaya port interface card has a total of 64 DSP resources. Each DSP resource can support a single G.711 communications interface, for a maximum total of 64 simultaneous conversations (1 DSP resource = 1 active gateway channel). The G.729/A protocol format uses compressed voice (8 Kbps) and requires two DSP resources per IP call requiring a circuit switched connection to a non-IP peripheral, for a maximum of 32 simultaneous conversations. The conversion process for G.729/A is more processing intensive than uncompressed G.711 (64 Kbps). The actual number of available channel connections to the Avaya Definity local TDM bus will change continually based on the type of voice codecs that are implemented for concurrent IP calls requiring TDM bus connectivity. The number of IP Media Processor Interface cards required in a system design will depend on IP call requirements for TDM bus connectivity, or transcoding. The NEC IP line interface card, known as the IP PAD, has 32 on-board compressors that perform the gateway function for IP stations requiring circuit switched connections to non-IP ports. NEC designed the card to have a maximum of 32 compressors because it occupies a single card slot that is limited to 32 TDM bus backplane connections.

Taking into account gatekeeper signaling and gateway resource capacities, the actual number of equipped and active IP endpoints that can be supported by an IP port interface card to maintain an acceptable QoS level is based on the number of IP endpoints, customer traffic patterns, and engineering requirements. The more likely a gateway channel will be used per IP station call, the lower the acceptable ratio between peripheral endpoints and channels. For example, if there are very few IP stations equipped in a converged IP-PBX and most ports are traditional TDM/PCM peripherals, a call generated by an IP telephone more likely will require a channel for connectivity to the local TDM bus to talk with a non-IP station/trunk. If a converged IP-PBX system is equipped with a very high percentage of IP telephones, as compared with traditional analog or digital telephones, and a significant percentage of off-premises calls are handled over an IP WAN, the acceptable ratio between IP endpoints and gateway channels will be larger. A converged IP-PBX system design using separate interfaces for gatekeeper signaling transmission and gateway functions theoretically can be equipped with no gateway interface cards, assuming there is no requirement for any IP call to use TDM bus resources. This scenario is almost impossible to envision in a current customer environment, unless it is a small satellite location with all IP stations and no local central office trunking is IP networked to a main PBX system.

IP port interface circuit cards typically support functions beyond their gateway responsibilities. Some of these functions are administration and maintenance support, echo cancellation, silence suppression, DTMF detection, conferencing, and improved voice quality through dynamic jitter buffers. Basic administration/maintenance functions include support of station registration and initialization, downloading firmware changes to desktop terminals, and diagnostics of errors and alarm conditions at the port endpoint. *Echo cancellation* technology reduces the effects of echo heard by a caller when on an active voice call. If the echo transmission signal is delayed, the resulting echo will be noticeable to the caller. An echo canceller monitors caller speech; when an echo occurs, a signal is generated, transmitted, and sent back to the caller to cancel the echo. *Silence suppression* is the detection of the absence of audio on the bearer communications channels. Another term for silence suppression is *voice activation detection*. When no voice packets are detected, the gateway bearer communications channels are released from a call and made available for voice packets transmitted by another caller.

The *jitter buffering* function is very important for maintaining voice QoS levels. The time for voice packets to be transmitted and received between endpoints is known as *delay*. The "end-to-end" delay time consists of two network delay elements, fixed and variable. Jitter is the difference between the two delay values of two voice packets in the transmission across the network. Fixed network delay may include propagation delay of signals between the two endpoints, voice encoding delay, and the voice packeting time for the IP voice codec. Fixed network delay times can be calculated and corrected for, but variable delay times present a different problem. Variable packet delays can be caused by network traffic congestion and serialization delays on network interfaces. The quality of voice communications is degraded if there is a variation in the arrival of voice packets at the receiving endpoint. Network congestion can occur any time and cause variations in arrival times. To compensate for delay variations, the IP station card equipped with jitter buffers turn delay variations into a constant value so that voice can be transmitted and played out smoothly. Digital signal processing (DSP) algorithms can be programmed for different buffer times based on the expected voice packet network delay. The algorithm can review each packet's timestamp included in the RTP header of the voice packet, calculate expected delays, and adjust the jitter buffer size accordingly. Extra buffer time can be programmed to account for variable packet delays and smooth out the packet flow. If packet delay exceeds the total jitter buffer time, the packet is dropped. Loss of one packet in a voice call does not significantly affect the quality of the call. If variable delay of voice packets is underestimated and many packets are dropped, the resulting voice call quality will suffer.

IP station equipment supported by an IP station interface card may include an IP telephone terminal instrument or a PC client softphone. Converged IP-PBX systems can support local IP stations that are physically located on the customer premises or remote at off-premises locations. Control signaling to and from the IP endpoint is over LAN/WAN facilities. A remote IP station can use private line WAN carrier facilities or PSTN services (ISDN, DSL, dial-up) for gatekeeper control signaling. A centrally housed IP port circuit card provides gateway function to non-IP ports.

Making IP Voice Calls

In a converged IP-PBX system, a call begins when the IP telephone goes off-hook and the PBX common control complex is alerted via a control signaling path between the IP telephone and the common control complex functioning as a gatekeeper. The signaling is transmitted to and from the LAN via an IP port interface card, dedicated Ethernet TCP/IP interface card, or integrated Ethernet connector. Dial tone packets or a control signal will alert the IP phone to activate a dial tone signal, an indication to the station user that the IP phone is ready for use. As the calling number digits are dialed, the common control complex will send multiple signals to the desktop, such as ring back or other call progress tones. If the call is being placed to a non-IP station port, such as an analog telephone, or a PSTN trunk circuit is required for an off-premises call, the following steps will occur:

1. A local TDM bus talk slot will be assigned
2. A gateway bearer communications channel will be assigned

Converged IP-PBX System Design

3. Voice communication signals will be transmitted over the LAN to the IP port circuit card
4. The packeted voice signals will be reformatted, buffered, and transmitted across the gateway channels to the local TDM, and the call will continue until one party releases (disconnects) the connection.

For purposes of the call, the IP telephone appears to the common control complex and switching network as a PCM peripheral endpoint for voice transmission and feature activation operations. Converged IP-PBX systems typically use the same proprietary control signals for IP telephone support as those used for digital telephones.

If an IP station user is placing an intercom call to another IP telephone, the common control complex will direct the IP telephone to start sending voice packets directly to the IP address of the called IP telephone. The common control complex also will direct the called party's IP telephone to send voice packets directly to the IP address of the calling party's IP telephone. The direct audio communications path between the two IP telephones, using only LAN facilities, is often referred to as an IP-PBX peer-to-peer LAN switched connection or direct IP. No traditional PBX circuit switched connections are used for the call.

Direct IP connections will be established automatically between two IP endpoints if several conditions are satisfied:

- Both IP endpoints are administered to allow direct IP connection
- No TDM connections are required for either IP endpoint, and a point-to-point LAN connection is available
- The IP endpoints are in the same network region or in different network regions that are interconnected via LAN/WAN facilities
- The two IP endpoints share at least one common voice codec in their voice codec lists and the internetwork region connection management voice codec list
- The IP endpoints have at least one voice codec in common, as shown in their current codec negotiations between the endpoints of the IP-PBX

If any of these conditions are not satisfied, the call may require TDM connectivity.

A direct IP connection established for an existing call may be torn down if circuit switched TDM connectivity is required during the call. Conditions that may require TDM connectivity, based on the IP-PBX system, are:

- Additional parties are conferenced onto the call, including IP endpoints
- A PBX signaling tone or announcement needs to be inserted into the connection
- The connection is put on hold—music on hold

When the event requiring TDM connectivity is no longer in effect, a direct IP connection may again be established. The generic term for call connections that change from direct IP to TDM connectivity and back to direct IP is *null capability*.

The first generation of converged IP-PBXs required talk slots on the local TDM buses to support multiparty conference calls for calls among IP endpoints, even if all the parties were IP endpoints. Each conferenced party required a talk slot per TDM local bus to support the call. The manufacturers have future plans to support non-TDM conference calls among IP endpoints through enhanced versions of their current IP port circuit cards. Planned versions of IP port circuit cards will include conferencing circuits to support multiparty calls exclusively among IP endpoints, thereby eliminating the need for TDM switched connections. The IP endpoints may be internal IP telephones or IP trunk circuits connecting off-premises stations.

IP Trunk Ports

The first generation of converged IP-PBXs could support only private network IP trunk calls. VoIP calls are not currently used for local trunk connections to and from the central office. Off-premises network calls are routed across IP routers over the customer WAN and are typically under the control of a policy management server. A policy management server uses a customer-defined program of IP telephony policies to support VoIP QoS levels across the customer WAN. The server can extend QoS differentiation for content networking through network-based application recognition support. The customer-based policies are managed through interaction with a directory server to provide different network classes of service to different station users. Through the server, end-to-end service can be provisioned, and service level agreements are maintained. Access control policies can be applied to maintain network and application security levels.

Converged IP-PBX System Design

Converged IP-PBX systems support two types of outgoing IP trunk calls:

- IP station generated calls
- Non-IP station generated calls

For IP station calls, if the ARS program selects an IP WAN trunk circuit to route a private network call, the IP line port circuit card providing control signaling to the desktop station will direct the IP telephone to transmit voice packets to an assigned IP WAN router. The IP WAN router will then handle the networking operations as programmed. Incoming IP trunk calls from the WAN can be routed directly to an IP station based on the control signaling directions from the terminating IP-PBX system gatekeeper.

An outgoing IP trunk call placed by an IP station is commonly handled without local TDM connectivity, except in cases requiring a circuit switched connection. A direct IP trunk call not requiring TDM connectivity must satisfy a set of conditions similar to that listed for IP station calls. If any of these conditions are not satisfied, a TDM switched connection may be required to place or receive a trunk call.

In a converged IP-PBX system, a non-IP station can also place an IP trunk call if the ARS program selects a route using customer IP WAN facilities instead of traditional PSTN trunk circuits. An integrated IP trunk port interface card provides a gateway between the circuit switched TDM bus network and the LAN/WAN facilities. An external VoIP gateway server is not required, as do circuit switched PBXs that do not support integrated IP trunk interface capabilities.

The IP trunk card functions similarly to a DS1 trunk circuit interface card in support of private line tie trunk control signaling and bearer communications channels. It connects to a local TDM bus and packets/compresses PCM voice communications signals for transmission over the Ethernet LAN to an IP router for access to the customer WAN. The card can also modulate and demodulate incoming and outgoing fax calls. IP trunk cards typically support ISDN protocols, such as ISDN D channel, for private network signaling, H.323 signaling, and a standard VoIP protocol stack.

IP trunk cards also monitor QoS parameters, such as latency, jitter, and packet loss to ensure that acceptable voice communications quality is maintained throughout the call. A customer can define a delay threshold, such as 150 ms, for monitoring purposes; if transmission delay exceeds this level, the call is automatically rerouted over circuit switched PSTN trunk carrier facilities. Packet loss measurements also can be thresholded and monitored for call rerouting purposes. In all cases call rerouting is performed transparently and seamlessly to the call parties. There are several methods used by IP-PBX system designers to monitor QoS transmission delay parameters, including analysis of packet time stamps embedded in the RTP signaling stream or calculating delay based on a self-generated signaling ping across the network.

IP trunk circuit card gatekeeper and gateway functions are similar to those of the IP station card. IP trunk cards typically support 24 (T1 carrier interface) or 30 (E1 carrier interface) gateway channels. There is always a 1:1 ratio between the number of trunk channel equivalents and gateway channel connections. Some cards can be configured to support increments of eight channels to support the maximum T1/E1 carrier channel capacity. Like IP station cards, the most common voice codecs supported by IP trunk cards are G.711, G.723.1, and G.729/A. Although IP-PBX control signaling may be proprietary across a private network of like IP-PBX systems, to support feature transparency functions, almost all systems attempt to comply with the latest version of ITU-T H.323 standards. H.323 compliance supports network connectivity within a multivendor network of different IP-PBX systems.

An IP trunk circuit card can be used for private networking configurations as an alternative to other PBX circuit cards: analog E&M tie trunk, T1/E1 interface, and ISDN PRI interface. It handles outgoing IP trunk calls and is used for incoming IP trunk calls to non-IP stations. Incoming IP trunk calls directed by the terminating IP-PBX gatekeeper to non-IP stations, such as analog telephones, require TDM connectivity. The IP trunk circuit card supporting outgoing IP calls by non-IP stations performs a reverse gateway function for incoming IP calls to non-IP stations. Incoming IP call voice packets are decompressed, reformatted into PCM signals by the IP port interface card, and transmitted across available gateway channels onto a local TDM bus for circuit switched connections to the called non-IP station.

The first release of each supplier's converged IP-PBX system was designed with dedicated IP station and IP trunk interface cards, the same approach used for traditional TDM/PCM ports. Avaya was the first converged IP-PBX system designer to develop a universal IP port interface for IP stations and IP trunk circuit connections. The Avaya IP Media Processor Interface's gateway resources are available for station and trunk calls. Pooling the gateway resources for all types of IP endpoints provides greater flexibility in traffic engineering design and can reduce customer hardware costs. Alcatel's OmniPCX 4400 originally required different IP interface cards for station, trunk, and remote cabinet support, but the latest system release supports a universal INT-IP card that replaces the three dedicated INT-IP cards. A universal IP port interface card likely will be a design trend followed by most converged IP-PBX system suppliers.

To summarize, there are four categories of IP trunk calls:

1. IP station to IP station (IP trunk circuit cards are not usually required)
2. IP station to non-IP station (IP trunk circuit card required at terminating IP-PBX to perform gateway functions)
3. Non-IP station to IP station (IP trunk circuit card required at originating IP-PBX to perform gateway functions)
4. Non-IP station to non-IP station (IP trunk circuit card required at originating and terminating IP-PBXs)

The major benefit of an IP trunk call is potential cost savings due to the reduced requirement for private line carrier facilities. If the IP call is implemented with a codec using compressed voice packets, such as G.729/A, less bandwidth is required for the IP trunk call than when placed over circuit switched PSTN carrier facilities. Although the LAN/WAN requires continual traffic engineering to maintain an acceptable QoS level and may experience less than optimal QoS levels due to higher than anticipated transmission delays, a customer balances lower-quality voice communications service with reduced telecommunications expense. If QoS levels exceed a defined benchmark, IP calls can be overflowed to PSTN carrier facilities.

Some customers first test IP trunk calls for only the last category of calls, when the only non-IP station equipment placing and receiving calls are facsimile terminals, before using IP networking for real-time voice calls. The ARS software can be programmed to select IP trunk routes for only certain types of calls between specific station users. For example, conference calls typically require a very high QoS level and may be set up only with PSTN trunk circuits. Call types classified as less important may use an IP trunk route as the first choice, accepting the possibility that a diminished QoS level occurs during the call.

Dispersed Common Equipment over LAN/WAN Infrastructure

Support of IP endpoints, stations, and/or trunks may not be the only trademark of a converged IP-PBX system design. A PBX system that supports neither IP stations nor trunks can be considered an IP-PBX if the system design includes geographically dispersed port cabinets/carriers using an IP LAN/WAN infrastructure for control signaling from the common control complex and voice communications between dispersed port interface equipment. Using a LAN/WAN infrastructure to support customer communications requirements across one or more customer premises locations based on a single IP-PBX system offers several potential key benefits:

- Single system image (numbering plan, features, systems management)
- Reduced networking costs between customer locations
- Scalable system expansion

There are several types of converged IP-PBX system designs in this category. They include upgrades of traditional circuit switched PBXs and IP-PBX systems whose original design was based on a LAN/WAN infrastructure to support desktop and networking communications requirements. The latter system designs were introduced by nontraditional PBX suppliers that entered the market during the late 1990s.

Upgraded Circuit Switched PBX

Circuit switched PBXs that are upgraded to support IP telephony can be designed to support dispersed common equipment for single or multiple customer premises configurations. It may be necessary to disperse common equipment across a single customer premises if it is a large campus environment with multiple buildings or even a very large single building complex. Multiple customer premises requirements can be satisfied by a variety of design options, but a single system solution is often the preferred choice. If an existing customer WAN is installed and can handle voice calls, there is a potential to save significant transmission service expenses that would normally be allocated to dedicated circuit switched trunk facilities.

Converged IP-PBX System Design

Nortel Networks implemented the first use of IP telephony by a traditional circuit switched PBX system. It supported remote communications requirements with the use of WAN links instead of more traditional T1/E1 trunk carrier circuits. In the early 1980s, Nortel Networks also was the first PBX supplier to offer a remote port cabinet option, remote peripheral equipment (RPE). In 1999, the supplier announced a remote port carrier using IP connectivity, the 9150 Remote Office Unit. A centrally located Nortel Networks Meridian 1 equipped with an ITG port interface board (Reach Line Card) can support a remote 9150 by using available WAN trunk facilities between the two locations. The remote unit has a port capacity of 32 digital Meridian 1 telephones, a TDM backplane to support local switching between stations (reducing dependency on the Meridian 1's center stage switch network), and an integrated IP gateway. Voice codecs supported include G.711, G.726, and G.729/A. The 19-inch carrier unit also has a 10Base-T Ethernet port connector for LAN connectivity and ISDN BRI trunk circuit interfaces for local trunk calls. Key QoS parameters on call sessions between the Meridian 1 and 9150 are monitored, included packet loss, jitter, and end-to-end packet delay (latency). If threshold parameters are exceeded during a call in progress, the call can be dynamically and transparently rerouted over ISDN BRI trunk circuits to the Meridian 1. With G.729 (8 Kbps) encoding, a single ISDN BRI B channel supports up to eight simultaneous calls back to the Meridian. Calls can be transitioned back to the IP WAN when the network can support acceptable QoS levels. If the WAN link to the Meridian 1 is lost, the remote 9150 supports limited call processing functions, including local switch connections, access to ISDB BRI trunk circuits, and basic station features (transfer, paging access).

Ironically, the 9150 unit does not directly support IP stations. IP stations remote to the Meridian 1 can be supported via an ITG line card in a centrally located IPEM port cabinet over a LAN/WAN link. Communications between a remote IP telephone and a digital telephone supported by the 9150 unit must be handled across the Meridian 1 center stage switch network at the main location. The remote IP station is totally dependent on the LAN/WAN link for all telephony services, and not even dial tone is supported when the link fails.

Other PBX suppliers soon followed Nortel's lead by offering IP remote cabinet/carrier options on their circuit switched PBXs. For example, the Avaya R300 Remote Office Communicator is designed for branch offices that support centralized Definity/IP 600 IP Communications Server features—essentially everything available at headquarters—over the WAN, to offices with fewer than 25 employees. The R300

Remote Office Communicator was designed to lower customer networking costs by converging voice and data onto a single WAN facility. An IP port interface card in the centrally located Definity/IP 600 system transmits call processor control signaling to the remote location and functions as a gateway for calls between the central and remote locations. The R300, an Avaya-customized Ascend Communications R3000 remote office concentrator, has built-in data networking and routing capabilities including firewall, VPN, and security features, with optional remote access concentration capabilities to improve network efficiency. The R300 has integrated port circuit interfaces that support up to 24 digital and two analog stations, WAN options such as full and fractional T1/E1 and BRI, a bidirectional 10/100 Ethernet port, local analog and digital T1-carrier trunking, E911, PPP data routing to the corporate LAN, and dial tone even during WAN failure or power failure. The unit supports local switching among TDM peripherals (stations, trunks). IP telephones may be supported at the remote location when using the R300 Ethernet port connection for signaling connections over the WAN back to the IP-PBX common control complex.

The Nortel Networks and Avaya remote IP carrier options are similar in concept (remote station/trunk support, IP control signaling over a WAN link, gateway interfaces at the main location) but somewhat unique in their sum capabilities (port capacities, types of stations supported, survivability features, data networking functions). Both IP options could, in theory, be used for dispersed communications services across a campus location, instead of remote locations, like the Siemens offering originally known as the Fiber Loop Exchange (FLEX) option. First available on the supplier's Hicom 300H platform (currently upgraded to HiPath 4000), the option consisted of a port cabinet shelf that could be installed remotely from the main system complex and a dark fiber cabling link to support IP-based control signaling and voice communications channels. Available in a redundant mode, the optical fiber cable interfaces to the Hicom 300H common equipment through an optional integrated gateway board, the HiPath HG 3800. The gateway card provides connectivity for the Hicom 300H hub location to remote shelves by using Fast Ethernet fiber over dedicated single or multimode fiber to campus locations within a 20-mile radius of the host location. The remote port cabinet shelf, designated the HiPath AP 3300, is designed with an integrated gateway interface circuit.

The newer HiPath 4000 platform can also support remote ports over a traditional LAN/WAN infrastructure. This option requires a HiPath HG 3570 gateway interface card, housed in a HiPath 4000 host system, with

Converged IP-PBX System Design

IP tunneling to a HiPath AP3300 remote port cabinet or an AP 3500 remote 19-inch rack mount carrier at a remote location. A HiPath HG 3575 is housed in the remote AP 3300 cabinet or AP 3500 carrier. The gateway cards are equipped with 10/100 Base-T Ethernet connectors and are used to transmit call processor control signaling from the centralized HiPath 4000 common control complex to the remote cabinet/carrier. Gateway resources convert PCM signals to IP packet format for LAN/WAN transport and provide TDM bus connectivity at the host and remote locations. The AP 3300 cabinet and AP 3300 carrier are equipped with TDM switch network backplanes to support local switching for station and trunk ports. In case of LAN/WAN failure, a dial-up modem connection over PSTN trunk circuits can provide gatekeeper control signaling support to the remote cabinet/carrier.

The Alcatel OmniPCX 4400 distributed processing, switching, and cabinet design easily lends itself to an IP LAN/WAN infrastructure for control and communications signaling. A fully configured system can support a cluster of 10 control cabinets and associated expansion port cabinets by using a variety of networking options, including TDM/PCM, IP, ATM, and frame relay. Each control cabinet/port cabinet group can be remote from the others. A control cabinet also can support a remote expansion cabinet or a single remote IP station. The IP networking option requires an INT-IP card to support gatekeeper control signaling between a centralized control cabinet (with call processor) and remote port cabinets, between control cabinets configured across a LAN/WAN, or to a remote IP station. The INT-IP card also supports gateway functions between dispersed circuit switched port cabinets configured across a LAN/WAN or between dispersed IP stations and a centralized port cabinet. The INT-IP interface card, designed with a 10/100 Base-T Ethernet connector, supports peer-to-peer LAN switching between IP endpoints, provides tone generation and signaling channel processing, and supports several voice codecs (G.711, G.723.1, and G.729/A).

Distributed Modular Design

A category of converged IP-PBXs was originally designed to leverage a customer's existing data communications LAN/WAN infrastructure. These are systems based on distributed, modular design architecture for call processing, switching, and port interfaces. Traditional analog telephones were used as the primary voice terminals (at least in the early system releases, until IP telephones were supported), and advanced system features and functions were available through CTI-based PC telephony. IP-PBX systems that best illustrate this IP-PBX system category are from Sphere Communications and Shoreline.

Sphere's Sphericall Enterprise Softswitch solution consists of several major system components:

- **Sphericall Manager**—Host platform for Sphericall Softswitch call control software; includes remote management and monitoring
- **PhoneHub**—MG carrier for up to 24 analog or CLASS stations from IP or ATM networks
- **COHub**—MG carrier for T1/E1 connections to PSTN or PBXs from IP or ATM networks; Q.sig, ISDN, CAS, and international protocols are supported
- **BranchHub**—Remote or small office (6 × 12) analog trunk and station MG carrier to IP or ATM networks; six lines of power failure transfer
- **VIM**—Remote office or campus extension carrier, ATM IAD for T1/E1 or fiber connections to the MAN/WAN; downlinks for voice/video/data

A Sphericall system requires a single centralized Sphericall Manager to support customer premises PhoneHub and COHub carriers and remotely located BranchHub and VIM carriers. IP telephones are supported by the manager through direct control signaling over the LAN or across a WAN. Redundant managers can be configured for purposes of survivability. Individual manager, PhoneHub, and COHub carriers are interfaced to each other using an Ethernet LAN. The manager uses TCP/IP transmission to support call processing signaling to dispersed port interface carrier equipment. The BranchHub is remotely linked over a customer WAN. The PhoneHub and BranchHub carriers have TDM switching backplanes to handle local intercom calls; calls between dispersed station hubs (local, remote), IP telephones, and trunk hubs are handled over the LAN/WAN via integrated MGs in each port carrier. A Web browser systems management interface tool allows system administration support from a centralized management workstation or from multiple dispersed workstations.

The Shoreline system is based on a similar design concept, with one distinct difference. The Shoreline3 system is a completely distributed, modular voice communication solution, with no single point of failure, which is layered on top of the IP network. At the heart of the system is the standards-based Distributed Internet Voice Architecture (DIVA) software, which uniquely distributes call control intelligence to voice switches connected anywhere on the IP network. In addition, DIVA distributes voice applications, including voice mail and automated attendant, to servers across locations rather than centralizing applications at the network core. There are four types of ShoreGear voice switch:

- **ShoreGear-24**—A 24-port (16 telephone ports and 8 universal analog telephone or trunk ports) stackable or rack-mountable nonblocking voice switch with an integrated IP media gateway
- **ShoreGear-12**—Twelve universal port stackable or rack-mountable nonblocking voice switch with integrated IP media gateway
- **ShoreGear-T1/E1**—Digital trunk interfaces to the central office; it can also be used as a VoIP gateway to other PBXs; alternatively, the ShoreGear-E1 can be used as a VoIP gateway to an existing PBX, thereby bridging legacy systems to the Shoreline system
- **ShoreGear-Teleworker**—Supports remote station users while maintaining full communications functionality

As the names imply, each port carrier has specific interface capabilities. Each ShoreGear voice switch has a local switching TDM backplane and a call processor that runs ShoreWare software to support fully distributed call control, voice applications, desktop applications (via CTI-based PC telephony), and management tools. ShoreGear voice switch carriers, equipped with Ethernet connectors, can be dispersed across a customer LAN/WAN infrastructure to support single- or multiple-location requirements. Voice QoS is monitored and maintained with dynamic jitter buffering and packet loss replacement. Voice codecs can be administered to support linear, G.711, ADPCM, and G.729/A compression formats, echo cancellation, and silence suppression. The distributed call control design provides a high level of local survivability in case of LAN/WAN link failure.

The primary design difference between the Sphere and Shoreline systems is that the Sphere is based on dedicated telephony call server (with optional redundant servers) and the Shoreline embeds call processing functionality and software in each port carrier. Both systems use circuit switched connections for intercom calls between stations interfaced to the same port carrier/hub unit. IP signaling format is used only for intercabinet communications, although the limited port capacity of each carrier/hub increases the likelihood that a significant percentage of calls will be handled across the LAN. Each system is based on an incremental modular expansion design and is ideally suited for a network of numerous small locations. A single-location customer configuration with significant port requirements will require a large number of port carrier/hubs. The interface capacity of each port carrier/hub is comparable to a single port interface circuit card in a traditional PBX system. In a large customer configuration, the limited port capacity of each carrier/hub increases the complexity of the network design necessary to support basic port-to-port communications because the QoS level of the customer LAN/WAN is a factor for most premises calls.

There are several important benefits to a dispersed LAN/WAN infrastructure design, including ease of expansion; single-system image across multiple customer locations, including unified dialing plan and feature-transparent operation; toll bypass using private WAN facilities; dynamic bandwidth use of network transmission resources; and centralized administration.

There is often confusion regarding the classification of IP-PBXs into different system design categories. The Sphere and Shoreline systems are sometimes categorized as client/server IP-PBXs because they lack a traditional common control and switching network complex and are heavily dependent on a LAN/WAN infrastructure communications signaling. Both designs are based on circuit switched networking within each port carrier and retain many of the characteristics of a traditional PBX. The Shoreline system can be viewed as a network of mini-PBX systems that uses an IP-based infrastructure to link the multiple systems. The Sphere system is based on a LAN-connected common control complex that supports a cluster of circuit switched port cabinets interconnected over an IP network. Instead of a multicarrier port cabinet capable of supporting dozens of port circuit cards, these systems have substituted a network configuration of port interface carriers/hubs, each one the equivalent of a single port interface card. There are two disadvantages to this approach: more complex hardware equipment is required to support large single-location port requirements and there are limited shared equipment resources. For example, an integrated center stage switching complex is sometimes more advantageous than a complex network of LAN switches. A centralized power supply to support a large system configuration can also be advantageous. As usual, there are advantages and disadvantages associated with every PBX system design.

CHAPTER **10**

Client/Server IP-PBX System Design

The first IP-PBX systems were based on a client/server design. When most telecommunications managers hear the term IP-PBX, they usually envision a client/server design, although the converged system design is gaining in popularity and likely will dominate the market for the next few years. An IP-PBX system based on client/server design fully uses and depends on a LAN/WAN switching network infrastructure for call control and communications signaling. Like converged IP-PBX systems, client/server designs are not standard or uniform across manufacturers, although the competing models share some common design elements.

The term *client/server* is borrowed from the world of data communications. It describes an IP-PBX system that does not use a traditional PBX common control complex and integrated circuit switched network or traditional common equipment hardware (port cabinets and port interface circuit cards). A client/server data communications design specifies a data processing topology in which a personal computer (client) depends on a centralized computer (server) for applications software and database management functions. For many years the traditional PBX system design was compared with a mainframe computer because all call control and switching functions were centralized and desktop terminals (teleprinters, CRTs) lacked processing functions of their own. As enterprise voice communications systems evolved toward distributed and dispersed modular design topologies, similar to the concurrent evolution of minicomputer and personal computer networks, the term client/server was used more and more to describe the improved PBX system design. The first IP-PBX systems more closely conformed to data communications and processing client/server design topology; hence, the adoption of the term.

The common control complex of a client and server IP-PBX is based on a telephony call processing server that transmits and receives control and status signals to/from LAN-connected peripheral endpoints, known as clients. The telephony server's primary role is as a gatekeeper to clients for call setup and teardown functions and to manage and control communications bandwidth requirements for each call. In IP-PBX terminology, a telephone terminal is also referred to as a client. The IP telephone client depends on the telephony server for dial tone, call routing, and desktop feature/function implementation, similar to the relation between an analog/digital telephone and a traditional PBX common control complex. The major distinction between a traditional circuit switched PBX system and a client/server IP-PBX is that the LAN/WAN infrastructure, not an internal network of TDM buses, is the primary switching network.

Some current and planned client/server IP-PBX models may be equipped with optional port cabinet/carrier equipment with integrated TDM bus backplanes for circuit switched connections, but only calls between ports connected to the same port cabinet/carrier are circuit switched; all other calls depend on the LAN/WAN infrastructure for transport and switching operations. This type of system also may be classified as a converged IP-PBX, because TDM/PCM circuit and IP packet switching is supported. Arguments can be made for categorizing it in either classification, but if the design primarily depends on IP packet communications, and circuit switching is secondary or optional, the system is best defined as a client/server design. Design differences between individual IP-PBX models, as illustrated by client/server designs with optional circuit switched port carrier equipment, make it difficult to definitely classify any PBX system into one category or another.

There are several hardware layer elements common to all client/server IP-PBX models:

- **Call processing layer**—Gatekeeper/telephony call server
- **Client layer**—Voice terminals and other communications devices
- **Applications layer**—Messaging, contact center
- **LAN/WAN infrastructure**—Ethernet switches, IP routers, telephony gateways

Differences between client/server IP-PBX models are based on how each design element is configured within each layer and between layers. There are also differences in how some features and functions are provisioned, and whether optional application services are integrated into the overall system design or provided through nonproprietary third-party equipment. Ethernet switches and IP routers are designed to industry standards, and products from different manufacturers are usually interchangeable at the infrastructure layer, except for a few select IP-PBX models that may require proprietary LAN/WAN solutions as part of their overall system design or integrate LAN/WAN interfaces into their hardware design.

The remainder of this chapter focuses on call telephony servers and gateways. IP telephones, the client layer of the architecture hierarchy, and LAN/WAN design issues are discussed in separate chapters elsewhere in this book.

Telephony Call Server

IP-PBX telephony call servers typically support the following system functions and operations:

- Gatekeeper for IP peripheral endpoints
- Call processing and feature provisioning
- Systems management, maintenance, and diagnostics

The telephony call server in a client/server IP-PBX system design may be one of two types:

- Proprietary, closed server cabinet
- Third-party, off-the-shelf server

A proprietary, closed server cabinet bundled with the IP-PBX's generic software program often supports one or more integrated advanced application software solutions and/or includes an integrated CTI Applications Programming Interface (API) for third-party software solutions. It may also include a variety of integrated telephony gateway interfaces for PSTN trunk circuit and proprietary port cabinet/carrier connections. The server cabinet also may include an SMDI for third-party voice VMSs.

There are several IP-PBX systems that do not include a telephony call server as standard hardware equipment. For these systems an IP-PBX manufacturer recommends one or more third-party, off-the-shelf servers that conform to a series of technical specifications (processor type, operating system, clock rate, main memory, disk drive capacity) to satisfy call processing and reliability requirements. External gateways and application servers are required to work behind a distributor- or customer-provided server to support non-IP interface requirements.

Looking at several of the leading client/server IP-PBX manufacturers reveals little commonality among the competitors. When Cisco Systems acquired the Selsius Systems IP-PBX solution, the product offering was based on a third-party server platform. Shortly after Cisco's acquisition, the product was redesigned and available only with a proprietary server cabinet, known as the MCS 7835. Cisco's AVVID IP Telephony System is currently available with the MCS 7835, a lower priced, reduced port capacity MCS 7825, or a third-party server from Compaq or IBM. The third-party servers are certified by Cisco and offer customers a more flexible hardware design choice. The 3Com NBX system originally was designed as a proprietary, closed server platform, and remains such.

3Com, after acquiring NBX Corporation, modified its SuperStack 3 Ethernet switch chassis to support call processing boards and the same telephony gateways available on the original NBX system, and currently offers it as the server platform for its large system SS3 NBX model.

The circuit switched PBX manufacturers who have entered the client/server IP-PBX market have taken different design paths. For example, Siemens offers two server platforms for its HiPath 5000 client/server IP-PBX system: the HiPath 5500 runs on a third-party server based on a Windows NT or Windows 2000 O/S platform, and the smaller HiPath 5300 is based on proprietary closed server cabinet design with an embedded Windows NT O/S. The Siemens-provided HiPath 5300 hardware includes an integrated H.323 gatekeeper that supports client authentication and registration and bandwidth management. It also provides address resolution for calls based on user name, e-mail, or phone number and proxy functions for unavailable endpoints. A Web-based systems management tool with an intuitive GUI is bundled into the generic software program.

The 5300 call processor is based on an Intel Celeron 433-MHz microprocessor with 256 MB of SDRAM memory. There is a 20-GB (IDE) hard drive and a CD-ROM drive. It has a 99.99 percent reliability level, with an MTBF of 60,000 hours. It is very compact at $88 \times 440 \times 380$ mm. The module is installable in a standard 19-inch rack mount carrier.

Nortel's Succession CSE 1000 is currently based on a proprietary server cabinet, although there are plans to offer a third-party server option. The heart of the Succession call server is a Succession Controller Card (SSC). The SSC has two dual-port IP daughterboards that support the system's MGs. The call server links to the MGs by a 100BaseTx LAN connection or direct point-to-point connections. Also residing on the SSC are two Flash drives that perform system software operations and store customer data. The first drive is the primary Flash drive that contains Succession CSE 1000 system data and the first copy of the customer data required to load the system. The primary Flash drive is programmed with system software before shipment. The second drive is the backup Flash drive that stores files the user can change, such as configuration data and the second copy of the customer database.

The Mitel Networks MN 3300 ICP is based on a proprietary server carrier. The 3300 ICP Controller carrier is based on a Motorola 8260 microprocessor and runs on a VxWorks O/S with a 11-GB hard drive. The main control processor is responsible for Ethernet to TDM gateway functions, real-time operations for the generic system software, integrated voice mail, Web browser systems management, and MITAI applications gateway (a TAPI CTI interface). There are three port interfaces for the printer, alarm, and low-level maintenance devices. The carrier has an integrated tone card and Conference board that supports up to 64 channels (eight per conference group).

Some of the proprietary, closed server cabinet platforms integrate several client/server design layers into a single hardware cabinet. Although there are many manufacturers that market a wide range of telephony gateway servers, including a few that also manufacture PBX systems, several of the proprietary, closed server cabinets are designed with integrated T1/E1 digital trunk telephony gateways for PSTN trunk circuit access, such as Mitel Networks' MN3300 ICP, Siemens's HiPath 5300, and 3Com's NBX/-SS-3.

Server Operating System

IP-PBX call telephony servers currently run on a variety of operating systems. For example, the Cisco Systems and Siemens client/server IP-PBXs models use Windows NT or Windows 2000 platforms. Windows-based operating systems were a natural choice for many IP-PBX designers because of their ubiquity in the marketplace. Windows NT was the operating system of choice for most of the converged IP-PBX ICS solutions listed above. Nortel Networks, 3Com, and Mitel Networks servers use VxWorks. There are several advantages to using VxWorks as an operating system for a telephony communications system. VxWorks was originally designed as a real-time operating system (RTOS) for industrial grade applications. It is a modified version of the UNIX operating system and can easily be adapted to support the many multitasking requirements of an IP-PBX system. As of this writing, several IP-PBX suppliers are planning to use or are evaluating Linux as an operating system. Alcatel's OmniPCX Office, designed for the small systems market, was the first converged IP-PBX to use Linux. Alcatel plans to make available a Linux-based call telephony server option in their large OmniPCX 4000 system. Avaya's evolving Definity ECS platform is being upgraded to support a Linux call server platform. There are several reasons to use Linux as an operating system for a server-based IP-PBX design, as Alcatel and Avaya were the first to discover.

The server of a client/server IP-PBX must handle not only basic telephony call processing functions but also additional systems management (configuration, maintenance/diagnostics, call costing), advanced applications (voice and unified messaging, contact center), gateway interface, and firewall functions. The features and functions supported by the server influence the hardware and software design. The cited server services are available on standard Windows NT/2000 and UNIX/Linux server platforms. If each platform is equally capable of supporting the required services, the cost of the solution becomes a deciding factor between the operating systems. Linux is not currently subject to license fees and does not require vast processing resources to run the typical IP-PBX services. Linux supports telephony applications running on a typical IP-PBX system, and is available at an attractive cost to the system designer and developer. The free availability of Linux allows source files to be accessed easily.

An IP-PBX must perform most processing operations in real-time. Linux responds to events in what is known as *soft real-time* because some of its kernel operations cannot be pre-empted. This means that a high-priority process cannot interrupt a low-priority process, although the high-priority process must take precedence. Although Linux does not handle task in real-time, software extensions are available to achieve an acceptable response time, such as Real-Time Linux (RTLinux) and Real-Time Application Interface (RTAI). Alcatel uses the RTLinux extension in its OmniPCX Office system and will use it in its larger OmniPCX 4400 converged IP-PBX model when its optional server call control design is available. Real-time applications for RTLinux and RTAI can be developed using the portable operating system (POSIX).

Although the Linux operating system has not been used by the first generation of client/server IP-PBXs, the potential cost savings of using freeware is too attractive for most system designers to ignore as an alternative operating system to Windows when developing the next generation of systems. Any modifications to Linux that improve its use as an operating system for telephony systems will be made available to the entire telecommunications community as part of the no-fee licensing agreement and provide a further incentive for others to use it.

Server Redundancy

There are three major functional elements at the call processing layer that can be addressed in terms of redundancy: main CPU, memory, and power. It is simpler to address redundant memory and power issues first. Call telephony server memory systems may be based on a redundant array of independent disks (RAID) design or disk mirroring. Instead of using a single memory disk, a set of memory disks are used to store copies of the same piece of data in different physical locations. Information is written simultaneously to two different disks for two purposes: protection against component failure and possible improvement in system performance. Another redundant server option duplicates internal power supply modules to protect against power failures or errors. RAID/disk mirroring and duplicated power supplies may slightly increase the cost of a server but are usually available with most client/server IP-PBX models to increase system survivability and reliability levels.

There are two methods to provide call processing redundancy:

- Duplicate processor/memory board(s) in the server cabinet
- Multiple server configuration

At the time of this writing, the only client/server IP-PBX system available in a single server cabinet design with a fully duplicated call processor board is the Cisco Systems 7750 ICS. The system contains a system processing engine (SPE) 310 card that executes system software, including the Cisco ICS System Manager, and embedded fault-management services. The Cisco CallManager generic software program running on the SPE 310 delivers intelligent call processing to support IP telephony services and applications in the ICS 7750. A redundancy option is available by provisioning additional Cisco CallManager cards.

The second method, using multiple servers, can be based on two different design concepts

- Dedicated back-up server
- Multiple active servers

The Cisco IP Telephony System was the first to offer a redundant call telephony server option for a client/server IP-PBX design. The Cisco system can be configured with one or both redundant design options. Cisco can configure a system with a dedicated back-up server for a single active server for systems with port requirements of up 2,500 IP stations. The active CallManager Server replicates its database to the back-up server. IP client endpoints are programmed to signal the primary active server for registration and all call processing activities, but can also be programmed to signal the secondary back-up server if the primary server fails to acknowledge signal requests. In a client/server IP-PBX system, station equipment monitors the state of the main call processor to determine which call telephony server is available for any and all call

Client/Server IP-PBX System Design

processing activities. This situation is the reverse of a traditional circuit switched PBX, where the common control complex monitors the state of its peripheral endpoints. If the primary active control processor in a circuit switched PBX system fails, the internal system diagnostics program automatically activates a seamless switchover to the secondary back-up call processor (if available). The station port has no role in this process.

TABLE 10-1 Cisco CallManager 3.2 Cluster Guidelines

Required Number of IP Phones within a Cluster	Recommended Number of Cisco CallManagers	Maximum Number of IP Phones per Cisco CallManager
2,500	Three servers total: ▪ Combined publisher/TFTP ▪ One primary Cisco CallManager ▪ One backup Cisco CallManager	2,500
5,000	Four servers total: ▪ Combined publisher/TFTP ▪ Two primary Cisco CallManagers ▪ One backup Cisco CallManager	2,500
10,000	Eight servers total: ▪ Database publisher ▪ TFTP server ▪ Four primary Cisco CallManagers ▪ Two backup Cisco CallManagers	2,500

Very large customer port requirements require multiple CallManager servers networked in a cluster configuration because of individual server port capacity limits. Since Cisco CallManager Release 3.0(5), a cluster can contain as many as eight servers, six of which are capable of call processing. The other two servers can be configured as a dedicated database publisher and a dedicated Trivial File Transfer Protocol (TFTP) server, respectively. The publisher server makes all configuration changes and produces CDRs. The TFTP server facilitates the downloading of configuration files, device loads (operating code), and ring types. A dedicated publisher server and a dedicated TFTP server are recommended for large systems. For smaller systems, the function of database publisher and the TFTP server can be combined. Table 10-1 lists guidelines for scaling devices with Cisco CallManager clusters. The Cisco recommendations provide an optimum solution. It is possible to reduce the amount of redundancy and, hence, use fewer servers. For small systems, the database publisher, TFTP server, and Cisco CallManager back-up functions can be combined.

The maximum number of registered devices per Cisco CallManager is 5,000 in the case of the MCS7835, including a maximum of 2,500 IP telephones, gateways, and DSP devices, such as transcoding and conferencing resources. In the event of failure of one Cisco CallManager within the cluster, the maximum number of registered devices remains 5,000 per Cisco CallManager in the case of the MCS7835.

The servers are configured in a mesh network topology to better respond dynamically to changes in the network. Although customer database information (station user profiles, group assignments, call routing tables) does not significantly change over short periods, IP telephones may be constantly changing locations and reregistering with different servers in the network cluster. It is important that new registration data is stored quickly. In case of server failure or network problems, a fully meshed topology allows peripheral endpoint devices to locate and register with back-up servers.

In a multiple server cluster configuration, a publisher server protects against the possibility of server failure and loss of database information. The function of the publisher server is to provide read and write access for database administrators and for the CallManager Server nodes themselves. It is recommended that a dedicated server be used as the publisher server instead of using an existing CallManager server for the function. Use of a separate server prevents database updates from affecting the real-time processing performed by the CallManager server as part of call processing. The publisher server maintains a TCP connection to each CallManager server in the network cluster. When there are administrative changes to a CallManager server database, the publisher server replicates the changes on each CallManager server. The publisher server also serves as the central storage database for all CDRs written by all CallManager servers in the cluster. If the publisher server is not available due to system failure or network problems, the CallManager servers store the CDRs locally and replicate them to the publisher server when it becomes available. In the Cisco system an IP telephone can be programmed to signal up to three servers for gatekeeper, call processing, and feature support. One server is always designated as the primary server, and two others can be assigned as the secondary and tertiary servers. If the primary server fails or there are signaling path problems between the primary server and IP endpoint, the secondary server can be used for telephony services. The IP phone maintains active TCP sessions with its primary and secondary Cisco CallManagers. This configuration facilitates switchover in the event of failure of the primary Cisco CallManager. Upon restoration of the primary server,

Client/Server IP-PBX System Design

the device reverts to its primary Cisco CallManager. When the primary server is not available, the secondary server assumes primary status and the tertiary one assumes secondary status. The original tertiary server will assume primary server status only when the original primary and secondary servers are not available. The back-up secondary and tertiary servers may be active or dedicated standby servers.

One problem with using active servers as secondary/tertiary servers for redundancy is that station port capacity rules apply to all active servers regardless of their multifunctional mode. Another limitation to this redundancy option is a requirement for all clustered servers to be configured within a contiguous customer premises LAN configuration. CallManager servers cannot be networked over WAN facilities with IP network routers; they must be part of a localized cluster. The redundant multiple active server design is sometimes referred to as *N+M* because there are multiple active servers and multiple back-up servers. This provides the very high level of call processing redundancy typically provided by circuit switched PBX common control complexes.

The distributed architecture of a Cisco CallManager cluster provides the following primary benefits for call processing:

- Spatial redundancy
- Resiliency
- Availability
- Survivability

If multiple location communications is required, CallManager Server clusters or a single server must be configured at each location and networked as discrete systems. Intercluster communication is provided according to H.323 protocol standards and permits a subset of the features to be extended between clusters. The set of features transparent across clusters are:

- Basic call setup
- G.711 and G.729 calls
- Multiparty conference
- Call hold
- Call transfer
- Calling line ID

A Cisco alternative solution to a multiple cluster network configuration to support remote communications requirements with local call processing support is Survivable Remote Telephony (SRS) Telephony technology for all Cisco IOS software platforms that support voice. SRS Telephony comprises network intelligence integrated into Cisco IOS Software. This service can act as the call processing engine for IP phones in the branch office during a WAN outage. SRS Telephony is a capability embedded in Cisco IOS software that runs on the local branch office access router. SRS Telephony automatically detects a failure in the network and, using the Cisco Simple Network Automated Provisioning (SNAP) capability, initiates a process to intelligently autoconfigure the router to provide call processing back-up redundancy for the IP phones in that office. The router provides call processing for the duration of the failure, thus ensuring that the phone capabilities stay up and operational. Upon restoration of the WAN and connectivity to the network, the system automatically shifts call processing functions to the primary Cisco CallManager cluster. Configuration for this capability is done once in the Cisco CallManager at the central site, simplifying deployment, administration, and maintenance.

The Nortel Networks Succession CSE 1000 is based on an optional distributed processing design, although the standard design is based on a centralized call server. The call server is a centralized resource for database management and call processor functions, but each configured MG can be equipped with a local call processing controller card, the same card installed in the call server. If connectivity is interrupted between the call server and a survivable MG, there will be automatic switchover to survival mode. Once in survival mode, the succession MG itself, via the call processor on its MGC card, will serve the call processing function for all port interface cards. Although the primary functions of the MG are to provide control signaling interface connections to system peripherals and support a gateway function for communications between IP and non-IP peripherals, the module cabinet can also function as a call processing server for directly linked peripherals.

Any survivable IP succession MG can run in stand-alone mode in the event of a link failure or a catastrophic outage of the call server. If a system has multiple-succession MG links, a customer may choose any or all of the remotes to be survivable, depending on business requirements. If the call server or the IP link fails, the survivable succession MGs will automatically restart. During this time, the succession MGs will attempt to register with the call server. If no connection can be established, each survivable succession MG will go into survival mode, act as a stand-alone PBX, and service all users in each survivable succession MG. If the IP link fails to a particular survivable succession MG, then the succession MG and the (optional) succession MG expansion associated with that succession MG would go into survivable mode, thereby restoring service to all the users in both chassis. When an MG is in survivable mode, the IP telephone station users will hear a different dial tone and see a notice on their display field that the telephone is operating under local mode service.

Database synchronization between the call server and succession MGs is performed during a system data dump and system start-up. The database used by the survivable succession MGs during survival mode is an identical copy of the database configured at the call server. Customers, whose service is being provided by a succession MG in survivable mode, are notified by special dial tone and telephone display information. In addition, during survival mode, service changes at the succession MGs are possible but cannot be data dumped and must be re-entered once the system returns to normal mode.

The succession call telephony server performs functions comparable to those of the Cisco CallManager, publisher, and TFTP servers. Although all three functions can run on a single server, Cisco recommends that the CallManager generic software run on a dedicated server. The Nortel dispersed design eliminates the need for dedicated back-up servers. Succession MGs can be local or remote from the centralized call telephony server. A remotely located survivable MG protects against WAN failures to ensure continued local call processing service. It effectively functions like the Cisco Systems SRS Telephony service.

Multifunction Server Design

Although the main function of the IP-PBX call telephony server is to support gatekeeper and basic telephony services (dial tone and call routing), several system suppliers have bundled it with one or more advanced functions and applications. The following are some examples of client/server IP-PBXs that use a single server cabinet design to support features and functions beyond basic telephony.

The 3Com NBX family has a variety of call processing features and applications supported by its centralized main server cabinet, including voice mail, automated attendant, hunt/call groups, CDR, CTI, and PC-based visual voice mail/email clients (IMAP4). For example, the high-end SS3 NBX V5000 Chassis call processor can support up to 72 auto-attendant ports, 1,000 voice mailboxes, and 400 hours of message storage. The SS3 NBX chassis is backward compatible with 3Com NBX

25 and 100 systems. The call processors support up to 750 devices, including telephones, voice mail ports, and PSTN lines. The call processors use the VxWorks operating system to maximize reliability in mission-critical applications. Dual 10/100-Mbps uplink ports deliver resilient failover protection, and dual-load–sharing redundant AC inputs boost reliability. The SS3 NBX V5000 chassis features four chassis slots, two resilient 10/100 switched uplink ports, universal AC power connection, and a built-in 3Com RPS uplink port. The call processors range from 250-device capacity single-power supply (3C10201) and dual-redundant power supply (3C10202) versions to a dual-redundant power supply, 750-device unit (3C10203); all three offer resilient 10/100 uplink ports, universal AC power connections, and simple Web browser-based administration.

The Mitel Networks 3300 ICP supports a range of fully integrated, advanced Mitel Networks enterprise applications, including voice mail; speech-enabled auto attendant and unified messaging; PDA and PC-based application integration; and optional and emerging VoiceXML, contact center, and customer relationship management (CRM) applications. These applications are run from the same server processor.

Mitel Networks' MN 3100 also supports a variety of services, including messaging. The MN 3100 ICP is a client/server IP-PBX that also can be classified as an integrated communications system (ICS) platform solution because its single carrier cabinet design supports voice communications (IP-PBX), data communications (IP router, Ethernet switch), and messaging application services. An ICS platform is defined as a communications system design that not only supports basic voice call processing requirements but also satisfies enhanced applications such as messaging, computer telephony, call center, and/or data communications networking without external system hardware. All of the system features and functions are supported "in skin," leading to an all-in-one shorthand description.

The Cisco 7750 ICS also falls into the ICS platform category. The system's SPE 310 board is an application server card that runs call processing, system management, and voice applications including voice mail, automated attendant, unified messaging, interactive voice response, automatic call distribution, and Web-based contact-center applications. The Cisco SPE 310 card offers customers the flexibility to add support for a broad range of Web-based communications applications as their business and communications needs grow.

Each of these IP-PBXs is based on a client/server design and can be classified as an ICS platform system. In addition to some advanced voice

Client/Server IP-PBX System Design

options, the Mitel Networks MN 3100 and Cisco 7750 ICS offerings also integrate voice and data communications capabilities. Both products are targeted at customers who are looking for a truly converged single system solution for their voice and data networking needs.

If advanced telephony features and applications are not integrated into the telephony call server, they can be supported with a dedicated application server configured on the LAN/WAN, but under the management and control of the telephony call server. This is a design concept used by most circuit switched PBXs using proprietary solutions or third-party CTI application server options. In many cases, the same application server and software solution used behind a manufacturer's circuit switched PBX system is also used with the client/server IP-PBX design. For advanced call center requirements, the Nortel Networks Symposium Call Center Server, originally designed for the Meridian 1 PBX platform, is compatible with the Succession CSE 1000 IP-PBX. The Siemens HiPath ProCenter Suites server works behind the Hicom 300H and HiPath 4000 PBX models and can be configured to work behind the HiPath 5000 IP-PBX models. The Mitel Networks 6100 Contact Center Solution, designed for the SX-2000 Light PBX, can interwork with the MN 3300 ICP system. The Nortel, Siemens, and Mitel contact center servers are not stand-alone systems but are functionally dependent on the IP-PBX call telephony server for control signaling support.

Application servers are also used for messaging applications. The Siemens Xpressions server that supports VMS and UMS behind the manufacturer's Hicom 300H and HiPath 4000 PBXs is also available as an option with the Siemens HiPath 5000 IP-PBX models. HiPath Xpressions unifies voice, fax, and e-mail into a single mailbox that can be accessed from any PC or touch-tone phone. Station users can listen to voice messages, listen to e-mails, and listen to fax messages from any analog DTMF, digital, or IP telephone on or off site. It's available in a VMS configuration and upgradeable to full unified voice, fax, and/or e-mail messaging. The solution's Windows NT server design is based on open standards and integrates with other IP-ready mobility and collaboration solutions in the Siemens HiPath MobileOffice portfolio. Nortel's CallPilot server provides similar capabilities working behind the Succession CSE 1000.

3Com's NBX Call Center is a server application that supports up to 25 agents, two supervisors, one administrator, and one database manager. The software manages call flows in real time, makes changes dynamically to optimize customer responsiveness, and handles up to 100 customer queues with the Intelligent Call Routing feature. The supervisor

GUI supports sophisticated real-time monitoring of alarms, ports, queues, and call status, allowing supervisors to drag and drop queue assignments to prevent abandoned calls. Alarms can be used to signal critical thresholds. Components include IBM servers, Sybase database software, and Apropos application software. The application server works behind the main system's call telephony server, the NBX or SS3 chassis equipment.

Telephony Gateways

IP telephones, including PC client softphones, communicate directly with the call telephony server over a customer LAN/WAN infrastructure. Proprietary port circuit cards housed in proprietary port carriers are not required for signaling between the IP desktop and the common control controller, unlike converged IP-PBX designs. Non-IP stations and trunk circuits require telephony gateway interfaces to support server control signaling and voice communication transmissions. Telephony gateways for analog telephones and other 2500-type compatible communications devices, such as facsimile terminals, may be provided through a variety of design methods:

- Integrated call telephony server gateway interfaces
- Desktop gateway modules: proprietary, third party
- Gateway servers/interfaces: proprietary, third party

Several proprietary, closed call telephony servers have integrated gateway interfaces for PSTN digital T1/E1 trunk circuits. The gateway interfaces usually support ISDN BRI or PRI services over the T1/E1 trunk circuits. The Mitel MN 3100, 3Com NBX, and Siemens HiPath 5300 systems have integrated PSTN digital trunk gateway interfaces. For example, the 3Com NBX's integrated analog line card connects up to four conventional (loop start) PSTN telephone lines, and the T1/PRI trunk card connects to a standard T1 circuit. The HiPath 5300 BRI gateway interface card supports four BRI ports (8×64-Kbps channels); the PRI gateway interface card includes a T1 carrier interface.

Mitel Networks uses a different approach to support non-IP peripherals on its MN3300 ICP. The 3300 ICP includes an analog services unit (ASU), and a network services unit (NSU), but also supports traditional Mitel SuperSet digital telephones by a link to a peripheral equipment (PE) cabinet. An ASU supports four analog trunks and 16 stations (including MOH, paging, and PFT); an NSU supports four T1 digital trunk interfaces. Up to four ASU and four NSU carriers are supported per controller carrier. What is unique about the system is that the call server also provides control signaling to an SX-2000 Light PE cabinet. Supporting the traditional PE cabinet protects a customer's substantial investment in the installed base of proprietary Mitel SuperSet voice terminals.

Mitel intends the MN 3300 system to be a migration vehicle for its large installed base of SX-2000 system customers and allows customers to link existing PE cabinets to the new call telephony server through one of two options: direct optical fiber cable connection or T1 trunk interface. Customers who want a centrally located call telephony server and PE cabinet can use the optical fiber link. The DTI can support remote PE cabinets. The 3300 ICP was the first client/server IP-PBX design to support common equipment originally designed for a circuit switched PBX system. All communications traffic between digital telephones is handled internally by the PE cabinet's integrated circuit switched TDM backplane. Calls between PE endpoints and other endpoints (IP telephones and ASU and NSU ports) are handled across the integrated controller gateway channels.

Desktop gateway modules may be proprietary or industry-standard H.323 equipment. The most common desktop gateways support 2500-type communications devices, such as analog DTMF telephones and facsimile terminals. The desktop communications device links directly to the gateway module and converts analog signals to IP format for control and communications signaling. For example, 3Com NBX analog devices are available as single-port stand-alone units and four-port chassis-based cards. The single-port ATA unit also includes an additional Ethernet port that allows an analog device and an Ethernet device to share the same Ethernet LAN cabling. The multiple-port NBX analog terminal card features four analog (FXS) ports. The units connect to a wide variety of industry-standard analog devices and fax machines and provide support for door phones, paging systems, and other applications that may require analog connectivity.

The gateways may be proprietary to an IP-PBX system, like the Siemens HiPath AP 1100 (available in one- and four-port interface models), or third-party products available from a large list of suppliers. For example, Ericsson markets a downsized version of its Webswitch IP-PBX for use as an H.323 gateway module. 3Com, a major enterprise data communications equipment supplier, is another IP-PBX supplier marketing desktop gateway modules, including those that support H.323 and SIP standards.

Another type of desktop gateway module is an add-on adapter that converts a proprietary digital PBX telephone into an IP-compatible voice terminal. A few client/server IP-PBX manufacturers, including Siemens and Nortel Networks, offer this as an option to upgrade installed digital telephones originally designed for use behind their circuit switched PBXs. The same adapters can support IP desktops behind the manufacturer's converged IP-PBX system solutions.

Gateway servers and interface modules/boards that are not fully integrated into the call telephony server or used as desktop devices are proprietary to a manufacturer's IP-PBX or conform to industry standards, such as H.323 or MGCP, and used as OEM solutions. One example of a proprietary solution is the Mitel Networks MN 3300 ICP gateway carrier that supports traditional analog trunks (loop start) and digital trunks (DASSII, DPNSS, QSig, Euro ISDN, and BRI) for connection to the PSTN and for connecting multiple sites or systems together. This allows multiple 3300 ICPs to be clustered or networked between multiple sites over IP or traditional TDM infrastructures to support up to 40,000 users. The MN 3300's call telephony server carrier supports the trunk gateway interface carrier.

The Cisco Systems IP Telephony system, when originally designed as the Selsius System, used desktop modules for support of non-IP communications devices and trunk circuits. The redesigned product supports analog station, analog trunk, and digital trunk interfaces with proprietary circuit boards that are housed in Cisco Catalyst 6000 Ethernet switch carriers. Three different modules are used for analog connections: 24-port analog station FXS (H.323 or MGCP), analog trunk circuit FXO (H.323 or MGCP), and analog E&M tie trunk (H.323 only). The FXS module supports fax relay, which enables compressed fax transmission over the IP WAN. An alternative to the FXS module is the standalone Cisco VG248 analog gateway module that supports 48 fully featured analog phone lines as extensions to the Cisco CallManager system. It is housed in a compact 19-inch rack-mount chassis, and its high-density gateway can be used for analog phones, fax machines, modems, and speaker phones. Digital PSTN trunk interfaces are supported by a limited-capacity stand-alone T1 adapter module or a Catalyst 6000 T1 and services module that provides eight T1 ports (192 channels) or DS0 voice trunks. The module supports voice trunk protocols such as ISDN Primary Rate Interface (PRI) and in H2 CY '00, channel-associated signaling (CAS). The module's DSP resources can also be programmed for call conference bridge services and voice codec transcoding applications, instead of digital trunk gateway interfaces.

Client/Server IP-PBX System Design

The Nortel Succession CSE 1000 MG module supports a variety of non-IP interfaces, such as analog station, analog trunk, and digital trunk. Each MG module has three IPE card slots and can support an expansion module for four additional slots. The first Succession CSE 1000 release is limited to a maximum of four MGs (28 card slots, maximum). Table 10-2 summarizes the many MG IPE cards and interfaces supported by the Succession MG.

TABLE 10-2
MG IPE Cards and Interfaces Supported by Succession MG

Gateway	Component	Interface	Maximum Capacity/ Gateway
MG	Digital trunk card1	1–T1/PRI2	3 per gateway
MG and gateway expansion	ITG line card	96 IP phones	2 per gateway or gateway expansion
MG and gateway expansion	Analog line card	16 analog telephones	3 per gateway or 4 per gateway expansion
MG and gateway expansion	Analog trunk	8 analog trunks CO, DID, RAN, music-on-hold	3 per gateway or 4 per gateway expansion
MG and gateway expansion	E&M trunk card	4 2-wire 4-wire tie trunks	3 per gateway or 4 per gateway expansion
MG and gateway expansion	Call Pilot 201i server card	UMS application	1 per gateway or gateway expansion; requires two consecutive card slots
MG and gateway expansion	E mobility 802.11 wireless IP gateway	24 wireless IP telephones	2 per gateway or gateway expansion; requires two consecutive card slots
MG and gateway expansion	Reach line card for remote office 9150, 9110, 9115, and IP adapters	16 or 32 ports for remote office digital phone connections	3 16-port cards per gateway or 4 per gateway expansion; 1 32-port card per gateway or 2 per gateway expansion
MG and gateway expansion	Integrated recorded announcement card	Multichannel recorded announcements	3 per gateway or 4 per gateway expansion

continued on next page

TABLE 10-2

MG IPE Cards and Interfaces Supported by Succession MG (continued)

Gateway	Component	Interface	Maximum Capacity/ Gateway
MG and gateway expansion	Integrated conference bridge card	16 or 32 port "meet me" conference bridge	3 per gateway or 4 per gateway expansion
MG and gateway expansion	Integrated call assistant card	Auto-attendant, voice processing	3 per gateway or 4 per gateway expansion
MG and gateway expansion	Integrated personal call director	"One number, follow-me application"	3 per gateway or 4 per gateway expansion

1. Requires 1 clock controller daughterboard per gateway with digital trunk card.
2. PRI requires a D-channel daughterboard card.

What is different about the Succession CSE 1000 as a client/server IP-PBX system is its requirement for IP station line circuit cards to provide call signaling to IP phones behind the call telephony server. The ITG line card supports the following functions in the Succession CSE 1000 system:

- Provides a bearer channel gateway between the IP domain and the Meridian TDM domain. Up to 24 simultaneous bearer channels can be supported.
- Provides a terminal proxy server (TPS) for up to 96 IP telephones (any combination of IP phones and soft clients). Provides a TFTP server for the IP phone to download its firmware.
- Provides a registration service to register and authenticate the IP telephones.
- All signaling to the IP telephones is via the TCP protocol.
- Provides a registration service to register and authenticate the IP telephones.
- All signaling to the IP telephones is via the TCP protocol.
- Sets up of media channels between the IP telephones, the audio channels on the ITG card itself, and other IP devices.
- Provides a pool of media channels, with codecs and echo cancellers that are used when the IP phone needs to be connected to the Succession CSE 1000 TDM switch fabric.
- Provides an arbitration mechanism when multiple servers have access to one IP phone.

Client/Server IP-PBX System Design

Nortel plans to support peer-to-peer LAN switching between IP peripheral endpoints without the need for ITG line cards to perform call setup and tear-down procedures, although the cards can be used for gateway channels to communicate with non-IP communications devices. The primary role of the Succession MG is very similar to the role of Nortel's circuit switched Meridian 1 IPEM port carriers: housing interface cards to support control signaling and communications transmission to all system ports. Call connections between non-IP ports linked to the same MG are handled through an integrated TDM switch network. Calls between non-IP ports linked to different MGs are handled over the LAN using gateway channels on the ITG line interface card.

The Mitel Networks, Cisco System, and Nortel Networks client/server IP-PBX gateway options are proprietary to the systems and fully integrated into the telephony server or proprietary hardware. Most of the gateways are necessary to support non-IP communications devices and PSTN trunk interfaces and will remain necessary for the foreseeable future. The only present client/server IP-PBX system configurations that do not require any gateway options are those systems servicing small, edge locations networked to a tandem or main PBX system. Only a relatively small line size system could function as a satellite location within a private network without a need to support at least one analog communications device, if only a facsimile terminal, and/or local PSTN trunk circuits.

Summarizing Client/Server IP-PBX Design Issues

It is apparent that client/server IP-PBX designs differ across models. There are even major design differences within one supplier's product family. For example, the Cisco Systems 7750 ICS all-in-one design is radically different from the supplier's larger MCS 7825/7835 multiple server cluster design; Siemens offers client/server models based on a closed, embedded Windows NT server or a customer-provided server option. Client/server IP-PBX designs may be far simpler than traditional circuit switched PBXs and converged IP-PBX platforms, but there is a sufficient number of design variables to create major differences between system models.

Figure 10-1 shows how three layers (call telephony server, gateways, and applications) of a client/server IP-PBX can be designed and integrated into the system architecture. The choices available to a designer include a proprietary or a nonproprietary call telephony server, integrated or external advanced applications support, integrated or external gateways. Gateways and application servers may be a mix of proprietary or third-party solutions. Designers may select proprietary components for quality control and development of feature/function options capabilities not supported by third-party solutions. Third-party components may reduce system costs and provide customers with more design flexibility in their purchase decisions.

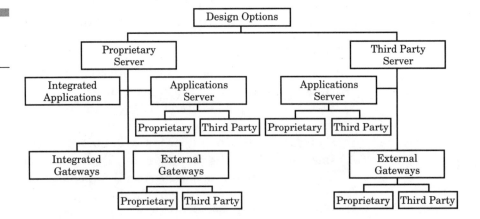

Figure 10-1
Client/server IP-PBX design.

Several IP-PBXs originally designed using third-party call telephony servers were redesigned with customized servers because distributors and customers often failed to use a third-party server with the technical specifications as recommended by the supplier. Several manufacturers that currently offer a proprietary call telephony server model plan to migrate their systems to a less expensive third-party server solution, but not until all the system design bugs and problems are solved after one or two system generations.

If issues related to proprietary or third-party components are not part of the customer-buying equation, there are other important design and performance criteria to be evaluated before a purchase decision is made. System reliability and survivability are as important when evaluating a client/server IP-PBX as they are in a circuit switched PBX design—port capacity, traffic handling, and call processing power. The following is a summary checklist of design performance issues for evaluating a client/server IP-PBX (many unique to an IP telephony communications system):

Client/Server IP-PBX System Design

- **System redundancy**—Fully duplicated or shared back-up components for call processing, memory, and power functions
- **System port capacity**—Stations (IP, non-IP) and trunk interfaces (analog, T1/E1, IP)
- **System traffic handling capacity**—Distribution of gateway channels and conference circuits
- **System call processing**—BHCC
- **Supported call control protocols and interfaces**—H.323, SIP, MGCP, SGCP, MEGACO, etc.
- **Voice codec support**—G.711, G.723.1, G.729(A,B), GSM, etc.
- **QoS support**—DiffServ, 802.1p/Q, COS, TOS, IP precedence, RSVP, dynamic jitter buffer, packet loss replacement
- **Standard messaging system interfaces**—AMIS-A, VPIM, LDAP, IMAP

The first client/server IP-PBX systems were shipped less than 5 years ago, and in that time the design and performance capabilities have changed significantly. It will take several more years until the product design stabilizes, and its reliability is comparable to current circuit switched PBXs, which have been shipping for more than 25 years. Although a new generation of data communications systems seems to appear every year or two, new voice communications system platforms take many years to evolve.

CHAPTER 11

LAN/WAN Design Guidelines for VoIP

There is no standard definition for QoS as it applies to real-time voice communications carried over an Ethernet LAN or IP WAN. As applied to a circuit switched PBX, QoS means consistent, reliable service delivery of control and communications signals in support of customer needs. This definition also can be used for LAN QoS in support of IP telephony. To enable LAN QoS requires all network elements, at all network layers, to work together to support a required level of traffic and service.

An IP-PBX by definition is not a virtual circuit switched communications system, like a traditional PBX, but rather a system that uses an IP network infrastructure. An IP network makes more efficient use of available bandwidth resources than does a circuit switched PBX and is designed to support the "bursty" nature of data communications traffic rather than the continuous traffic flow of real-time voice communications. IP networks can adapt to changing traffic conditions, but the level of service can be unpredictable. When used to support an IP-PBX system, the IP network must be properly designed and engineered to support the unique real-time traffic requirements of voice as opposed to less stringent data communications requirements.

QoS techniques manage bandwidth according to different application demands and network management settings but cannot guarantee a service level if resources are not available and allocated. Reserving resources for voice communications can seriously affect other network traffic. A priority for QoS network designers has been to ensure that best-effort traffic is available after resource allocations have been made. QoS-enabled high-priority voice applications must not harm lower-priority data applications.

The Internet was based on a dumb network concept with intelligent endpoints to transmit and receive datagram packets flowing through a series of network routers. IP does not deliver reliable service over the Internet: packets can be dropped by routers and are retransmitted as necessary. The service mechanism can assure data delivery, but not timely delivery. This "best-effort" service may be adequate for data networking services, but it is not good enough for voice communications.

Audio and video traffic demands sufficient bandwidth with low-latency requirements when used in two-way communications. A major challenge for network planners is to design a LAN infrastructure that satisfies an acceptable QoS level that PBX system users have grown accustomed to for their voice communications applications. A newly installed IP-PBX system in a green field location provides an ideal situation, but if a network is already installed and operating, introducing IP telephony-grade QoS should not disrupt existing services and applications.

LAN/WAN Design Guidelines for VoIP

LAN QoS levels fluctuate over time due to unanticipated changes in customer usage patterns and traffic flow. If QoS is degraded for short periods, it may significantly affect IP telephony services in ways noticeable by all system users, even if data communications services appear satisfactory. There are several reasons QoS can change:

- Temporary excessive network usage
- Insufficient link capacity
- Insufficient switch/router resources
- Traffic flow peaks
- Traffic flow interference
- Improper use of resources

Several basic control methods can be employed to manage QoS levels to ensure the higher grade of service level required by real-time voice communications:

- Reserving fixed bandwidth for mission-critical voice communications applications
- Restricting network access and usage for defined users or user groups
- Assigning traffic priorities
- Designating which kinds of traffic can be dropped when congestion occurs

There are several high-level decisions facing network planners and managers regarding the type of QoS-based network to be designed and operated. The network planner must decide whether network users are involved in the QoS functions or whether the network is in total control of QoS functions. If a network user has knowledge of QoS functions and a limited degree of QoS control, the network QoS is said to be *implicit*. If network QoS functions are predetermined and only the network administrator can program changes when needed, the network QoS is said to be *explicit*.

Another planning issue is whether QoS is soft or hard. Network QoS is said to be soft when there is no formal guarantee that target service levels will be met, even if QoS functions are implemented. Hard network QoS is a guarantee of service at a predefined level of QoS. Hard network QoS is usually available only with connection-mode transport, such as ATM constant bit rate (CBR) service.

Network QoS is also manageable by network design, by installing the necessary physical resources to support target service levels. IP-PBX system voice quality and availability can be determined by the physical

LAN infrastructure and available cable bandwidth. Cisco Systems, a leading IP-PBX system supplier and the dominant supplier of data communications systems, has developed and published an IP telephony network planning guide. The Cisco guide is a planning tool for its CallManager IP-PBX system customers, but it is also useful as a network design guide for customers who plan to install and operate any converged or client/server IP-PBX system.

Fundamental LAN Planning Guidelines

The Cisco guide recommends a detailed analysis of the following LAN elements:

- LAN/campus topology
- IP addressing plan
- Location of TFTP servers, DNS servers, DHCP servers, firewalls, network address translation (NAT) gateways, and port address translation (PAT) gateways
- Potential location of gateways and call telephony servers
- Protocol implementation including IP routing, Spanning Tree, VTP, IPX, and IBM protocols
- Device analysis including software versions, modules, ports, speeds, and interfaces
- Phone connection methodology (direct or daisy chain)

According to the Cisco guide, the significant LAN topology issues are:

- Available average bandwidth
- Available peak or burst bandwidth
- Resource issues can may affect performance including buffers, memory, CPU, and queue depth
- Network availability
- IP phone port availability
- Desktop/phone QoS between user and switch
- Network scalability with increased traffic, IP subnets, and features
- Back-up power capability

LAN/WAN Design Guidelines for VoIP

- LAN QoS functionality
- Convergence at Layers 2 and 3

IP addressing issues that should be reviewed are:

- Phone IP addressing plan
- Average user IP subnet size use for the campus
- Number of core routes
- IP route summary plan
- DHCP server plan (fixed and variable addressing)
- DNS naming conventions

Potential considerations with IP addressing include:

- Route scalability with IP phones
- IP subnet space allocation for phones
- DHCP functionality with secondary addressing
- IP subnet overlap
- Duplicate IP addressing

The locations (or potential locations) of servers and gateways are important to ensure that service availability is consistent across the LAN infrastructure and for multiple sites. Gateways and servers in the review should include:

- TFTP servers
- DNS servers
- DHCP servers
- Firewalls
- NAT or PAT gateways
- Call telephony server
- Gateway location

After determining the location of these network elements, the following issues should be analyzed:

- Network service availability
- Gateway support (in conjunction with the IP telephony solution)
- Available bandwidth and scalability
- Service diversity

IP telephony scalability and availability issues will be affected by protocols in the network. The following areas for the protocol implementation analysis are:

- IP routing including protocols, summarization methods, non-broadcast media access (NBMA) configurations, and routing protocol safeguards
- Spanning Tree configuration including domain sizes, root designation, uplink fast, backbone fast, and priorities in relation to default gateways
- HSRP configuration
- VTP and VLAN configuration
- IPX, DLSW, or other required protocol services, including configuration and resource usage

With regard to protocol implementation, the following issues should be reviewed:

- Protocol scalability
- Network availability
- Potential impact on IP telephony performance or availability

All network devices should be reviewed and analyzed to determine whether the network has the desired control plane resources, interface bandwidth, QoS functionality, and power management capabilities. The checklist for this process includes:

- Device (type and product ID)
- Software version(s)
- Quantity deployed
- Modules and redundancy
- Services configured
- User media and bandwidth
- Uplink media and bandwidth
- Switched versus shared media
- Users per uplink and uplink load sharing/redundancy
- Number of VLAN supported
- Subnet size, and devices per subnet

For establishing a network baseline, it is important that the following measurements be made to determine voice quality levels and potential problem areas:

LAN/WAN Design Guidelines for VoIP

- Device average and peak CPU
- Device average and peak memory
- Peak backplane use
- Average link use (prefer peak hour average for capacity planning)
- Peak link use (prefer 5 minute average or smaller interval)
- Peak queue depth
- Buffer failures
- Average and peak voice call response times (before IP telephony implementation)

Cabling questions that may help determine the readiness of the infrastructure for IP telephony readiness include:

- Does the building wiring conform to EIA/TIA-568A?
- Does your organization comply with National Electric Code for powering and grounding sensitive equipment?
- Does your organization comply with the more rigorous IEEE 1100-1992 standard for recommended practices of grounding and powering sensitive equipment?
- Does the organization have standards for data center and wiring closet power that include circuit distribution, available power validation, redundant power supply circuit diversity, and circuit identification?
- Does the organization use UPS and/or generator power in the data center, wiring closet, phone systems, and internetworking devices?
- Does the organization have processes to SNMP manage or periodically validate and test back-up power?
- Does your business experience frequent lightning strikes? Are there other potential natural disasters?
- Is the wiring to your building above ground?
- Is the wiring in your building above ground?

Network bandwidth consumption is required for each VoIP stream. In any conversation, two such streams are required: one in each direction. The required bandwidth per conversation will be based on several factors, but of primary importance is the codec used to digitize, compress, and convert an analog voice signal into IP format. The two codecs of most interest are G.711 and G.729A. G.711 is the TIA recommended codec to optimize IP telephony QoS because it reduces impairment of the voice signal across the network, but the signal is uncompressed and requires a high amount of bandwidth. To save on network transmission costs, G.729A is used for off-premises traffic because it uses a compression algorithm to reduce bandwidth requirements. Table 11-1 lists the basic bandwidth requirements for each codec at different sampling periods.

TABLE 11-1

Baseline Bandwidth Consumption

Codec	Sampling Period (MS)*	Voice Payload (bytes)	Packets per Second	Bandwidth per Conversation
G.711	20	160	50	64 kbps
G.711	30	240	33	64 kbps
G.729A	20	20	50	8 kbps
G.729A	30	30	33	8 kbps

Values are rounded to the nearest integer.
*The sampling period is the waiting period for collecting voice information before being forwarded.

In addition to the basic datagram, there is the additional overhead of headers including RTP headers, UDP headers, IP headers, and link headers. All of the media types listed in Table 11-2 assume RTP, UDP, and IP headers at 40 bytes per packet.

TABLE 11-2

Bandwidth Consumption with Headers

Codec	Ethernet 14 Bytes of Header
G.711 at 50 pps	85.6 kbps
G.711 at 33 pps	78.4 kbps
G.729A at 50 pps	29.6 kbps
G.729A at 33 pps	22.4 kbps

Bandwidth allocations can be reduced by using RTP header compression and VAD. RTP header compression reduces the size of the IP/UDP/RTP header from 40 bytes to 2 bytes. VAD reduces the bandwidth requirement by approximately one-half because bandwidth is allocated only to the talking party. The VAD values listed in Table 11-3 can be misleading because the media streaming is not deterministically interrupted 50 percent of the time: VAD-controlled media streaming is a function of the measured audio levels at the sending end, which are

LAN/WAN Design Guidelines for VoIP

influenced by factors such as background noise. For this reason, caution should be used when reducing the bandwidth requirements.

TABLE 11-3 Bandwidth Consumption with cRTP and VAD

Codec	Ethernet 14 Bytes of Header
G.711 at 50 pps	85.6 kbps
With cRTP	70.4 kbps
With VAD	42.8 kbps
With cRTP and VAD	35.2 kbps
G.711 at 33 pps	78.4 kbps
With cRTP	67.7 kbps
With VAD	39.2 kbps
With cRTP and VAD	33.9 kbps
G.729A at 50 pps	29.6 kbps
With cRTP	14.4 kbps
With VAD	14.8 kbps
With cRTP and VAD	7.2 kbps
G.729A at 33 pps	22.4 kbps
With cRTP	12.3 kbps
With VAD	11.2 kbps
With cRTP and VAD	6.1 kbps

Network performance and capacity planning help to ensure that the network will consistently have available bandwidth for data and VoIP traffic and that the VoIP packets will consistently meet delay and jitter requirements. Cisco recommends the following six-step process for network capacity and performance planning:

1. Determine baseline existing network use and peak load requirements
2. Determine VoIP traffic overhead in required sections of the network based on busy hour estimates, gateway capacities, and/or CallManager capacities
3. Determine minimum bandwidth requirements

4. Determine the required design changes and QoS requirements based on IP telephony design recommendations and bandwidth requirements (overprovide where possible)
5. Validate baseline performance
6. Determine trunking capacity

Factors Affecting QoS: Packet Loss and Latency

Packet Loss

The two major problem areas that affect IP telephony QoS are packet loss and delay. The two QoS impairment factors are sometimes interrelated.

Packet loss causes voice clipping and skips, often resulting in choppy and sometimes unintelligible speech. Voice packets can be dropped if the network quality is poor, the network is congested, or there is too much variable delay in the network. Poor network quality can lead to sessions frequently going out of service due to lost physical or logical connections. To avoid lost or late packets, it is necessary to engineer the IP telephony network to minimize situations that cause the problem, but even the best-engineered system will not stop congestion-induced packet loss and delay. To combat this problem, it is recommended that a buffer be used on the receiving end of a connection. Buffer length must be kept to a minimum because it contributes to end-to-end network delay. Dynamic receive buffers that increase or decrease in size can be used to handle late packets during periods of congestion and avoid unnecessary delays when traffic is light or moderate.

Packet problems that occur at the sending end of a connection can be handled by methods such as *interleaving* and *forward error correction* (FEC). Interleaving is the resequencing of speech frames before they are packeted. For example, if each packet has two frames, the first packet contains frames 1 and 3 and the second packet contains frames 2 and 4. If a packet is lost, the missing speech frames will be nonconsecutive and the gaps will be less noticeable to the receiving party. FEC is a method that copies information from one packet to the next packet in the sequence. This allows the copied data to be used in the event a packet is lost or late.

Different methods are used at the receiving end of the connection. Unlike a circuit switched network, a packet switched network breaks communications signals into small samples, or packets of information. Each packet has a unique header that identifies packet destination and provides information on reconstruction when the packet arrives. Packets travel independently across the LAN/WAN and can travel by different routes during a single call. Packets can be lost for two primary reasons: dead-end routes and network congestion. Network congestion can lead to packet drops and variable packet delays. Voice packet drops from network congestion are usually caused by full transmit buffers on the egress interfaces somewhere in the network. The packet is purposely dropped to manage congested links. As links or connections approach 100 percent use, the queues servicing those connections become full. When a queue is full, new packets attempting to enter the queue are discarded. This can occur on an Ethernet switch or IP network router. Network congestion is typically sporadic, and delays from congestion tend to be variable in nature. Egress interface queue wait times or large serialization delays cause variable delays of this type.

DSP elements in most current voice codecs can correct for up to 30 milliseconds of lost voice. If the voice payload sample is no greater than this loss time, the correction algorithm is effective, if only a single packet can be lost during any given time. There are several methods that can compensate for lost or long-delayed packets. It is not practical to search for a lost packet to try to retrieve it. A preferred option is to conceal packet loss by replacing lost packets with something similar. One approach is to replay the last ordered packet in place of the lost one. This is a simple solution that is acceptable for rare packet loss, but a more complex solution is required for situations of frequent packet loss.

Several techniques are available for replacing a lost packet. One technique is to estimate the information that would have been in the packet. This concealment method generates synthetic speech to cover missing data. The concealment technique should have spectral characteristics similar to those of the speaker. This is relatively easy for a CELP-type codec such as G.729A because the speaker's voice signals are modeled during the encoding process. It is a more difficult process if a waveform codec such as G.711 is used, because the amplitude of the waveform is coded rather than making assumptions about how the sound was produced. G.711 codec packet loss concealment requires more complex processing algorithms and greater memory requirements and adds to system delay. A waveform codec, such as G.711, compensates for this; it can rapidly recover from packet loss because the first speech sample in the first good packet restores the speech to the original, whereas CELP-based codecs require a few frames to catch up.

The concealment process requires the receiver codec to store a copy of the synthetic packet in a circular history buffer that calculates the current pitch and waveform characteristics. With the first bad packet, the contents of the buffer are used to generate a synthetic replacement signal for the duration of the concealment. When two consecutive frames are lost, repeating a single pitch can result in harmonic artifacts, or beeps, that are noticeable when the erasure lands on unvoiced speech sounds, such as s or f, or rapid transitions, such as the stops p, k, and d. Concealment algorithms often increase the number of pitch periods used to create replacement signals when multiple packets are lost. This results in a variation of the signal and creates more realistic synthetic speech.

There must be a smooth transition between synthesized and real speech signals. The first good packet after an erasure needs to be merged smoothly into the synthetic signal. This is done by mixing synthesized speech from the buffer with the real signal for a short time after the erasure period.

Packet loss can become noticeable when a few percentages of the packets are dropped or delayed, and it begins to seriously affect QoS when the percentage of the lost packets exceeds a certain threshold (roughly 5 percent of the packets). Major problems also occur when packet losses are grouped together in large packet bursts. The methods for dealing with packet loss must be balanced against adding delay packet transport between connected parties. Delay issues and solutions are covered in the next section.

Latency

Delay, commonly referred to as latency, is the time delay incurred in speech by the IP telephony system. It is usually measured in milliseconds from the time a station user begins to speak until the listener actually hears speech. One-way latency is known as *mouth-to-ear latency*. *Round-trip latency* is the sum of the two one-way latencies comprising a voice call. Round-trip latency in a circuit switched PBX system takes less than a few milliseconds; PSTN round-trip latency is usually tens of milliseconds but almost always less than 150 milliseconds. Based on formal Mean Opinion Score (MOS) tests, latency at or under 150 milliseconds is not noticeable to most people. Latency up to 150 milliseconds receives good to excellent MOS scores ranging between 4 and 5 (1–5 scale) and provides for a satisfactory IP telephony QoS experience. One hundred fifty milliseconds is specified in the ITU-T G.114 recommendation as the maximum desired one-way latency to achieve high-quality voice. Switched network latency above 250 milliseconds, more common for international calls, becomes noticeable and receives fair MOS scores. Latency above 500 milliseconds is annoying and deemed unsatisfactory for conducting an acceptable conversation.

Latency in an IP telephony network is incurred at several nodal points across the voice call path, including the IP telephony gateways at the transmitting and receiving ends of a conversation. Latency is cumulative, and any latency introduced by any component in an IP telephony system will directly affect the total latency experienced by the station users.

The gateway network interface for an IP peripheral voice terminal may be an integrated component of the telephone instrument or an external device, such as a desktop IP adapter module, or embedded on a port circuit interface card housed in a PBX port carrier. The network interface in a gateway includes any hardware or software that connects the gateway to the telephone system or network. The typical network interface frames and converts digitized audio PCM data streams into the internal PCM bus for transport across a DSP. There is usually very little latency induced in this process, with typical maximums well below 1 millisecond. The DSP function is more complex because it involves compression or decompression of speech, tone detection, silence detection, tone generation, echo cancellation, and generation of "comfort" noise. The entire DSP mechanism is known collectively as *vocoding*.

DSP operations depend on processing entire frames of data at one time. The side effect of processing data in frames is that none of the data can be processed until the frame is completely full. Digitized speech arrives at a fixed rate of 8,000 samples per second, and the size of the frame processing the data will directly affect the amount of latency. A 100-sample frame would take 12.5 milliseconds to fill, and a 1000-sample frame would take 125 milliseconds to fill. Deciding on the frame size is a compromise: the larger the frame, the greater the DSP efficiency, but with that comes greater latency. Each standard voice coding method uses a standard frame size. The maximum latency incurred by the framing process depends directly on the selection of vocoder. Table 11-4 shows the frame duration and size for the three most commonly available voice codecs used by current IP-PBX systems.

TABLE 11-4

Vocoder Framing

Voice Codec	Bandwidth (bits/s)	Frame Duration(ms)	Frame Size (bytes)
G.711	64,000	Flexible	Flexible
G.723.1	5,300–6,300	30	24
G.729a	8,000	10	10

A G.711 voice codec can be programmed for frame size specifications, and very small frame duration delays can be used. A typical G.711 programmed frame duration is 0.75 milliseconds. A G.723.1 voice codec results in far greater frame delay than a G.729A voice codec, with only a slight comparative bandwidth savings.

After the collection of an entire frame is completed, the DSP algorithms must be run on the newly created frame. The time required to complete the processing varies considerably but never exceeds the frame collection time; otherwise, the DSP would never complete processing one frame before the next frame arrived. A DSP responsible for multiple gateway channels would continually process signals from one channel to the next. The latency incurred due to the DSP process is usually specified as the frame size in milliseconds, although the actual total latency from framing and processing is actually somewhere between the framing size and no more than twice the frame size.

There are three other gateway processes that add to latency: buffering, packetization, and jitter buffer. Buffering can occur when the resulting compressed voice data frames are passed to the network. This buffering is done to reduce the number of times the DSP needs to communicate to the gateway main processor. In other situations, it is done to make the result of coding algorithms fit into one common frame duration (not length).

A multichannel gateway might be operating with different voice codecs on different channels. For example, a universal IP port interface card in a converged IP-PBX system may be handling G.729A off-premises calls across several gateway channels and G.711 premises-only calls across other gateway channels. For example, multiple G.711 frames may be collected for each G.729A frame, irrespective of the coding algorithm, to allow the transfer of one buffer per fixed period of 10 milliseconds.

As coded voice (compressed or noncompressed) is being prepared for transport across a LAN or WAN, it needs to be assembled into packets. This process typically is done by the TCP/IP protocol stack with UDP and RTP. The selection of these protocols improves timely delivery of the voice data and eliminates the overhead of transmission acknowledgments and retries. Each packet has a 40-byte header (combined IP/UDP/RTP headers) that contains the source and destination IP addresses, the IP port number, packet sequence number, and other protocol information needed to properly transport the data. After the IP header, one or more frames of coded voice data would follow.

An important consideration for voice coder selection is the decision of whether to pack more than one frame of data into a single packet. A G.723.1 voice coder (which produces 24-byte frames every 30 milliseconds) would have 40 bytes of header and 24 bytes of data. That would make the header 167 percent of the voice data payload, and a very inefficient use of bandwidth resources. The most common way to reduce the inefficiency of the IP packet overhead is to put more than one coded voice frame into each IP packet. If two frames are passed per packet, the overhead figure drops to 83 percent, but another frame period is added to the latency total. This is a trade-off dilemma of an IP telephony system. To avoid increased latency but reduce overhead, multiple voice frames across gateway channels can be transported in the same packet. When voice from another channel in the originating gateway is going to the same destination gateway, the data can be combined into a single packet. The standard H.323 protocol does not support this latency saving process, but proprietary solutions can implement it.

Jitter buffer latency is based on the variability in the arrival rate of data across the network because exact transport times cannot be guaranteed. Network latency affects how much time a voice packet spends in the network, but jitter controls the regularity at which voice packets arrive. Typical voice sources generate voice packets at a constant rate. The matching voice decompression algorithm also expects incoming voice packets to arrive at a constant rate. However, the packet-by-packet delay inflicted by the network may be different for each packet, resulting in irregular packet arrival at the gateway. During the voice decoding process, the system must compensate for jitter and does this by buffering one packet of data from the network before passing it to the destination DSP. Having these "jitter buffers" significantly reduces the occurrence of data starvation and ensures that timing is correct when sending data to the DSP. Without jitter buffers, there is a very good chance that gaps in the data would be heard in the resulting speech. Jitter buffering improves the speech quality heard by the receiving station user but incurs more latency. The larger the jitter buffers, the more tolerant the system is of jitter in the data from the network, but the additional buffering causes more latency. Most systems use a jitter buffer time of no longer than 30 milliseconds, although 20 milliseconds is the most commonly used time. Jitter buffer time is usually programmable by the system administrator. Jitter buffering can be programmed at the desktop gateway or at any network gateway node.

Beyond gateway latency, there is network latency. Network latency can occur at network interface points, router nodes, and firewall/proxy server points. Network interfaces are points at which data is passed between different physical media used to interconnect gateways, routers, and other networking equipment. Examples are the RS-232C modem and T1-interface connections to the PSTN or LAN/WAN links. For a connection to a relatively slow analog transmission circuit via a RS-232C modem, a delay of more than 25 milliseconds would occur; a T1-interface connection might incur a 1-millisecond delay; and a 100-Mbps Ethernet connection might incur a delay of less than 0.01 millisecond based on a 100-byte data packet. Table 11-5 presents latency as a function of link speed and packet size link speed at different network links.

TABLE 11-5 Latency as a Function of Link Speed and Packet Size Link Speed at Different Network Links

Serialization Delay as a Function of Link Speed and Packet Size Link Speed	64 Bytes	128 Bytes	Packet Size 256 Bytes	512 Bytes	1024 Bytes	1500 Bytes
56 kbps	9 ms	18 ms	36 ms	72 ms	144 ms	214 ms
64 kbps	8 ms	16 ms	32 ms	64 ms	128 ms	187 ms
128 kbps	4 ms	8 ms	16 ms	32 ms	64 ms	93 ms
256 kbps	2 ms	4 ms	8 ms	16 ms	32 ms	46 ms
512 kbps	1 ms	2 ms	4 ms	8 ms	16 ms	23 ms
768 kbps	0.640 ms	1.28 ms	2.56 ms	5.12 ms	10.24 ms	15 ms

Source: Cisco Systems

Routing latency can be incurred because each packet is examined for address destination and overhead headers before directing the packet to the proper route. The queuing logic used by many currently installed routers was designed without considering the needs of IP telephony.

There are problems resulting from the real-time requirement of voice communications. Many existing routers use best-effort routing, which is far from ideal for latency-sensitive voice traffic. The current IP routers support priority programming, the absence of which results in the router delaying all data during congestion situations, irrespective of the application. For example, routers supporting the IETF's RSVP allow a gateway-to-gateway connection to establish a guaranteed bandwidth commitment on the intermediate network equipment, which would dramatically reduce the variability in packet delivery and improve the QoS. Multi-Protocol Label Switching (MPLP) is another recent router programming tool that can reduce routing latency.

Network firewalls or proxy servers that provide security between the corporate intranet and Internet must examine every incoming and outgoing IP packet. This process can incur a sizable amount of latency, so their use is almost always avoided in IP telephony applications. Routers with packet filter features can support some network security without significant added latency. Stand-alone firewalls or proxy servers must receive, decode, examine, validate, encode, and send every packet. A proxy server provides a very high level of network security but can incur more than 500 milliseconds of latency. This is not a problem to the Web-browsing applications for which proxy servers were designed, but it is clearly unacceptable for real-time voice communications requirements. This is one reason using the relatively insecure Internet as a voice network is not yet practical.

When all latency elements are added up, one-way latency can seriously affect IP Telephony QoS. Table 11-6 lists potential one-way latency for a voice call.

TABLE 11-6 Potential One-Way Latency for a Voice Call

Latency Source	Latency (ms)
Framing	1 (G.711) to 30 (G.723.1)
Processing time	10 (worst case)
Buffering	0–5
Packetization	30 (two frames per packet)
Jitter buffering	2–30/buffer
Media access delay	2/hop
Network interface	1 (1.54-Mbps T1)
Routing	10–50 (router dependent)

If a G.723.1 voice codec and multiple jitter buffers are used, the latency example in Table 11-6 borders on the edge of an unsatisfactory call experience. For any voice call, several of the latency sources are known and fixed per call; however, several are variable but manageable. QoS control tools must be used to minimize latency or compensate for it.

QoS Controls

QoS controls can be segmented into several categories: traffic authorization, traffic modification, and traffic adaptation. Traffic authorization controls a station user's access to resources within a domain of control. Traffic authorization methods include admission control, eligibility control, and application control. These are forms of restriction that allow traffic only if a station user provides a password, the station user is on an access list, or the station user is permitted to do so by a policy management server. Traffic modification controls the type of traffic on the network through classification (segregating traffic into different classes), shaping (smoothing out traffic peaks to avoid overload situations), or policing (dropping traffic that doesn't respect policies). Traffic adaptation methods include protocol control, path control, user behavior, congestion avoidance, and congestion management.

There are several commonly used QoS mechanisms that are supported by most of current IP-PBX systems. The two most common class of service (CoS) mechanisms are IEEE 802.1p/Q tagging (Layer 2) and type of service (ToS) prioritization (Layer 3). Both provide prioritization but have their limitations. A better mechanism, developed by the IEEE's IEFT, is differentiated services (DiffServ), an advanced architecture of ToS.

802.1p/Q

The IEEE 802.1p standard for QoS prioritization is a specification defining 3 bits within the IEEE 802.1Q field in the MAC header (OSI Layer 2). The 802.1Q was designed originally to support VLAN operability and then extended to support traffic priorities. IEEE 802.1p adds 16 bits to the Layer 2 header, including 3 bits that can be used to classify priority (the tag). Frames with 802.1p implementation are called *tagged frames*. The standard specifies six different priorities, which do not offer extensive policy-based service levels. Typically, a NIC card in a LAN system sets the bits according to its needs, and Layer 2 switches use this information to direct the forwarding process.

If multiple LANs are interconnected by routers (Layer 3 switches), then the Layer 2 bits must be used to drive Layer 3 QoS mechanisms. The 802.1p/Q mechanism does not operate on an end-to-end basis in an internetwork but does provide a simple method of defining and signaling an end system's requirements within the entire network environment. The Layer 2 header is read only at the switch level—the boundary routers, where traffic congestion occurs—and cannot take advantage of prioritization based on 802.1p unless it is mapped to a Layer 3 prioritization scheme. Even though prioritization is achieved within the switched network, it is lost at the LAN/WAN boundary routers.

Another potential problem is installing a LAN switch supporting 802.1p in a network with non-802.1p switches, which could lead to instability: older switches would misinterpret the unexpected 16 bits specified by the standard. Implementing 802.1p in older networks could require a costly upgrade of all switches.

IEE standard 802.1D is also supported by some IP-PBX systems for traffic prioritization. IEEE 802.1D extends the concept of MAC bridging to define additional capabilities of bridged LANS: to expedite traffic capabilities in support of the transmission of time-critical information in a LAN environment and provide filtering services that support the dynamic use of Group MAC addresses in a LAN environment. IEEE 802.1D Spanning Tree Bridge Protocol is a widely used bridge standard for interconnecting the family of IEEE 802 standard LANs. In this standard, a shortest path spanning tree with respect to a predetermined bridge, known as a *root bridge*, is used to interconnect LANs to form an extended LAN. The spanning tree defines a unique path between a pair of LANs, but this path may not be a shortest path. Moreover, because only one spanning tree is used, some bridges and some ports may not be used at all.

ToS

ToS was first defined in the early 1980s but largely unused until recent IP traffic bottlenecks at the boundary routers required prioritization for better service levels. The IPv4 protocol always contained an 8-bit field, called the ToS field, originally intended for use in packet prioritization. The most recent version, called *IP Precedence*, is a control mechanism that provides end-to-end control of QoS settings. The ToS octet in the

Ipv6 header includes three precedence bits defining eight different priority levels ranging from highest priority for network control packets to lowest priority for routine traffic. Three of the ToS bits are used to flag sensitivity to delay, throughput, and packet loss. Many boundary routers and ToS-enabled Layer 3 switches read the precedence bits and map them to forwarding and drop behaviors. Devices use IP Precedence bits, if set, to help with queuing management.

Differentiated Services (DiffServ)

An evolving IETF QoS control mechanisms is known as DiffServ. DiffServ will not be based on priority, application, or flow, but on the possible forwarding behaviors of packets, called per-hop behaviors (PHBs). DiffServ is rule based and offers a control mechanism for policy-based network management. The DiffServ framework is based on network policies because different kinds of traffic can be marked for different kinds of forwarding. Resources can then be allocated according to the marking and the policies. The IETF Working Group is completing a series of standards that redefine Ipv6 ToS bytes, renamed the Differentiated Services Code Point (DSCP). The new byte indicates the level of service desired and maps the packet to a particular forwarding behavior (PHB) for processing by a DiffServe-compliant router. The PHB provides a particular service level (bandwidth, queuing, and dropping decisions) in accordance with network policy.

Under DiffServ, mission-critical packets could be encoded with a DSCP that indicates a high bandwidth, 0-frame–loss routing path. The DiffServ-compliant boundary router would then make route selections and forward the packets accordingly, as defined by network policy and the PHBs the network supports. The highest-class traffic would get preferential treatment in queuing and bandwidth, and the lower class packets would be relegated to slower service.

The DSCP is 6 bits wide, allowing coding for up to 64 different forwarding behaviors. The DSCP replaces the older ToS bits, and it retains backward compatibility with the 3 precedence bits so that non–DS-compliant, ToS-enabled devices will not conflict with the DSCP mapping.

There are currently two standard PHBs, expedited forwarding (EF) and assured forwarding (AF). EF has one codepoint (DiffServ value), minimizes delay and jitter, and provides the highest level of aggregate QoS. Traffic that exceeds the traffic profile is discarded. AF has four service classes and three drop-precedences for each service class (12

LAN/WAN Design Guidelines for VoIP

total codepoints). Excess traffic is not delivered with the same level of probability as traffic within the defined profile, and it may or may not be dropped. DiffServ assumes the existence of service level agreement (SLA) between networks sharing a border. The SLA establishes policy criteria and defines the traffic profile.

Other QoS control mechanisms include RSVP, subnet bandwidth management (SBM), and multiprotocol label switching (MPLS). RSVP was used by the first generation of client/server IP-PBXs but is deemed too complex, with too much overhead for many parts of the network. SBM is concerned with layer protocols above Layer 2 for internetworking between multiple LANs. MPLS is used primarily for private network routing applications, with limited appeal for premises-only communications applications.

Another approach to IP telephony is the use of VLANs. VLANs can provide more efficient use of LAN bandwidth, are used to distribute traffic loads, and are scalable to support high-performance requirements at a microsegment level. Traffic types, such as real-time voice and delay-insensitive data, can be segmented. IEEE 802.1Q is used as a VLAN packet tagging standard.

QoS Control Summary Points

In its IP telephony network planning guide, Cisco summed up the following core QoS principles for building a network infrastructure to support an IP-PBX system:

- Use 802.1Q/p connections for the IP phones and use the auxiliary VLAN for voice
- Classify voice RTP streams as EF or IP Precedence 5 and place them into a second queue (preferably a priority queue) on all network elements
- Classify voice control traffic as AF31 or IP Precedence 3 and place it into a second queue on all network elements
- Enable QoS within the campus if LAN buffers are reaching 100 percent use
- Always provision the WAN properly by allowing 25 percent of the bandwidth for overhead, routing protocols, Layer 2 link information, and other miscellaneous traffic
- Use low latency queuing (LLQ) on all WAN interfaces

- Use link fragmentation and interleaving (LFI) techniques for all link speeds below 768 kbps

A Final QoS Factor: Echo

Echo in a circuit switched telephone network is caused by signal reflections generated by the hybrid circuit that converts signals between a four-wire circuit (separate transmit and receive pairs) and a two-wire circuit (one transmit and receive pair). Reflected signals of the speaker's voice are heard in the speaker's ear. It is usually acceptable in the PSTN because the round-trip delays through the network are shorter than 50 milliseconds, and the echo is masked by the normal side tone every telephone generates.

In an IP telephony network echo becomes a problem because the round-trip delay through the network is almost always longer than 50 milliseconds. Echo cancellation techniques are required and always used. ITU standard G.165 defines performance requirements that are currently required for echo cancellers. The ITU is defining much more stringent performance requirements in the G.IEC specification.

TIA IP Telephony QoS Recommendations

The TIA has done extensive research and analysis to understand IP telephony voice quality. It used the ITU-T recommendation G.107 E-model to develop its own recommendations for optimizing IP telephony QoS levels, categorizing them by sources of potential speech impairment: delay, speech compression, packet loss, tandeming, and loss plan. The E-model consists of several models that relate specific speech impairment factors and their interactions with end-to-end performance. The resulting transmission rating (R) is:

$$R = Ro - Is - Id - Ie + A$$

- Ro, the basic signal-to-noise ratio, is based on send and receive loudness ratings and the circuit and room noise.

LAN/WAN Design Guidelines for VoIP

- Is is the sum of real-time or simultaneous speech transmission impairments, e.g., loudness levels, side tone, and PCM quantizing distortion.
- Id is the sum of delayed impairments relative to the speech signal, e.g., talker echo, listener echo, and absolute delay.
- Ie is the equipment impairment factor for special equipment, e.g., low bit-rate coding (determined subjectively for each codec and for each percentage of packet loss and documented in Appendix I to ITU-T G.113).
- A, the advantage factor, improves the R value for new services, such as satellite phones, to take into account the advantage of having a new service and to reflect user acceptance of lower quality for such services. Customers assume that the advantage factor will be reduced over time as service improves and users get accustomed to its benefits. It is not recommended to include a non-zero advantage factor for IP telephony because it is a replacement for existing services rather than a completely new service.

The specific recommendations, as summarized in TIA/EIA/TSB116, are:

- **Delay recommendation 1**—Use G.711 end to end because it has the lowest Ie value (equipment impair value) and therefore allows more delay for a given voice quality level.
- **Delay recommendation 2**—Minimize the speech frame size and the number of frames per packet.
- **Delay recommendation 3**—Actively minimize jitter buffer delay.
- **Delay recommendation 4**—Actively minimize one-way delay.
- **Delay recommendation 5**—Accept the [TIA's] E-model results, which permit longer delays for low Ie codecs, like G.711, for a given R value (transmission rating factor).
- **Delay recommendation 6**—Use priority scheduling for voice-class traffic, RTP header compression, and data packet fragmentation on slow links to minimize the contribution of this variable delay source.
- **Delay recommendation 7**—Avoid using slow serial links.
- **Speech compression recommendation 1**—Use G.711 unless the link speed demands compression.
- **Speech compression recommendation 2**—Speech compression codecs for wireless networks and packet networks must be rationalized to minimize transcoding issues.

- **Packet loss recommendation 1**—Keep (random) packet loss well below 1 percent.
- **Packet loss recommendation 2**—Use packet loss concealment (PLC) with G.711.
- **Packet loss recommendation 3**—If other codecs are used, then use codecs that have built-in or add-on PLCs.
- **Packet loss recommendation 4**—New PLCs should be optmized for less than 1 percent of (random) packet loss.
- **Transcoding recommendation 1**—Avoid transcoding where possible.
- **Transcoding recommendation 2**—For interoperability, IP gateways must support wireless codecs or IP must implement unified transcoder-free operations with wireless.
- **Tandeming recommendation 1**—Avoid asynchronous tandeming, if possible.
- **Tandeming recommendation 2**—Synchronous tandeming of G.726 is generally permissible. Impairment depends on delay, so long-delay digital circuit multiplication equipment (DCME) equipment should be avoided.
- **Loss Plan recommendation 1**—Use TIA/EIA/TSB122-A, the voice gateway loss and level plan.

Following the Cisco Systems and TIA recommendations and guidelines may prove to be a difficult task if aging network infrastructure is installed that cannot support most, if not all, of these QoS control mechanisms. It is obvious from the material covered in this chapter that a close working relationship between voice and data communications personnel is required to successfully implement and operate an IP-PBX systems. If IP telephony QoS is not comparable to the experience station users have grown accustomed to with their circuit switched PBX system, the new technology may be rejected as an enterprise communications solution, even if it offers potential cost savings and the benefit of new applications support. A green field location offers the best bet for a large IP-PBX system installation because it is easier to begin from scratch than to attempt a network upgrade while continuing to support ongoing communications operations.

The DiffServ model divides traffic into a small numbers of classes. One way to deploy DiffServ is simply to divide traffic into two classes. Such an approach makes good sense. If you consider the difficulty that network operators experience just trying to keep a best-effort network running smoothly, it is logical to add QoS capabilities to the network in small increments.

Suppose that a network operator has decided to enhance a network by adding just one new service class, designated as "premium." Clearly, the operator needs some way to distinguish premium (high-priority) packets from best-effort (lower-priority) packets. Setting a bit in the packet header as a one could indicate that the packet is a premium packet; if its a zero, the packet receives best-effort treatment. With this in mind, two questions arise:

- Where is this bit set and under what circumstances?
- What does a router do differently when it sees a packet with the bit set?

A common approach is to set the bit at an administrative boundary, such as at the edge of an Internet service provider's (ISP's) network for some specified subset of customer traffic. Another logical place would be in a VoIP gateway, which could set the bit only on VoIP packets.

What do the routers that encounter marked packets do with them? Here again there are many answers. The DiffServ working group of the IETF has standardized a set of router behaviors to be applied to marked packets, which are the PHBs. PHBs define the behavior of individual routers rather than of end-to-end services. Because there is more than one new behavior, there is a need for more than one bit in the packet header to tell the routers which behavior to apply. The IETF has decided to take the ToS byte from the IP header, which has not been widely used in a standard way, and redefine it. Six bits of this byte have been allocated for DSCPs. Each DSCP is a 6-bit value that identifies a particular PHB to be applied to a packet. Current releases of Cisco IOS software use only 3 bits of the ToS byte for DiffServ support. This is adequate for most applications, allowing up to eight classes of traffic. Full 6-bit DSCP support is under development.

One of the simplest PHBs, and one that is a good match for VoIP, is EF. Packets marked for EF treatment should be forwarded with minimal delay and loss at each hop. The only way that a router can guarantee this to all EF packets is if the arrival rate of EF packets at the router is limited strictly to less than the rate at which the router can forward EF packets. For example, a router with a 256-kbps interface needs to have an arrival rate of EF packets destined for an interface that is less than 256 kbps. In fact, the rate must be significantly below 256 kbps to deal with bursts of arriving traffic and to ensure that the router has some ability to send other packets.

The rate limiting of EF packets may be achieved by configuring the devices that set the EF mark (e.g., VoIP gateways) to limit the maximum arrival rate of EF packets into the network. A simple, albeit conservative, approach would be to ensure that the sum of the rates of all EF packets entering the network is less than the bandwidth of the slowest link in the domain. This would ensure that, even in the worst case, where all EF packets converge on the slowest link, the link is not overloaded and the correct behavior results.

In fact, the need to limit the arrival rate of EF packets at a bottleneck link, especially when the topology of the network is complex, turns out to be one of the greatest challenges of using only DiffServ to meet the needs of VoIP. For this reason, an approach based on integrated service and the RSVP is appropriate in those situations where it is not possible to guarantee that the offered load of voice traffic will always be significantly less than link capacity for all bottleneck links.

CHAPTER 12

PBX Cabling Guidelines

Telephony wiring dates back 125 years ago, to the days when Alexander Graham Bell was tinkering with the first telephone. Telephones traditionally have used loop current for voice communications and signaling transmission. For many years single-pair (two-wire) cabling had supported telephones working behind a PBX system, but system equipment innovations, beginning with the introduction of digital switching and stored program call control, forced changes in the cabling infrastructure during the late 1970s. The first generation of proprietary PBX telephones, first electronic and then digital, required multiple wiring pairs to support the more advanced features and functions available with the new technology. At the same time, the early data LANs required a wiring infrastructure of their own, based on coaxial cable. As customer premises voice networks and data networks evolved in the mid-1980s, issues such as a common infrastructure and increasing transmission bandwidth requirements needed to be addressed. The existing telephony wiring system, fine for voice but inadequate for data, needed a major overhaul.

In 1985, two standards committees began working on specifications for a generic telecommunications cabling system to support a mix of communications media (voice, data, video) in a multivendor environment. The TIA and the Electronic Industries Association (EIA) formed a joint committee known as the EIA/TIA 41.8 Committee. After 6 years of work, the TIA/EIA 568 standard was issued. TIA/EIA 568 is more formally known as the Commercial Building Cabling Standard and outlines specifications for a generic telecommunications cabling system. The American National Standards Institute (ANSI) also adapted this standard, so it is sometimes referred to as ANSI/TIA/EIA 568.

There is a corresponding series of specifications known as ANSI/TIA/EIA 569: Commercial Building Standard for Telecommunications Pathways and Spaces. The purpose of ANSI/TIA/EIA 569 is to standardize design and construction practices within and between buildings that support telecommunications equipment and transmission media. The standards are outlined for rooms or areas and pathways into and through areas where telecommunications media and equipment are installed. To simplify the implementation and administration of the cabling infrastructure, another series of specifications were developed, ANSI/TIA/EIA 606: The Administration Standard for the Telecommunications Infrastructure of Commercial Building.

In addition to the standards specified by the ANSI/TIA/EIA recommendations, the International Standards Organization (ISO) defined a generic cabling system recommendation known as ISO/IEC IS 11801. The ISO standard is intended for global usage and is broader in scope than the ANS/TIA/EIA standards for the North American market. The European version of ANSI/TIA/EIA standard is EN 50173 and is more similar to 568 than to the ISO standard.

Cabling System Fundamentals

A structured cabling architecture design is intended to accommodate telecommunications technology changes with minimal impact on any of the other cabling subsystems, such as, electrical cabling. The target life cycle of an average cabling installation is up to 20 years. It is expected that a few generations of telecommunications systems will be installed and replaced or upgraded. Another planning assumption is that networking and bandwidth requirements will certainly increase during the life cycle of the cabling system. The following are key factors used to specify networks and cabling, as identified by Avaya in its SYSTIMAX CSC guidebook:

- Usage patterns, including combined size and duration of peak loads for all applications
- Expected increase in bandwidth demands
- The number of users and anticipated changes in that number
- Location of users and maximum distances between them
- The likely rate of change in users' locations (churn)
- Connectivity with current and future devices and software
- Space available for cable runs
- Total cost of ownership
- Regulations and safety requirements
- Importance of protection against loss of service and data theft

PBX systems traditionally have been based on a star network topology. A star network topology includes many point-to-point links radiating from central equipment. The early LAN topologies were based on ring-and-bus network designs. A ring network topology has a continuous transmission loop that interconnects every device. The most familiar example of a ring network topology is the IBM token ring LAN. A bus network topology is a communications link that connects devices along the length of a cable. The original Ethernet LAN was based on a bus network topology.

Today's dominant LAN technology is based on Ethernet standards. The logical topology of an Ethernet LAN is a bus topology, but the physical topology of the network is a star. Ethernet workstations that connect to an Ethernet hub or switch communicate over a high-speed bus housed in a hub or switch, but these network nodes are connected in a clustered star network topology. The star topology favored by PBX systems and adapted by Ethernet LANs is now the accepted communications system network topology.

The first Ethernet LAN installations were based on coaxial cable used for the transmission medium. During the mid-1980s the cabling used by PBX systems, known as unshielded twisted pair (UTP), was adapted for Ethernet LANs. Telephony UTP cabling was classified by IBM's cabling system specifications as Category 3 and was used for 10Base-T Ethernet LANs operating at 10 Mbps. A 10Base-T Ethernet LAN used two pairs of Category 3 UTP cabling. A 100Base-T4 Ethernet LAN used four-pair Category 3 UTP cabling. A 100-Mbps Fast Ethernet, also known as 100Base-TX, used two-pair Category 5 UTP cabling. The 1000-Mbps (1 Gbps) Ethernet, 1000BASE-T, uses four-pair Category 5 UTP cabling. The 1000Base-TX, a lower cost alternative to 1000Base-T, uses the recently introduced Category 6 UTP cabling. PBX system telephony requirements can be satisfied with any of these UTP cabling types, making possible a single network cabling system infrastructure for voice and data communications applications.

In the SYSTIMAX SCS guidebook, Avaya lists the following considerations for choosing the type of customer network cabling:

- Maximum distance between network hubs and nodes
- Space available in ducting and floor/ceiling cavities
- The levels of electromagnetic interference (EMI)
- Likely changes in equipment served by the system and the way it is used
- Level of resilience required
- The required life span of the network
- Restrictions on cable routing that dictate cable bend radius
- Existing cable installations with potential for reuse

For the past two decades, most customers have used or installed two different cabling systems for telephony and data LAN applications. The evolution of the PBX system to an IP telephony platform will allow the large installed base of customers with installed circuit switched PBX systems to slowly phase out an infrastructure with two cabling systems and allow customers who are designing an entirely new converged voice/data network the opportunity to install a single cabling system. PBX systems installed before 1990 were implemented with Category 3 UTP, but more recent installations may have been based on Category 5 UTP, the same wiring used for data LANs. A newly installed communications system installation likely would be based on a generic cabling infrastructure using Category 5 UTP to provide for future needs.

A generic cabling system is a structured telecommunications cabling system capable of supporting a wide range of customer applications. Generic cabling can be installed before the definition of required applications because application-specific hardware (telephones, computers, etc.) is not part of the structured cabling design. Generic cabling can be enhanced through the use of *flood wiring*, which is the installation of sufficient cabling and telecommunications outlets in a work area to maximize the flexibility of the location for devices connected to the network. Many customers are currently installing four or six telecommunications outlets per work area, although the recommended minimum is two.

Cable Interference and Noise Issues

Electromagnetic flux is a potential problem that can disrupt network communications wherever there are active electrical and electronic devices. The selection of the right cabling and its network routing design is important to reduce communications interference problems. All network components, including the connectors and patch panels, must be designed to satisfactorily perform in the presence of external noise. Cable routing should conform to the manufacturer's recommendations and always avoid potential interference sources. Likely office building sources of EMI are lift motors (elevators), automatic doors, and air-conditioning units. The older the equipment, the more likely it will produce EMI. Closed metal conduits and ducting for the cabling system will provide extra protection against EMI sources that cannot be corrected or avoided. Balanced transmission over UTP cable offers strong protection against external noise. In EMI-sensitive or hostile environments, the only solution may be optical fiber cable that is immune to external noise.

There are regulations specified by the FCC (part 68 and part 15 subpart) that cover telecommunications network electromagnetic compatibility (EMC) with other electronic devices. Network system installers and users are responsible for conforming to EMC guidelines. Installers must ensure that cable specifications for routing and ducting eliminate interference problems. Some manufacturers provide warranties on the EMC performance of certified installations using their cabling.

In addition to the potential for interference from external electrical and electronic source devices, the active pairs in a multipair cable can interfere with each other. Interference between cable pairs is known as *crosstalk*. Crosstalk measurements may be performed with two methods: pair-to-pair and PowerSum. The pair-to-pair method measures only the maximum interference caused by any other single active cable pair. Near end cross talk (NEXT), the pair-to-pair measurement metric, is defined as the signal coupled from one pair to another in a UTP cable. It is called NEXT because it measures the crosstalk at the end where one pair is transmitting (and the transmitted signal is largest and, hence, causes the most crosstalk). Crosstalk is minimized by the twists in the cable, with different twist rates causing each pair to act as antennas sensitive to different frequencies so that signals are not picked up from neighboring pairs. Keeping the twists as close as possible to the terminations minimizes crosstalk. Far end crosstalk (FEXT) measures the effect of signal coupling from one pair to another over the entire length of the cable, and it is measured at the far end.

Another frequently cited measurement associated with crosstalk is the attenuation to crosstalk (ACR) ratio. Attenuation is the reduction in signal strength due to loss in the cable. ACR measures how much "headroom" the signal should have at the receiver. It is important that the signal strength at the receiver end be high enough for reception by the network hub/switch to pass through to workstation nodes or other hubs/switches. Ethernet LANs send very high-speed signals through the cable, and the attenuation varies with the frequency of the signal. Attenuation tests are performed at several wavelengths, as specified in the 568 standards. The test requires a tester at each cable end, one to send and one to receive. The loss between the ends is calculated, recorded, and compared with pass/fail criteria for UTP cable at Category 3, 4, and 5 frequencies.

Performance losses can be greater than indicated by pair-to-pair measurement if there are several active pairs in a multipair cable strand. For this reason, the preferred method of measuring crosstalk is known as PowerSum. It is based on the measurements taken when all pairs in a multipair cable are active. This is the more realistic crosstalk measurement for Fast Ethernet and Gigabit Ethernet LANs, where all pairs are used to carry signals, often simultaneously. PowerSum is the recommended method to use for cables with more than four wires.

Overview of ANSI/TIA/EIA 568

The original version of 568 has undergone several revisions. The first major revision, known as 568A, addressed bandwidth concerns, resulting in 100-MHz Category 5 cabling specifications. The most recent revision is known as 568B and provides for significant performance improvement and enhancements over the previous set of recommendations:

- Significant performance improvements over Class D (Category 5 cabling)
- A sequential naming system (Categories 5, 6, 7; Classes D, E, F)
- New cabling must be a strict superset of existing Class D
- New cabling must be backward compatible with existing modular RJ-45 connectors
- Category 5 and 6 components, working together, must offer at least Category 5 performance
- 100-meter topology must be supported
- Specification parameters must be formula based (for a more precise mathematical model)
- The meter frequency should be extended 25 percent to accommodate future DSP electronics

Engineering and architectural firms use the 568 standard during the planning phase of a commercial building's cabling system. With the use of 568 standards, a cabling system should support up to 50,000 individual station users and up to 10 million square feet of office space across a 2-mile long campus. Because the typical PBX system supports a few hundred stations users and most customers require far smaller space coverage, there are few problems when using the 568 standard. The 568 standard also will accommodate the evolution of the circuit switched PBX toward a client/server IP telephony design based on a LAN infrastructure.

The 568 standard is known as a structured cabling architecture. There are six major elements:

1. Entrance facility
2. Equipment room
3. Backbone cabling
4. Telecommunications closet
5. Horizontal cabling
6. Work area

Entrance Facility

The entrance facility is the building location where the outside cable plant interfaces with the building's backbone cabling. For PBX systems, the outside cable plant consists of the trunk network facilities of transmission service carriers, such as local telephone operating companies or long distance interchange carriers. The entrance facility is usually in a secure area that has multiple conduits through the building's exterior walls, through which cabling can be pulled by the transmission services carrier. A commonly used name for the entrance facility location is the *network demarcation point*. The network demarcation point is where the network carrier's service ends and the customer's enterprise communications system begins. Entrance facilities also may contain cabling, usually optical fiber, from other customer campus buildings. All customer entrance facilities should conform to ANSI/TIA/EIA 569 standards.

Equipment Room

The equipment room houses telecommunications equipment, such as PBXs, ACDs, messaging systems, and IVR systems, and data network communications equipment, such as routers and policy management servers. UPS equipment also may be housed in the equipment room for emergency power requirements. Large customer campus configurations may have multiple equipment rooms. This would apply for PBX configurations with distributed port cabinet designs, such as the Ericsson MD-110.

The equipment room can double as a telecommunications closet in small customer environments. The entrance facility, or demarcation point, may share the same building location as the equipment room because the distance between the trunk distribution frame and the PBX system is limited. Equipment rooms should be in a secure location, have adequate environment controls, such as heating and air conditioning, sufficient power supplies, and adequate space for equipment cabinets and racks. Some customers may use the equipment for system administration and monitoring stations. It is recommended that equipment rooms conform to ANSI/TIA/EIA 569 standards.

Backbone Cabling

The backbone cabling connects the building's telecommunications closets to the equipment room and entrance facility. It can also connect equipment rooms distributed across a campus environment. The recommended backbone cabling topology is a star originating from the equipment room.

The 568 standard recognizes four media options for the backbone cabling. Any of the following media can be used in combination:

- 100-Ω UTP multipair backbone cable not to exceed 800 meters (2,624 feet)
- 150-Ω Shielded twisted pair (STP) cable not to exceed 700 meters (2,296 feet)
- 62.5/125-μ multimode optical fiber not to exceed 2,000 meters (6,560 feet)
- Single-mode fiber optical cable not to exceed 3,000 meters (9,840 feet)

Although UTP cabling is the standard for most horizontal cabling applications, optical fiber cabling has become the medium of choice for backbone cabling applications. Optical fiber cable is preferable to copper wiring for several reasons:

- Longer loops
- Greater immunity to noise
- Higher transmission bandwidth
- Increased security (very difficult to tap)
- Elimination of electrical problems, such as short circuits or floating grounds

Telecommunications Closet

The main function of a telecommunications closet is the termination of horizontal cable. It houses telecommunications equipment that provides for mechanical terminations and/or cross connections for the horizontal and backbone cabling systems. It is a 568 requirement to have at least one telecommunications closet or equipment room per building; there is no specified limit on the number of telecommunications closets per building. It is a 569 requirement to have at least one telecommunications closet for each building floor, and additional closets should be added if the floor area to be served exceeds 1,000 square meters or the horizontal distance to the work area is longer than 300 feet. It is recommended that telecommunications closets conform to the 569 standard.

The equipment room houses the main distribution frame (MDF), and telecommunications closets can house an intermediate distribution frame (IDF). *Main crossconnect* is an old telephone company term describing the MDF; *intermediate crossconnect* was used in place of IDF. Old crossconnections are quickly being replaced with modern patch panels. A *patch panel*, formerly known as the crossconnect, is a facility enabling the termination of cable elements and their connections, primarily by means of patch cords or jumpers (hook-up wires). Patch cords are short multiwire stranded cables with connectors, such as RJ-45 plugs, on each end to establish connections on a patch panel. Patch cords provide an easier way to rearrange circuits without using the special tools required to install the older jumper wires (also known as hook-up wires) on old-style crossconnections with 66 punchdown blocks. In the main and intermediate crossconnections, the patch cord lengths should not exceed 20 meters, according to the 568 standard. Jumpers that bridge horizontal wiring in the telecommunications closet should not exceed 6 meters. Lengths in excess of these will be deducted for the maximum cable length. The 568 standard specifies that patch cords have stranded conductors.

Current patch panels consist of rack hardware with 110 punchdown blocks to directly connect the patch cords. Patch cord connectors are used instead of older crossconnections and jumpers when requirements call for UTP 5/5E/6 high-bandwidth transmission. Patch panels and patch cords provide superior management features to the older crossconnect technology. The recommended patch cord connector is a modular 8-pin connector more commonly known as the RJ-45 plug.

Horizontal Cabling

The horizontal cabling extends from the telecommunications closet to the telecommunications outlet at the station user's work area. Horizontal cabling is based on a star topology, and each work area should be connected to the telecommunications closet. This cabling design is sometimes referred to as *home-run wiring*. Multidrop connections between telecommunications outlets are not permitted. The maximum horizontal loop length is 90 meters, with an additional 10 meters allowed for work area cables, patch cables, and other telecommunications closet connections. The 90 meters + 10 meters distance restrictions are imposed to support the 100-meter restriction of Ethernet LAN technology over UTP. The recognized cables include:

PBX Cabling Guidelines

- Four-pair 100-Ω UTP cable
- Two-pair 150-Ω STP cable
- Two-fiber 62.5/125-μ optical fiber cable

There are several categories of UTP cabling as defined by the different standards organizations, including ANSI/TIA/EIA, ISO, and EN, that are covered in the 568 specifications. These categories are:

- **Category 3**—UTP cable and hardware whose transmission characteristics are specified up to 16 MHz
- **Category 4**—UTP cable and associated connected hardware whose transmission characteristics are specified up to 20 MHz
- **Category 5**—UTP cable and hardware whose transmission characteristics are specified up to 100 MHz
- **Category 5E**—UTP cable and hardware whose transmission characteristics are specified up to 100 MHz, but with improved NEXT as compared with Category 5 (Table 12-1)
- **Category 6**—UTP cable and hardware whose transmission characteristics are specified up to 250 MHz. This is the newest specification to provide for improved performance of existing parameters, plus some new defined parameters. Category 6 is intended as a replacement for Categories 5 and 5E.

TABLE 12-1

Comparison of Category 5 and Category 5E UTP Cable

Specification	Category 5	Category 5E
Frequency range (MHz)	1–100	1–100
Attenuation (dB)	24	24
NEXT (dB)	27.1	30.1
PowerSum NEXT (dB)	NA	27.1
ACR (dB)	3.1	6.1
PowerSum ACR (dB)	NA	3.1
ELFEXT (dB)	17	17.4
PowerSum ELFEXT (dB)	14.4	14.4
Return loss (dB)	8	10
Propagation delay (ns)	548	548
Delay skew (ns)	50	50

NA, not applicable

Tables 12-2 and 12-3 define horizontal UTP cable attenuation and NEXT for Category 3, Category 4, and Category 5 UTP cables.

TABLE 12-2

Horizontal UTP Cable Attenuation (per 100 M @20°C)

Frequency (MHz)	Category 3 (dB)	Category 4 (dB)	Category 5 (dB)
0.064	0.9	0.8	0.8
0.256	1.3	1.1	1.1
0.512	1.8	1.5	1.5
0.772	2.2	1.9	1.8
1	2.6	2.2	2
4	5.6	4.3	4.1
8	8.5	6.2	5.8
10	9.7	6.9	6.5
16	13.1	8.9	8.2
20		10	9.3
25			10.4
31.25			11.7
62.5			17
100			22

TABLE 12-3

Horizontal UTP Cable NEXT Loss (Worst Pair) at More than 100 Meters

Frequency (MHz)	Category 3 (dB)	Category 4 (dB)	Category 5 (dB)
1.5	53	68	74
0.772	43	58	64
1	41	56	62
4	32	47	53
8	27	42	48
10	26	41	47
16	23	38	44
20		36	42
25			41
31.25			39
62.5			35
100			32

UTP cable is the most commonly used transmission medium for horizontal cabling for voice telephony and data LAN applications. The wiring itself is typically constructed of four pairs of 22- or 24-gauge solid copper conductors. The majority of today's installed PBX cabling systems is based on 24-gauge UTP wiring, and the most often quoted station equipment loop lengths are based on 24-gauge wiring. The typical 24-gauge loop lengths for PBX telephone and module equipment are as follows:

- Analog voice terminal— 3,000 to 5,000 meters
- Digital voice terminals and data modules—1,000 to 1,500 meters
- Attendant consoles—1,000 meters
- ISDN BRI voice terminals and data modules—600 (S/T-interface) and 5,000 (U-interface) meters

Loop lengths using 22-gauge would increase slightly by a factor of about 10 percent.

Category 1 and 2 UTPs are not covered or recognized as part of the 568 standard. Category 7 specifications are not finalized but will have parameters specified to 600 MHz; it makes use of bulky and expensive STP cables. Category 7 will improve performance for existing parameters and define some new parameters including new connector interfaces.

Telecommunications Outlet

The telecommunications outlet at the work area can support a mix of voice, data, and video communications. The 568 standard specifies a minimum of two outlets (connectors) for each station user work area. The outlets should be configured as follows:

- One connector is supported by a four-pair 100-ohm UTP cable, Category 3 or higher
- The second connector is supported by one of the following:
  - Four-pair 100-Ω UTP cable (Category 5 or higher recommended)
  - Two-pair 150-Ω STP cable
  - Two-fiber 62.5/125-μ optical fiber cable

The color code for the 100-ohm UTP cable is specified as follows:

- **Pair 1**—White–blue and blue
- **Pair 2**—White–orange and orange

- **Pair 3**—White–green and green
- **Pair 4**—White–brown and brown

The amount of untwisting of individual pairs to terminate should be less than or equal to 0.5 inch for Category 5 or higher, and less than or equal to 1.0 inch for Category 4. The cable bend radius should not be less than four times the cable diameter.

The recommended wall jack to support Category 5 cabling is based on a RJ-45 connector as specified by 568 standards. Telephones traditionally have been connected with RJ-11 connectors. The POTS-based RJ-11 connector has six-position jack pin/pairs; the RJ-45 connector, originally designed and used for ISDN BRI equipment, is currently used for high-speed transmission requirements, such as 10/100-Mbps Ethernet LAN applications. There are differences between the 568A and 568B standards for pin colors. The older 568A standard is used extensively in older installations; the more recent 568B standard is less prevalent but is the new specification. The EIA 568A and 568B standards for the RJ-45 connector are shown in Figure 12-1. The primary differences between the 568A and 568B standards is a switching of colors between the second and sixth pin assignments and switching the Pair 2 and Pair 3 pinout positions (Figure 12-2).

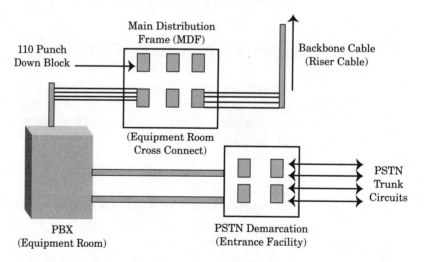

Figure 12-1
PBX cabling and distribution design (1).

Work Area

The work area is usually defined as the station user location and includes all of the necessary components to connect station equipment to

PBX Cabling Guidelines

Figure 12-1
PBX cabling and distribution design (2).

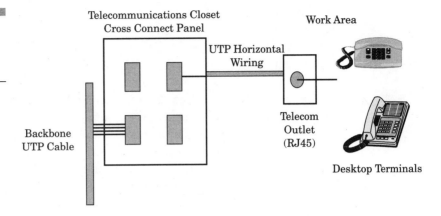

the telecommunications outlet. Station equipment can include telephones, modems, facsimile terminals, computers, and audioconferencing and videoconferencing equipment. Patch cables connect the station equipment to the telecommunications outlet. The patch cable and horizontal cable characteristics must match, although special interface adapters may be used. The special interface adapters, however, must be exterior to the telecommunications outlet. Some networks require application-specific electrical components on the telecommunications outlet of the horizontal cabling. These application-specific components should not be installed as part of the horizontal cabling but placed external to the outlet as needed in the work area.

Overview of ANSI/TIA/EIA 569

ANSI/TIA/EIA 569 is the Commercial Building Standard for Telecommunications Pathways and Spaces. The purpose of 569 is to standardize design and construction practices within and between buildings that support telecommunications equipment and transmission media. The standards are outlined for rooms, areas, and pathways into and through which telecommunications transmission media and equipment are installed. The standard is limited to the telecommunications aspect of building construction and design and does not cover safety aspects.

The specifications of 569 cover the following building elements:

- Entrance facilities
- Equipment room

- Backbone pathways
- Telecommunications closet
- Horizontal pathways
- Workstation

The entrance facility, equipment room, telecommunications closet, and workstation areas were described briefly in the preceding section on 568. The backbone and horizontal pathways are used for the corresponding cabling described above. Backbone pathways consist of intra- and interbuilding pathways. Intrabuilding pathways consist of conduits, sleeves, and trays. They provide the means for routing cables from the entrance facility to telecommunications closets and from equipment rooms to the entrance facility or the telecommunications closet. Interbuilding pathways interconnect separate buildings and consist of underground, buried, aerial, and tunnel pathways. Horizontal pathways are facilities for the installation of the telecommunications transmission media from the telecommunications closet to the telecommunications outlet at the workstation area.

The 569 specifications require a minimum of one telecommunications closet per floor, and that additional closets should be added if the floor area to be served exceeds 1,000 square meters or the horizontal distance to the work area is greater than 300 feet. At least one telecommunications outlet per workstation area is specified.

A very important area covered by 569 is labeling and color-coding specifications designed to simplify installation and maintenance of the cabling infrastructure.

Labels are divided into three categories: adhesive, insert, and other. Adhesive labels must meet UL requirements for adhesion, defacement, legibility, and exposure. Insert labels must meet UL requirements for defacement, legibility, and exposure. Other labels include special-purpose labels, such as tie-on labels.

The 569 color coding rules are:

- Termination labels at the two ends of the cable should have the same color
- Crossconnections between termination fields generally should have two different colors
- The color orange is used for the demarcation point
- Green identifies network connections on the customer side of the demarcation point

PBX Cabling Guidelines

- Purple identifies the termination of cables originating from common equipment
- White indicates the first level of the backbone media
- Gray indicates the second level of the backbone media
- Blue identifies the termination of station telecommunications media
- Brown identifies interbuilding backbone cable terminations
- Yellow identifies the termination of auxiliary circuits, alarms, security, and other miscellaneous circuits
- Red identifies the termination of KTSs
- White may be used to identify second-level backbone terminations in remote "non-hub" buildings

CHAPTER **13**

PBX Voice Terminals

The main station user interface to the PBX system is the telephone, but today's typical voice terminal bears little resemblance to the desktop instrument used a generation ago. Telephones are still regarded as a necessary communications tool, despite the growing popularity of the Internet and e-mail. Desktop computers currently may outnumber telephones in a growing number of offices, but real-time voice communications remains an important business priority. Telephones represent a significant percentage of the upfront purchase price of a PBX system and are the most primary variable cost element for any system installation. There are wide price variations among telephones, based on the technology platform (analog, digital, ISDN BRI, IP, mobile) and model (entry, standard, advanced, executive, etc.). Minimal function, industry-standard 2500-type analog telephones can be purchased for less than $50, but multiple-line appearance models with a few add-on options can break the $1,000 barrier. Due to these price variations, telephones may represent as little as 15 percent, or as much as 50 percent of PBX system costs. Desktop functions and performance correlate directly to the price of the telephone, and customers are often confused about which type of telephone might fully exploit the feature and function capabilities of their PBX system. The accompanying diagram illustrates the current price range of desktop PBX digital, BRI, and IP telephone instruments without module options (Figure 13-1).

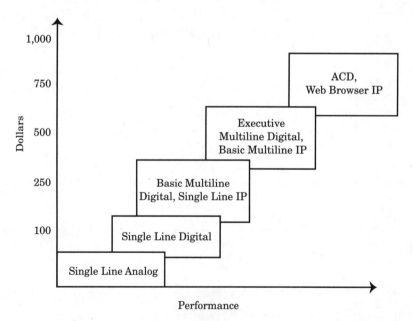

Figure 13-1
PBX telephone price ranges.

Voice Terminal Categories

PBX voice terminals can be classified into several basic telephone categories:

- 2500-Type analog telephone
- Digital telephone
- Mobile
- PC client softphone

Within the latter three categories are several subcategories. The distinguishing technical difference between an analog telephone and a digital telephone is that the latter terminal has an integrated codec that digitizes voice signals for transmission to the PBX system over the inside wiring system. The digital transmission format can provide a higher degree of feature and function performance at the desktop level, instead of at the PBX system common equipment. Before the design, development, and widespread availability of digital telephones, the first generation of stored program control PBX systems supported electronic telephones, sometimes referred to as hybrid telephones. Voice transmission between the desktop and the PBX system was analog, but the telephone design included integrated circuitry, sometimes a microprocessor chip, that could support multiple line appearances. Programmable feature/function buttons and limited function display fields were supported, but the performance capabilities were limited due to in-band signaling techniques between the port circuit card and the desktop. Digital telephones, supported by an out-of-band signaling channel, were capable of far greater performance potential.

Mobile telephones is a term used to categorize cordless, wireless, and cellular telephone handsets. Cordless telephones working behind a PBX system are likely to use digital radio transmission between the handset and the base station, but analog voice transmission is supported between the base station and the PBX port circuit card. The terms *wireless* and *cellular* usually describe telephones that do not require any local loop wiring, although the base station transceiver is hardwired back to a central switching system, such as a PBX system.

PC client softphones are based on a CTI platform from which the PBX common control complex functions as a server to the desktop terminal. With a client softphone option, communications control and signaling between the desktop and the PBX system is handled over a CTI link (desktop or client/server configuration) behind a traditional common control complex or LAN-based call server. The former type of softphone application requires an associated telephone instrument for voice communications transmission at the desktop. The latter is an example of a softphone based on an IP telephony platform, without a traditional desktop telephone. An IP-based softphone requires the PC client be equipped with a sound board with a combination microphone/speaker or a computer handset.

Analog Telephones

DTMF analog telephones that conform to the old Bell System 2500-type standard specifications are nonproprietary and can be supported by all PBX systems. However, the analog port circuit cards for each PBX system are proprietary. Transmission between the telephone and the PBX is analog based, with in-band signaling for all dialing and feature activation operations. Analog-based voice communications and signaling, transmitted over a 4-KHz transmission line, is carried over two-wire (single pair) UTP telephone wiring between the wall jack and the PBX system. The embedded signaling bandwidth limits support of integrated desktop features and functions, particularly display-based information. Analog telephones typically have a single line appearance, although a second virtual line may be supported for answering incoming calls after the original connection is placed on hold. Two line appearance analog telephones require multipair wiring and multiple port circuit card terminations. Analog telephone in-band signaling over a single wiring pair does not support multiple line appearances behind a PBX system.

Analog telephones may be equipped with a limited number of fixed feature buttons, such as hold, and an array of programmable speed dial buttons. The instrument may be equipped with a message indicator and limited function display field. Displays are usually limited to dialing and call duration information, time clock, and caller line ID (CLID) incoming call directory numbers. More detailed information such as name display, call diversion information, or feature/function menus commonly available in more sophisticated, and more costly, digital telephones are not available.

Some analog telephones have an integrated hands-free answer intercom speaker or a two-way simplex speakerphone. Almost all currently available desktop audioconferencing products are based on standard analog transmission standards between the desktop and the PBX system and supported with the same analog station port circuit used for 2500-type analog telephones. Audioconferencing products are typically equipped with a DTMF keypad and several fixed feature buttons, e.g., mute, and support full duplex speakerphone operation. The Polycom Soundstation is an example of an analog-based desktop audioconferencing product.

Digital Telephones

The second category of voice terminals is digital telephones. Digital telephones have a codec that digitizes analog voice signals at the desktop, using PCM as the encoding scheme. Excluding IP telephones, digital telephones are supported through an out-of-band signaling and control channel between the desktop and PBX port circuit card. There are several subcategories of digital telephones:

- Proprietary
- Universal Serial Bus (USB)
- ISDN BRI
- IP

Proprietary. Most digital telephones currently working behind a PBX system are proprietary, and work exclusively with the manufacturer's PBX system(s). Some manufacturers, such as Siemens and NEC, have designed their digital telephones to work across different product families of communications system (KTS/Hybrid and PBX systems). All current proprietary digital telephones are supported by a 2B+D communications/signaling transmission format between the desktop and port circuit card. Although the 2B+D transmission format is most closely associated with ISDN BRI services, the dual communication channel/dedicated signaling channel format was first implemented on a proprietary digital PBX telephone in 1980, before there were ISDN standards. The first digital PBX telephone was the Intecom ITE model, capable of supporting digital communications from desktop to desktop across the PBX cabling infrastructure and switching network, with an optional data module for supporting modemless data communications from the desktop. Each B channel can support 64-Kbps communications transmission for voice, data, or video signals. What makes a digital telephone proprietary is the D-channel signaling protocol. The protocol is proprietary and unique for each manufacturer's PBX system. Although the proprietary nature of the digital telephone's D-channel format restricts customer flexibility in selecting telephones for use behind their

PBX systems, it is used to support advanced features unique to each system, particularly display-based system capabilities and functions. Figure 13-2 illustrates a typical multiple line digital telephone instrument from Avaya.

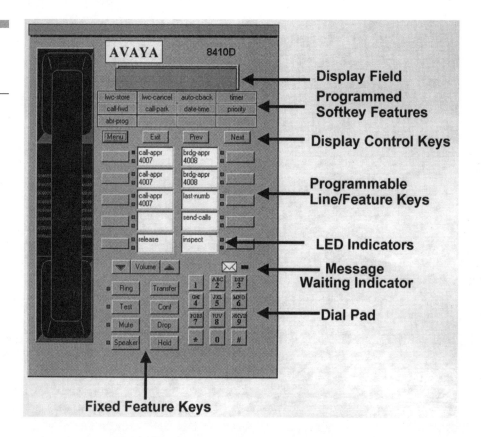

Figure 13-2
Typical multiple line appearance digital telephone.

USB. A special type of proprietary digital telephone is uses an integrated Universal Serial Bus (USB) interface. USB is an external bus standard capable of very high-speed transmission rates, up to 12 Mbps, and can support a wide variety of communications applications. The original intent of a digital telephone equipped with a USB interface port was for desktop CTI applications. The USB port passes signaling and control messages/commands between the voice terminal and a desktop computer. Each PBX manufacturer originally designed proprietary CTI API port interfaces for implementing desktop PC telephony applications. It was believed that the design, development, and adaptation of USB standards would stimulate the market for desktop CTI installations behind PBX systems, but few manufacturers incorporated the interface port in their telephone instrument designs.

A recent innovation using the USB link supports IP telephony. A digital telephone with a USB port eliminates the need for a RJ-11 jack interface between the phone and the PBX cabling system because communications transmission and signaling can be handled over the LAN infrastructure by using a desktop computer as the intermediary link to the LAN. The failure of CTI as a PBX station option has historically limited demand for USB telephones, but the emergence of ToIP communications may help create future demand for the option. Customers using a USB telephone in a ToIP installation can continue to enjoy the look and feel of a traditional telephone with a traditional keypad and handset for dialing and call answering operations while using the desktop computer to facilitate feature/function access (using CTI client software). Frequent computer processing and software problems, a major reason most station users are reluctant to use a softphone, do not affect most USB telephone functions or operations because the computer serves only as a physical connection to the LAN. Of all the major PBX system manufacturers, Nortel Networks has been the most active in promoting use of its USB telephone model for ToIP applications behind its IP-PBX communications systems.

ISDN BRI. A third category of digital telephones is ISDN BRI. The ISDN BRI telephone communications link to the desktop is 2B+D, but unlike proprietary sets, its D-channel signaling format conforms to National ISDN (NISDN) specifications. ISDN BRI telephones have limited access to some proprietary PBX features, and the level of display information is also limited compared with proprietary digital telephones. Although ISDN BRI telephones have been available since the early 1990s, support of ISDN BRI telephones offers customers some benefits not available with proprietary digital instruments, such as passive bus operation and bonding of the two communications bearer channels. The former is the ability for a single ISDN BRI 2B+D communications link to support up to eight desktop telephones, each with its own directory number. Although only two telephones can be active simultaneously, customers can save money on port circuit hardware and cabling when these telephones are used in low-traffic environments. The latter application supports high-speed transmission to the desktop, up to 128 Kbps, for data or H.320 video communications applications.

IP. The fourth digital telephone category, and the most recent, is an IP telephone. IP telephones are included in the digital category because voice signals are digitized with standard 8-bit coding and 8-KHz sampling before a VoIP audio codec converts the digitized sample to IP format. IP telephones may conform to VoIP protocol and signaling formats standards, but each IP-PBX system uses proprietary signaling bit data in support of unique features and display characteristics. The accompanying photograph of the Cisco 7960 illustrates a current generation IP telephone (Figure 13-3). Although most IP telephones with Web browser capabilities have similar feature capabilities, the look and feel of each supplier's telephone models is distinct. The accompanying photographs of three supplier's high-end IP phone models illustrates this (Figure 13-4).

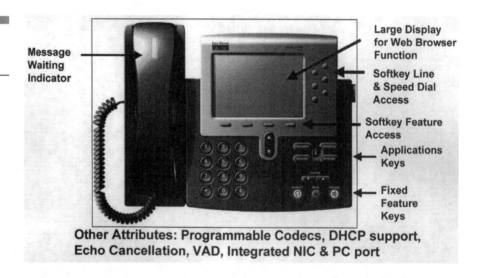

Figure 13-3
IP phone attributes: Cisco 7960.

Figure 13-4
Contrast in styles: IP telephones with Web browser displays.

IP Telephone Design Basics

All manufacturers base their IP telephone on proprietary design schematics and circuitry, but there are common design elements across the unique terminals. IP telephone basics include:

- User interface
- Voice interface
- Network interface
- Processor complex and associated logic

The accompanying diagram illustrates the internal design elements of an IP telephone instrument (Figure 13-5.)

Figure 13-5
IP telephone design.

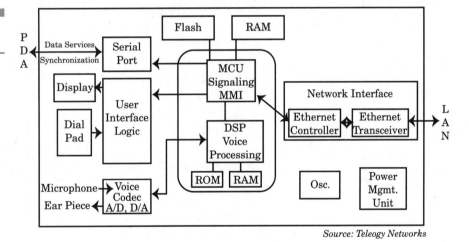

Source: Teleogy Networks

The user interface elements provide four classic telephone user function interfaces: keypad for dialing numbers; a variety of keys for line and feature access; a display for user prompts, caller feedback, messages, and other call processing information; serial interface to allow communications to an external device, such as a PDA, to allow synchronization of telephone information; speed dialing; and customer programming. An audible indicator (ringer) is also included to announce incoming calls.

The voice interface converts input analog voice signals into 8-bit digital word bit samples. Speech signals are sampled at an 8-KHz rate to create a 64-Kbps digital bit stream to the processor by using a standard

PCM codec. Voice signal compression and IP encoding functions are performed by processor complex elements. The processor complex performs voice processing, call processing, protocol processing, and network management software functions. The complex consists of a DSP for voice-related functions and a MCU for the remaining control and management functions. The DSP and MCU each have associated memory. DSP memory usually includes RAM and ROM elements; MCU memory usually includes RAM and Flash elements. The Flash memory element supports software upgrades.

The network interface allows the transmission and reception of voice packets to and from the telephone terminal based on 10BaseT or 10/100BaseT Ethernet running TCP/IP protocols. Some IP telephones may be equipped with multiple RJ-45 Ethernet connector ports and an integrated Ethernet hub/switch to support connections to the customer premises LAN and desktop PC clients. Newer IP telephones also may be designed with a USB connector port.

Basic IP telephone software modules include a variety of user interface drivers (display, keypad, ringer, user procedures), voice processing modules, telephony signaling gateway modules, network management modules, and system service modules. The voice processing software modules include a PCM interface unit; a tone generator (call progress tones, in-band DTMF signaling digits); a line echo canceler unit (ITU G.168-compliant echo cancellation on sampled, full-duplex voice port signals); an acoustic echo canceler for terminals equipped with a speakerphone; VAD; voice codec unit (compression and packeting of the 64-Kbps digital stream received from the station user based on a variety of algorithms, such as G.711, G.723.1, G.729/A, etc.); packet playout unit (compensation for network delay, jitter, and packet loss); packet protocol encapsulation unit (based on RTP, which runs directly on top of the UDP); voice encryption (to ensure privacy); and a control unit (coordinates the exchange of monitor and control information between the voice processing module and the telephony signaling and network management modules).

The telephony signaling gateway subsystem in an IP telephone performs the basic functions for call setup and teardown procedures. Software modules used by this subsystem include call processing, address translation and parsing, and network signaling. The most widely implemented network signaling standard used by currently available IP-PBX systems is H.323 protocol. H.323 is an ITU standard that defines several signaling and protocol specifications for multimedia communications between LAN-based terminals and network equipment. The main H.323

standards used for VoIP in an IP telephone are H.225–Call Signaling Protocol (based on Q.931), H.245–Control Protocol; RAS Protocol; and RTCP. An emerging network signaling standard not currently used by any commercially available IP-PBX, but being planned for by most suppliers, is SIP. SIP is the protocol developed and promoted by the Internet Engineering Task Force (IETF) and is forecasted to be widely implemented in network hosted services, such as IP-Centrex, and may eventually replace H.323 as the primary signaling protocol used by premises communications systems.

Distinct IP Telephony Features/Functions

Integrated port interfaces. Compared with a legacy digital telephone, an IP telephone can be designed and equipped to provide several unique feature/function capabilities. An IP telephone design attribute not available with traditional digital telephones is the integration of a multiport Ethernet hub/switch to allow multidevice sharing of a single connector port to the Ethernet switched network. Most current IP telephone models are equipped with two Ethernet port connectors: one connector for the Ethernet network and one for a desktop PC client. Mitel Networks has indicated that its next-generation models will have another external connector port to support two Ethernet devices external to the telephone. An integrated Ethernet port interface reduces telecommunications outlets, inside wiring, and Ethernet switch port requirements. Cisco Systems was the first supplier to incorporate an integrated Ethernet switch into its IP telephones in its 7900 series. Mitel Networks followed Cisco's approach by including integrated Ethernet switch ports in its second generation of IP telephones. Avaya, still marketing its first generation of IP telephones, offers its IP telephones with an integrated Ethernet hub.

The difference between an IP telephone with an integrated switch or hub may not be important to most customers, but providing a high level of voice-grade communications to the desktop is of primary importance. Voice communications QoS at the desktop can be supported using a variety of methods, such as Ethernet LAN 802.1 p/Q, or CoS programming (by switch or hub port). For example, each Cisco 7900 IP telephone internal Ethernet port can be programmed for different classes of service; the default service level of the voice port is a 5 and the data port is a 0. The system administrator can override the default service levels, if required, by an individual desktop station user. IP telephones with an internal Ethernet hub must include customized software to prioritize voice communications.

Besides Ethernet port connectors, current IP telephones may also support peripheral data devices through a USB port or infrared interface to a PDA. A USB port theoretically can be used for a variety of devices, such as printers, scanners, or digital cameras. There are several reasons to link a PDA through an infrared interface, including dialing from the directory or programming. Mitel Networks has introduced an IP telephone model with a docking station interface for a PDA. The PDA likely would function as the instrument's display field, and provide data download capabilities for call processing and handling applications.

Ethernet power distribution. IP telephones, like PC clients, require power. Traditional PBXs power analog and digital telephones use internal power supplies to distribute power over inside telephony wiring. Converged (IP-enabled circuit switched) IP-PBX cannot distribute power across integrated IP gateway circuit cards to the LAN; neither can the LAN-connected telephony servers used in client/server IP-PBX designs. The first generation IP telephones were powered with an AC/DC transformer connected to a local AC power outlet. Each IP telephone required its own transformer and a dedicated UPS for emergency power support. Although an IEEE subcommittee had been working on its recommended standard for in-line power over an Ethernet LAN, IEEE 802.3af, Cisco could not wait and developed its own proprietary solution. Other proprietary solutions soon followed from other IP-PBX system suppliers, including 3Com and Alcatel. Third-party solutions, from suppliers such as PowerDsine, are available and work with IP telephones from other leading IP-PBX suppliers, such as Avaya and Siemens. In-line power options are currently priced at $50 to $100 per Ethernet port, but prices are expected to decline over time.

An Ethernet switch is equipped with an integrated or external power patch module, and power is distributed directly only to IP telephones, supported by the switch. Power is transmitted over unused Ethernet cabling wire pairs to only those Ethernet ports identifying themselves to the switch as IP telephone devices. IP telephones identify themselves to the LAN switch during an automatic self-discovery installation method or through manual programming by the system administrator. The Ethernet switch queries the IP telephone as to how much power is required or assumes a default power level.

Some of the basic specifications of IEEE 802.3af are:

- DTE power shall use two-pair powering, where each wire in the pair is at the same nominal potential and the power supply potential is between the two pairs selected.
- The power detection and power feed shall operate on the same set of pairs.
- The DTE power maximum voltage shall not exceed the limits of SELV per IEC 950.
- For DC systems, the minimum output voltage of the source equipment power supply shall be at least 40 V DC.
- For DC systems, the source device shall be capable of supplying a minimum current of at least 300 mA per port.
- The solution for DTE powering shall support mid-span insertion of the power source.
- 802.3af systems shall distribute DC power.

Until the IEEE 802.3af standard is finalized, IP telephones will continue to be powered by available in-line power options or local AC power transformers.

Compressed voice. Traditional digital telephones are designed with codecs that digitize analog voice signals into digital format using 8-bit word encoding and 8-KHz sampling, resulting in 64-Kbps digital transmission over inside wiring and across the internal PBX switching network. IP telephones can compress voice signals for lower transmission rates and decreased bandwidth requirements. The most common digital encoding schemes currently used for voice transmission over Ethernet and IP WAN networks are G.711 (64 Kbps), G.723.1 (5.3 to 6.3 Kbps), and G.729/A (8 Kbps). G.711 is traditional PCM (no compression), but the two other codec specifications use compression algorithms. The total bandwidth used for voice transmission with IP transmission protocol is greater than the noted transmission rates; about 16 Kbps of additional transmission bandwidth is required because an IP destination address and overhead signaling bits are added to the voice datagram packets. Compressed voice transmission creates an overhead delay factor that may affect the quality of a conversation, but the trade-off is the potential for more efficient use of expensive off-premises network transmission resources. A T1 carrier circuit that typically supports a maximum of 24 voice-grade channels can support an equal or greater number of voice channels, with sufficient available bandwidth for concurrent data communications transmission, if voice is encoded using G.729/A compression. Using an IP telephone for voice compression eliminates the need for stand-alone telephony gateway equipment linking a traditional PBX system and an IP router. Calls placed from an IP telephone can be routed directly across a LAN and WAN without IP telephony servers.

Other IP telephone functions that reduce transmission bandwidth requirements are VAD and silence suppression. VAD detects voice communications signals entering the handset mouthpiece (microphone), and silence suppression signals the onset of "silent" voice transmission. A telephone call usually has a high percentage of silence during a conversation between parties, often as much as 50 percent of total talk time. A circuit switched connection is highly inefficient because much of the time there is no voice activity, but 8-bit words of "silence" are transmitted. With VAD and silence suppression, an IP telephone can reduce bandwidth transmission requirements because packets are not continually transmitted when no one is talking. When there are no voice communications signals picked up by the IP telephone microphone, a special signaling packet is transmitted to the destination IP address indicating the beginning of a silent period, when no new voice packets are being transmitted between the two endpoints. When voice activity resumes, another signaling packet is forwarded to inform the destination IP address that incoming voice packets are now on their way, effectively ending the period of silence. VAD and silence suppression packets are transmitted only when someone is actually talking, resulting in fewer packets and more efficient use of network resources.

Web browser. The most significant feature difference between a legacy digital telephone and an IP telephone is the integration of an embedded Web browser and pixel-based display monitor. The first question most people ask about Web-enabled IP telephones is: "Why do I need a telephone with Internet access if I have a PC?" The manufacturers of Web-enabled IP telephones are quick to point out that their product should not be considered a replacement for a fully functional PC client, but as a supplemental communications device for access to information when data processing is not required. These new IP telephones are best described as network communications portals that combine telephony functions with access to network information servers.

Thin client IP telephones have many of the internal design attributes of a computer: CPU, memory, operating system, applications software, and embedded communications protocol stacks. The RTOS of the thin client IP telephone may be proprietary, as in the Cisco Systems 7940/7960 models or the popular VX Works RTOS used by the Siemens optiPoint 600. Avaya's 4630 IP telephone was the first Web browser

PBX Voice Terminals

model with a color display and touch screen control. The use of color can greatly enhance the functionality and ergonomics of the desktop instrument, particularly when displaying graphic information or photographs. Touch screen control, instead of cursor control buttons, provides point-and-click mouselike activation of features and menu selection. A telephone with touch screen control is not new; industry veterans may recall the Northern Telecom M3000 digital telephone introduced in 1985.

General desktop applications using an integrated Web browser include:

- Access to directories external to the IP-PBX system directory database
- Messaging (voice, text, fax)
- Web page information screens
- Personal calendar
- Conference planning
- Transportation schedules and reservations
- Financial data (real-time stock quotes, investor information)

The accompany diagram of the Avaya 4630 IP telephone with a color touchscreen display illustrates the various applications supported by an IP telephone with an integrated Web browser interface (See Figures 13-6, 13-7, 13-8, and 13-9.)

Figure 13-6
Screenphone applications.
Source: Avaya.

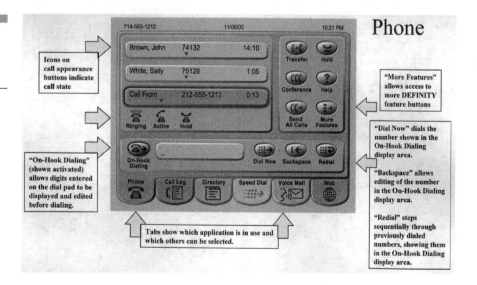

Figure 13-7
IP screenphone applications.
Source: Avaya.

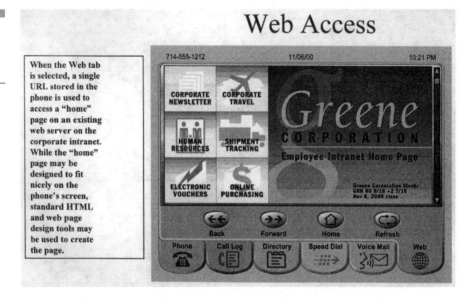

When the Web tab is selected, a single URL stored in the phone is used to access a "home" page on an existing web server on the corporate intranet. While the "home" page may be designed to fit nicely on the phone's screen, standard HTML and web page design tools may be used to create the page.

Figure 13-8
IP screenphone applications.
Source: Avaya.

Directory

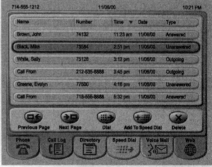

Call Log

PBX Voice Terminals

Figure 13-9
IP screenphone applications. Source: Avaya.

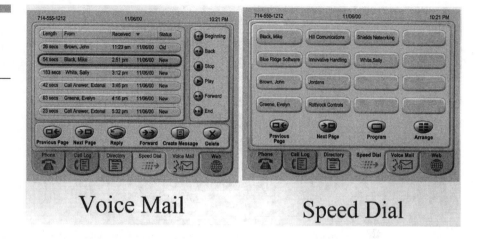

Voice Mail   Speed Dial

Using a telephone for e-mail or calendar access may seem strange if a personal computer is only inches away on the desktop, but it can be quicker and easier with the telephone. Telephones are always "on," and information access is immediately available at a touch of a button. Booting up a desktop computer is getting longer and longer, as each release of Windows becomes more and more complex and the number of programs loading grows even larger. Many companies have several antivirus programs that run a series of system and memory checks before the computer is ready for use. The reliability level of a telephone has proved to be at least an order of magnitude greater than desktop computers, and it is less likely that the telephone will freeze due to program interactions or some other operating system glitch.

The Web browser feature can be especially useful in vertical markets where voice station users do not normally have a desktop computer. The healthcare, retail, and hospitality sectors are characterized by a significant number of stations users who have voice-only instruments at their disposal. For example, many nursing stations still have dumb CRT terminals for information access. In the retail sector, most point-of-sale (POS) terminals have no Web server access. In hotels, guest rooms have telephones, and Ethernet ports, but no computers. There is also a sizable number of installed telephones across all industry sectors with no nearby PC client. Many telephones are not located on a desktop shared by a computer: lobby telephones; cubicle telephones; conference room telephones; and wall-mounted telephones in hallways, cafeterias, or locker rooms. An IP telephone with a Web browser can be used as an information kiosk in public locations, such as shopping malls, bus terminals, or airports.

Mobile. There are three subcategories of mobile telephones for use behind a PBX system: cordless, premises wireless, and cellular. PBX cordless telephones can be proprietary or standard 2500-type analog. Proprietary cordless telephones are supported by proprietary PBX port circuit cards and have a unique signaling and control channel that allows for multiple line appearances and full PBX feature access and performance (including display-based information). Usually using spread spectrum technology and operating in the 900-MHz frequency range, a proprietary cordless telephone can often be used as a substitute for desktop models. A growing number of circuit switched PBX systems supports this option, including Avaya, Nortel, Siemens, NEC, and Toshiba. Analog cordless telephones, the same type commonly used for residential applications, appear to the PBX system as 2500-type telephones and offer limited feature/function access but a degree of station user mobility not offered by fully wired desktop models.

Premises wireless handsets are included as part of a premises wireless telephony option working behind the PBX system. The wireless handsets for these systems are proprietary to each system's controller cards and base station transceivers. Base station coverage is limited in terms of geography and traffic handling. Most base stations support radio transmission ranges of about 50 to 150 meters, and between 2 and 12 simultaneous conversations per coverage cell. The wireless handsets closely resemble consumer cellular telephones, with several notable differences. Several manufacturers market wireless handsets with multiple line appearance buttons, fixed and programmable feature/function keys, and multiline displays that provide station users with information and data comparable to those of desktop digital telephone models. The high cost of a premises wireless handset and the infrastructure required to support coverage and traffic has limited the appeal of wireless telephony options, despite the ability of the station user to stay in touch with the PBX system regardless of location within the customer premises.

The first generation of premises wireless handsets was based on traditional circuit switching TDM/PCM standards. The recent introduction of wireless IP telephony solutions allows customers to use the existing LAN infrastructure to support distributed base stations. IP-PBX systems can interface directly to the wireless LAN infrastructure, but an MG is required for work behind a circuit switched PBX system. A leader in wireless IP is Symbol Technologies, whose Spectrum 24 wireless LAN system supports a wireless IP handset for use behind a PBX system. The Spectrum 24 uses spread spectrum frequency hopping within the 2.4- to 2.5-KHz band for transmission between access point transceivers and handheld communications devices. Data rates up to 2 Mbps per channel are supported. Each access point serves as an Ethernet bridge and can support wireless transmission coverage up to 2,000 feet in open environments and up to 180 to 250 feet in a typical office or retail store environment. Symbol's NetVision Phone system provides enterprise voice communications capability and allows for integration into an existing PBX system (via a gateway) for premises and off-premises communications. The system includes NetVision Phones, access points, and a telecom gateway (third party). Each access point typically can support between 12 and 16 active clients and up to 10 voice-only conversations. There is a voice prioritization algorithm at the access point and client levels to minimize voice transmission delays. Fast roaming and load balancing support hand-offs between access points. Access point pinging detects and tracks station devices. The NetVision Phone is based on the ITU H.323 standard and converts analog voice signals into compressed digital packets (G.729/A 8-bit sampling rate, 160 bytes per packet) that are sent via the TCP/IP protocol over standard data LAN networks with the CSMA/CA wireless access protocol. TCP/IP addressing is used to tie to an extension number or a name directory. Several dialing mechanisms are supported:

- Direct entry of complete or partial IP addresses
- Direct entry of an "extension" number
- Speed dial operation via speed dial keys
- Recall/redial of a previous number
- Using a name directory internally mapped to an IP address
- Pressing the Send button begins the keypad dialing process

NetVision is a single line telephone, with a second "virtual line appearance" to support two concurrent conversations (one line is active and the other is in the hold mode). Intercom calls are supported between the phones over the LAN infrastructure, including broadcast capability to any number of phones. A multiline display field provides for incoming CLID services, and fixed function keys are used for one-button feature access. Symbol also offers a NetVision Dataphone for use with Spectrum 24. This telephone handset has an integrated Web client for accessing applications and databases and bar code scanning capability. Proprietary versions of NetVision telephones are used by Nortel Networks and Mitel Networks behind their IP-PBX systems. The NetVision IP wireless telephony system interfaces to the IP-PBXs via port interface gateway line cards. The accompany diagram illustrates the integration of the wireless NetVision handsets into an Ericsson MD-110 PBX configuration (Figure 13-10). The NetVision terminals are typical of IP wireless handsets that are designed for enterprise mobile applications.

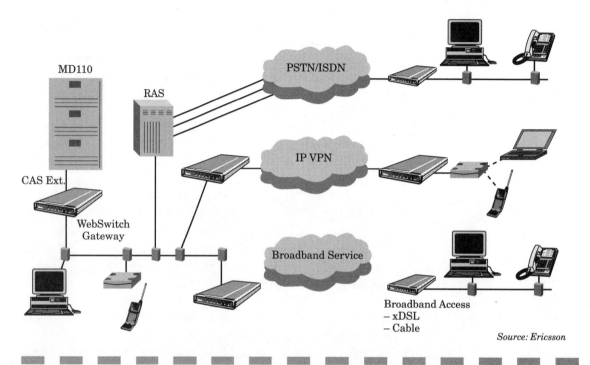

Source: Ericsson

Figure 13-10 Virtual IP telephony extensions.

Premises cellular is the third mobility communications option. The same cellular handset used with network cellular services, such as Sprint PCS, AT&T Wireless, and Cingular, can also interwork with a PBX system for premises mobile communications requirements. The first premises cellular options required an on-site mobility server and cell transceiver that linked to a local carrier's network. The mobility server provided an interface between the PBX system and the premises cellular infrastructure to support control signaling and feature support to cellular handsets while the station user was on the customer premises. This mobile communications option had several drawbacks, including cost (mobility servers and transceivers are expensive for limited numbers of subscribers) and network compatibility. The premises transceiver could link to only one cellular carrier service, such as TDMA or GSM. All premises subscribers required a cellular handset that worked with the same network carrier service. Although some business customers supplied their employees with a cellular handset and had a low-cost contract with a single service provider, the more likely scenario was that PBX station users had a great variety of cellular handsets supported by different network service carriers. A better solution was needed than an expensive cellular infrastructure linked to a single service provider.

Ericsson, a leader in mobile communications networks, developed a more cost-effective and flexible premises cellular option. The MD-110 Mobility Extension option is based on an integrated interface circuit card housed in the PBX's port carrier that can support a cellular handset with the use of any type of service standard from any local carrier. An ISDN PRI trunk circuit link is used to network the PBX system to the cellular network. Dialing procedures from the cellular handset will be in line with the terminal's existing network service procedures, plus fully support the MD110-procedures, including station features (via voice prompts) and network call routing. The Ericsson Mobility Extension option is carrier service provider and transmission/encoding independent.

PC client softphone. The final category of PBX telephones is the PC client softphone. There are several categories of softphones. The first generation of softphones was based on CTI desktop applications using first-party (desktop telephone API link) or third-party (client/server configuration) call control. The CTI-based softphone requires a telephone instrument (analog or digital) for voice transmission to/from the desktop. An IP softphone is a PC client functioning as the voice terminal using an integrated microphone/speaker option to support LAN-based voice transmission, with signaling and control to and from a telephony server over the LAN/WAN infrastructure. For implementation of either softphone, a station user accesses and implements PBX features (dialing, call answering, call coverage, call processing) using a keyboard and/or mouse control for a GUI computer screen. Communications solutions using PC client software tools offer station users many advantages over traditional telephone instruments, with a limited number of feature/line keys and relatively small noncolor display fields. The accompanying diagram is an illustration of the Nortel Networks i2050 soft client phone (Figure 13-11). Some suppliers also offer customized client keyboards with integrated handsets for use as a softphone. The accompanying photograph is a Siemens optiKeyboard designed for use with its family of softphone client solutions (Figure 13-12.)

Figure 13-11
IP softphone: Nortel Networks i2050.

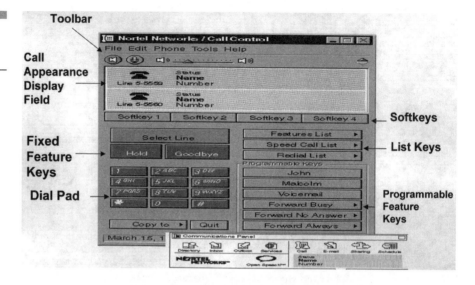

Figure 13-12
Siemens optiKeyboard.

Market demand for CTI-based softphones has been very weak. Many station users prefer to depress traditional telephone buttons to access features rather than interact with a GUI-based computer screen to perform drag, point, and click operations. Telephone instruments also offer a far greater degree of reliability than PC hardware/software and are not affected to the same level as AC-powered desktop computers by local power problems. A major problem associated with first-party control CTI softphones was the requirement of a relatively expensive digital telephone equipped with an API link to the desktop computer. Third-party control client/server CTI configurations could be implemented with a lower-priced analog telephone, but station user functionality is severely affected when the desktop computer fails or is not performing properly. The primary market for desktop CTI has been among call center customers because the current ACD agent position depends heavily on desktop computer equipment and GUI-based interactions, and the cost of the solution is not significant compared with overall contact center expenses.

The emergence of IP-PBX systems may spur demand for PC client softphones because the cost of the solution may be far less than that of a high performance IP telephone. There likely will be great resistance to IP softphones from most station users who have grown comfortable with traditional telephone instruments, but the many potential benefits of the new solution may stimulate market demand.

Desktop Voice Terminal Attributes

A telephone is more than just a handset and a dialpad. The appearance and operation of current digital telephones more closely resembles the cockpit panel of the Concorde than yesteryear's rotary dial telephone. The performance and price range of PBX telephone models vary from a standard single line analog telephone used for basic dialing and answering intercom operations at $25, to sophisticated digital, display-based multiline featurephones used for complex call center applications at $750. Adding one or two available options to the latter model can easily increase the final price tag to more than $1,000. The following is a descriptive checklist of PBX desktop telephone attributes developed to assist customers in their selection of the proper voice terminal type and model best equipped to satisfy their communications needs.

Loop length. Loop length is simply defined as the length of the telephony wiring cable between the wall jack and the PBX common equipment cabinet. Most PBX systems can support analog telephone loop lengths of at least 10,000 feet when using 24 AWG UTP cable. The PBX system loop length range for proprietary digital telephones is about 3,000 to 5,000 feet, 24 AWG. PBX system ISDN BRI loop length for S/T interface models is about 1,800 feet, 24 AWG; U-interface models can be supported with lengths up to 16,000 feet, 24 AWG. IP telephones have a 100-meter loop length limit to an Ethernet switch.

Number of wiring pairs. Analog telephones and most proprietary digital telephones are currently supported with one twisted wiring pair. ISDN BRI telephones use single pair for U-interface models and multipair (typically three pair) for S/T-interface models. IP telephones usually require only two pairs, although Category 5 cabling provides for four pairs.

Call control/audio coders (IP telephones only). IP telephones may be designed using proprietary call control standards, or one of many published standards specifications, such as H.323, SIP, or MGCP. Audio coders are used to convert PCM signaling into compressed IP formatted signaling. Commonly available protocol stacks are G.711, G.723.1, and G.729/A. In addition to TCP/IP, other pertinent IP telephone protocol capabilities include File Transfer Protocol (FTP), SNMP, and SNTP. Compatibility with Microsoft NetMeeting is another important attribute.

Ethernet auto-sensing/DHCP (IP telephones only). IP telephones are designed with integrated Ethernet NIC (RJ-45) ports. When an IP telephone is plugged into the Ethernet network, an IP address can be assigned automatically with a DHCP server instead of by manual programming.

Voice activity detection/silence and echo suppression (IP telephones only). IP telephones, unlike proprietary or ISDN BRI digital telephones, transmit voice packets only when there is voice activity. When a station user is not speaking into the mouthpiece, there are no transmitted packets. This feature reduces station user LAN bandwidth requirements. Echo suppression techniques improve the quality of voice conversations.

Integrated Ethernet hub/switch (IP telephones only). IP telephones equipped with an integrated Ethernet hub or switch have port interfaces to support adjunct Ethernet desktop equipment, such as a PC client. With the hub/switch option, both desktop devices can share a common RJ-45 wall jack connection. The voice and data terminals can share the same IP address or have dedicated IP addresses. QoS for voice transmission to and from the desktop for a hub connection is based on customized software or firmware. QoS voice transmission for switch connections at the desktop can be based on standard packet marked LAN switch QoS solutions, such as 802.1p/Q or DiffServe.

PBX Voice Terminals

Programmable line appearances/feature buttons. Standard analog telephones support a single line appearance. Digital telephone models are available that support a single line appearance or multiple line appearances. Digital telephone models support a wide range of programmable line appearances, although most station user requirements are limited to a few lines. Most manufacturers offer models capable of supporting up to 16 line appearances, and some available models can support more than 30 line appearances for heavy call coverage operations requiring dedicated bridged line appearances for individual stations. Each programmed line appearance is a distinct PBX system directory number. Some PBX systems support multiple call appearances (multiple programmable buttons supporting the same directory number) instead of distinct line numbers.

Programmable buttons not used for line/call appearance usually are used for one-button feature access or speed dialing. Some single line analog telephones may be equipped with programmable buttons for feature/speed dialing operations. Some digital telephones may be equipped with a limited number of programmable line/call appearance buttons but have additional feature/speed dialing programmable buttons. Programmable line/feature buttons usually have associated LED or LCD indicators to show the status of a line appearance or feature.

Fixed feature buttons. Many, but not all, digital telephones have several fixed feature buttons for simplified access to the most popular station user features, such as hold, transfer, release, last number redial, and conference. The fixed feature buttons may or may not have status indicators. Some manufacturers color code their buttons by function category, for example, Nortel Networks telephones have blue buttons to signify buttons or keys used for talk (line keys, speakerphone), red buttons for hold, and orange buttons for the release (goodbye).

Softkey buttons. Digital telephone softkeys ,rather than fixed or programmable buttons, are used for feature and function access. A few softkeys can be used as an alternative to 10 or 20 dedicated feature buttons. Softkey features are usually context sensitive based on station user status: on-hook, off-hook, and on-line. Features are displayed above an associated softkey and change base on station user status. Softkeys can be used for directory access, station programming, and advanced applications such as message system access. Display menu keys, including scroll and select keys, are used to move the display cursor and activate a feature.

Application feature buttons. Several current digital telephones have fixed buttons for features and applications such as call log (incoming and outgoing calls), directory access (LDAP-based), language services, ringing tones and patterns, or display formatting. IP phones with integrated thin client Web browsers have a fixed application button to access a display menu of Web services. Digital telephones designed for ACD agent positions usually have fixed buttons for ACD-specific features such as Supervisor Alert and Work Code.

On-hook dialing. The on-hook dialing feature allows a station user to dial a telephone number without lifting the handset. A built-in speaker allows the station user to hear dial tone, ringing, and answer before using the handset. This feature operates independently of a two-way speakerphone function.

Hands-Free Answer Intercom (HFAI). The HFAI function allows a station user to answer an incoming intercom call without lifting the handset. The caller is "announced" over an integrated speaker. The called party then goes off-hook to use the handset to talk.

Speakerphone. A built-in speakerphone allows a station user to dial telephone numbers, answer incoming calls, and speak to the connected party without lifting the handset. A speakerphone button activates and deactivates the function. A simplex speakerphone supports talk in one direction at a time; duplex speakerphone operation supports two-way conversation.

Some suppliers also make available a more fully functioned audiconferencing unit that interfaces to the telephone instrument.

Mute. The mute button deactivates the mouthpiece or speakerphone microphone and places the telephone is a mute state, thus suppressing outgoing talk. The mute feature is often used during audioconferencing calls to suppress background sounds or comments. Some digital telephones have a "smart" mute button that automatically senses whether the station user is talking through a handset, speakerphone, or headset.

Headset interface. An integrated port interface supporting a headset is usually standard on telephone models used for ACD agent positions, but there is increasing demand for the attribute to support office applications. Some digital telephones may support two headset interfaces to allow a third party to listen to a conversation.

Alphanumeric display. Digital telephones may be equipped with an alphanumeric display field to provide the station user with call information (dialed number, outgoing/incoming trunk identification incoming caller number, calling name display, call diversion information, call duration, and charging), feature/function menu associated with context-sensitive softkey operation, station programming guide, and time and date. Most current digital telephones have multiple line displays capable of supporting 40 to 80 alphanumeric characters per line. Several digital telephone models have very large display fields of 5 to 10 lines, support self-labeled programmable buttons (eliminating paper labels), and may exhibit programming icons. IP telephones with large display fields also may be programmed to download Web server pages.

Analog telephone display capabilities are usually limited in size and function. Typical analog telephone display fields are one or two lines, with usually fewer than 40 alphanumeric characters, and display the following types of information: dialed number, CLID, time and date, and call duration. Name displays are not supported on analog telephones.

Adapter modules. There are different adapter modules for digital telephones, primarily for proprietary models. Adapter modules for ISDN BRI and IP models are currently limited. Support of adapter modules, sometimes referred to as *add-on modules*, is not available across all models of a particular product line. Many entry-level digital telephone models do not support module options. Some digital telephone models may include a module capability as standard, although most of the following modules are usually optionally priced.

DATA MODULE. A data module supports DTE for modemless circuit switched data communications applications. The data module capability may be integrated into the digital telephone design, a plug-in module, or an external standalone device with links to the telephone and DTE. Most data modules are designed with a RS-232C interface; other data interfaces, such as RS-449, also might be available. Some digital telephone data modules are also used for CTI link applications for first-party call control at the desktop computer. This module option is available only for proprietary and ISDN BRI digital telephones.

CTI API MODULE. An API module is required for first-party call control CTI PC telephony applications. Some manufacturers use the data communications module for CTI links, and a few have a dedicated CTI link. USB telephones have an integrated port interface that can be used as a

CTI link. This module option is available only for proprietary and ISDN BRI digital telephones.

ANALOG COMMUNICATIONS DEVICE ADAPTER. An analog communications device adapter supports an adjunct analog communications device, such as an analog telephone, modem, fax terminal, or audioconferencing equipment, by using the second bearer communications channel of the digital transmission link between the PBX common equipment and the digital desktop. Almost all current PBX systems allow the system administrator the option of assigning a unique directory number to the analog communications device instead of programming an extension behind the digital telephone. The digital telephone and analog communications device can be simultaneously active. This module option is available primarily for proprietary digital telephones but may become available on a select number of IP telephones during the next few years. March Networks (Mitel Networks) has an option on its newest-generation IP telephones to support an adjunct audioconferencing unit. The accompanying photograph illustrates the Mitel Networks Audioconferencing unit interfaced to the S5520 (a modified SuperSet 4025) (Figure 13-13).

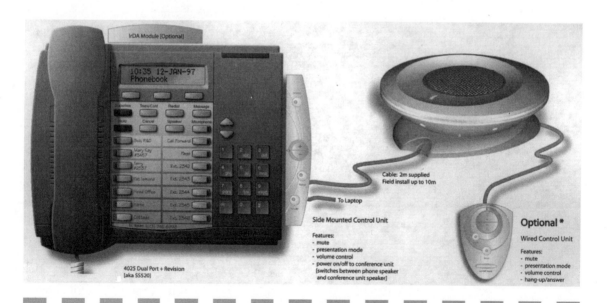

Figure 13-13 Mitel Networks SuperSet 5520 with audioconferencing option.

DIGITAL COMMUNICATIONS DEVICE ADAPTER. A digital communications device adapter supports an adjunct digital telephone by using the second bearer communications channel of the digital transmission link between the PBX common equipment and the digital desktop. Both desktop digital telephones have distinct directory numbers and can be simultaneously active. This option is unique to proprietary digital telephones.

ISDN BRI ADAPTER MODULE (PROPRIETARY ONLY). An ISDN BRI adapter module converts proprietary digital telephone transmission protocol at the desktop to conform to ISDN BRI specifications. The module option is used primarily to support H.320 videoconferencing applications for desktop computer terminals connected to the digital telephone or high-speed 128-Kbps data communication transmissions across the PBX switching network.

IP ADAPTER MODULE (PROPRIETARY ONLY). An IP adapter module is used to upgrade an existing proprietary digital telephone in support of IP telephony applications, precluding the need to replace the voice terminal with a native IP telephone. The adapter module may be a plug-in or an external gateway device.

ADD-ON BUTTON MODULE. One or more line/feature button (key) modules can be supported by select digital telephones to increase the number of available programmable line appearances or feature buttons at the desktop. Some digital telephones can support up to three add-on modules and support an additional 60 programmable line/feature buttons. This module option is used primarily for desktop station users responsible for large group call coverage. It is currently available with proprietary digital telephones but may soon be available with IP telephones.

DISTANCE EXTENDER MODULE (PROPRIETARY ONLY). A distance extender module increases the standard loop length for a proprietary digital telephone in support of campuslike system installations or remote teleworking applications, assuming right of way for cabling. The module may require local AC power.

SECURITY MODULE. A security module encrypts voice communications for privacy applications. Both telephones must be equipped with the same module equipment.

Power Requirements

Traditional PBX voice terminals are supported with standard in-line power over the telephony cabling system. The common equipment generates the current. Some desktop voice terminal equipment options may require local power using an AC power transformer unit. The first generation of IP telephones was powered locally with an AC power transformer. The second generation of IP telephones is currently supported with in-line power generated at the local switching closet.

CHAPTER 14

PBX Networking

The features, functions, and architecture designs supporting calls across public or private network carrier facilities determine PBX networking capabilities. The two primary objectives of PBX networking features and functions are to minimize telecommunications service expenses and provide management and control over all off-net and on-net calls.

Off-premises calls placed over the PSTN or private network are managed and controlled by ARS. Synonyms for ARS include LCR, route optimization, and MERS. Calls placed to station users configured within a private network configuration may be routed on-net to premises-based station users dispersed across a single customer location or off-premises to locations classified by any of the following categories:

1. Remote location with a single station user or a small station user group without local PBX common equipment
2. Remote location with small, intermediate, or large station user group with local PBX common equipment cabinets/carriers
3. Location with full function PBX system

Public Networking

Automatic Route Selection (ARS)

All station user calls, direct distance dialed and private network, are routed to trunk groups that have access to exchange carrier facilities terminating in a central office. The many types of port circuit interfaces and trunk facilities were described in Chapter 6 (Traditional PBX Common Equipment). The software feature controlling access to the trunk groups is ARS. PBX ARS features and functions support control and routing of calls over public network carrier facilities and across private networks. ARS is used to select among various types of trunks:

1. Local CO, analog or digital
2. Foreign exchange (FX)
3. Wide area telecommunications service (WATS)
4. Tie line (private line); analog and/or digital
5. ISDN PRI service circuits

When the ARS software program routes calls to public network carrier facilities, it is referred to as *off-net routing*; when calls are routed to private network carrier facilities, it is referred to as *on-net routing*. Off-net calls are based on the public network dialing format and terminate in the public network. On-net calls are based on a private network dialing format and usually terminate within the private network, although some calls may be routed off the private network and onto public network facilities, if the called station is not configured as part of the network. The private networking feature used to describe the latter routing option is known as *tail end hop off* (TEHO). On-net routing programs support tie-trunk protocols, proprietary private network protocols, and industry-standard private network protocols.

The ARS feature is activated when a station user dials the access code for an off-premises call requiring a trunk circuit, followed by the telephone number to be called. Public network calls in North America are usually dialed beginning with the digit 9, which alerts the common control complex that an outside line (trunk) is required to place the call. Calls placed over traditional private tandem networks are usually dialed with a customer-selected digit, such as 8, to distinguish the call from public network calls. Customers with traditional private tandem networks create a unique numbering plan for on-net calls. An *intelligent private network* call is placed by dialing a station directory number within the network that matches the local directory number of the dialed station. Intelligent private networks do not require a separate private network numbering plan because the entire network of PBXs operates as a single homogeneous network for most system features and functions, including the uniform dialing plan. Small network configurations may be supported using a four- or five-digit numbering plan; larger networks may require more dialed digits for on-net calls because a limited digit dialing plan cannot support the number of station users.

ARS routing table rules determine whether the call is routed off- or on-net. For off-net calls the ARS program determines which trunk group the call is routed to for network access. PBX system administrators rank trunk groups from lowest to highest cost to ensure that calls are sent over the least costly route available to the caller, based on customer's class of service or restriction level. Public network route selection is based on data in the dial plan databases. The ARS software analyzes and compares the dialed digits with the digit string patterns in the ARS dial plan. If there is no database match, the call is blocked. Blocked calls may be the result of call restriction levels for individual callers. It is common today for system administrators to block many outgoing 900 calls, or direct-dialed international calls, except for select station users. Restriction features may limit an individual station user's dial capabilities to internal calls or calls to very select off-net locations, such as local exchange calls.

If there is a database match, the system may request an account code before continuing, or the ARS feature can immediately define the call route and determine which digits should be sent over the network while the call processing software seizes an available trunk according to the station user's network class of service level. If all trunk circuits are busy for the lowest-cost trunk group, the call may be routed to the next highest-level trunk group based on call costs, or the call may be queued until a trunk circuit is available in the originally selected trunk group.

The trunk circuit seized by the PBX system is determined by analysis of the ARS routing table and associated trunk tariff database. The ARS routing table is made of routing patterns that map to call routes for the dialed CO location code. CO codes are defined by specific country and city locations. For each dialed CO code, there may be one or more routing patterns, and each routing pattern may have one or more call routes. More than one pattern of dialed digits can translate to the same routing pattern. The facility restrictions level (FRL) feature determines access to a select call route within a routing pattern. For any particular call route, multiple trunk groups may be used to handle the call, if the customer subscribes to more than one exchange carrier service. The lowest-cost route is usually programmed as the preferred route if the trunk circuits are available.

The most common ARS feature capabilities are:

1. Area code/office code restriction (toll restriction)
2. Alternate route selection (route advance)
3. Time-of-day routing
4. Day-of-week routing
5. Trunk queuing
6. Digit analysis and manipulation
7. Call screening

ISDN Features

ISDN evolved from the original analog-based telephony PSTN that provided end-to-end connectivity to support a wide variety of services to which users had access through a limited set of standard multipurpose customer interfaces. The major attributes of ISDN include:

- End-to-end digital connectivity
- Access and service integration

PBX Networking

- User control of service features
- Standardized user-to-network interfaces

The CCITT (later the ITU) originally defined two major, globally standard interfaces: BRI and PRI. The BRI provides a station-direct interface to ISDN network facilities or ISDN-compatible customer premises equipment. The PRI provides an interface for switch connections, such as PBX-to-PBX. Both interfaces have similar format structures: one D-channel and multiple B-channels. The B-channel, or bearer channel, is a 64-Kbps transmission path for basic and primary rates and carries voice, data, and image or video communications traffic to and from the network or switch. The D-channel, or data channel, uses 64 Kbps for primary line and 16 Kbps for basic lines and carries packeted signaling and control information across the interface. The D-channel signaling is out-of-band because it supports calls on the separate B-channels. In North America, the PRI has 23 B-channels and 1 D-channel (23 B+D); outside of North America PRI has 30 B-channels (30B+D). ISDN services are carried over T1/E1 transmission circuits using the DS1 format structure: (24/32) 64-Kbps channels. Globally, the BRI has 2 B-channels and 1 D-channel (2B+D).

The ITU standard established for D-channel signaling is the Q.931 message-oriented signaling protocol. This protocol allows the use of the same access lines for many different services and the introduction of new services without requiring users to replace existing ISDN-compatible communications equipment. ISDN is defined internationally as the physical interfaces between communications equipment and networks and the signaling protocols exchanged by the elements.

Several PBX network features that work with ISDN PRI services, such as ANI and Station Identification (SID), CBCSS, non-facility associated signaling (NFAS), and D-channel backup.

ANI/SID

ANI is delivery of the originating calling party's 10-digit billing telephone number to an ISDN subscriber's premises communications system and/or desktop terminal equipment. When a telephone call is made from an equal access CO, the ANI is passed from the local exchange carrier network to the interexchange carrier network for transport across Common Channel Signaling System 7 (CCSS7) packet switched facilities to the destination local CO exchange. The ANI is included in the call setup message using Q.931 message format over the D-channel. ANI is delivered with ISDN BRI or PRI trunk services.

ANI is an interexchange carrier service often confused with local exchange carrier CLID service. CLID is a local access transport area (LATA) service feature that is one of the commonly available customized local access signaling system (CLASS) features. ANI is a service feature offering that was originally developed in support of long distance telephone calls placed to inbound call center operations, with a requirement for ISDN trunking to the customer premises. PBX/ACD systems receiving the ANI can use the information for call screening, analysis, decision making, and routing procedures, including a database lookup procedure to match the ANI with a customer file. ANI also can be used to identify the geographic location of the calling party, because the area code is included in the digit string. ANI is no longer used exclusively for call center applications but is considered an important element for enhanced call screening functions at the system or desktop level. Another useful ANI benefit is preidentifying the calling party for network security purposes.

An ISDN feature similar to ANI is SID. SID is usually more useful than ANI because the feature delivers the originating caller's telephone number behind a PBX system. ANI is the trunk billing number of the PBX trunk circuits and does not identify individual callers provisioned behind the switching system. SID can distinguish station users to a greater degree than ANI.

ANI and SID are useful features for a variety of customer applications that do not include formal in-bound call center systems. Collecting ANI/SID data allows PBX customers to better track and analyze incoming long distance calls by geographic area (regional and local). Collecting and storing ANI/SID in-bound call data in a call log database can support improved out-bound customer service and telemarketing operations. PBX customers can use a variety of network carrier services to route incoming calls to different PBX systems by using ANI/SID data to load-balance calls across locations for increased call handling efficiency and performance.

CBCSS

CBCSS is a PBX network feature that allows a single ISDN PRI services trunk group to carry calls to multiple network carrier services or facilities or carry calls using different interexchange carriers. CBCSS uses the same routing tables as those used by ARS/AAR. Without CBCSS, each trunk group must be dedicated to a specific carrier service or facility. Implementing the CBCSS feature allows a variety of services to use a single trunk group. The services are specified on a call-by-call basis. This optimizes trunking efficiency because traffic is distributed fully over the total number of available trunks, regardless of peak time period service requirements. Examples of services typically requiring dedicated trunk groups without using CBCSS are in-bound and out-bound WATS; direct long distance dialing; VPN; digital data services including ATM, frame relay, and IP; digital video services; presubscribed common carrier operator international 800 calls; and other user-defined services.

CBCSS allows services to share the same trunks to reduce total trunk requirements. The probability of denied feature and service access due to blocked trunk access is also reduced. Network engineering is simplified because trunk engineering analysis can use total traffic data instead of analysis by a per-service basis. By dynamically changing the mix of trunk circuits accessing different services and facilities, the system can function more efficiently.

CBCSS customer programming tools allow administrators to dynamically assign and reassign individual trunk members in a trunk group access to different services and facilities based on pre-set schedules (time of day, day of week), real-time traffic loads, or on demand.

The NFAS feature allows an ISDN PRI DS1 interface D-channel (signaling channel) to transmit signaling information for B-channels on ISDN PRI DS1 facilities other than the one containing the D-channel. A single D-channel can carry signaling information for numerous B-channels on different DS1 carrier facilities, thus providing a more economical interface between the PBX system and the ISDN network. This means that a customer can configure a single D-channel to support more than the standard 23 B-channels available on a facility-associated signaling ISDN PRI DS1 carrier circuit. A single D-channel can therefore provide signaling for 50, 100, or more B-channels based on the software programming limitations of a PBX system supporting the NFAS feature.

PBX systems implementing NFAS also can support D-channel backup. If a D-channel signaling link fails, a backup D-channel transports the signaling. The feature requires that one D-channel be administered as the primary D-channel and a second be administered as the secondary D-channel. When a transition from one D-channel to the other occurs, all stable calls (calls already answered) are preserved. Some messages may be lost, resulting in a loss of call-related information, but the calls themselves will be maintained.

Private Networking

If the ARS feature determines that a placed call is to be routed on-net, the call is handled over private network facilities. There are many private network configurations, ranging from support of a single station user working at home, to linking of dozens or hundreds of PBX systems and locations scattered across the globe. Private networking options can be classified according to the following criteria:

1. Number of PBX systems in the network design
2. Number of station users at each network location
3. Locations of station users relative to a local PBX system
4. Available transmission carrier facilities
5. Traffic capacity requirements between network locations
6. Feature transparency requirements across the network
7. Survivability requirements for each network location
8. Operations, management, maintenance, and service procedures

Single System PBX Network: On-net Multilocation Support

The simplest network design consists of a single PBX system. There are several single system design configurations based on the location of the system's station users. These configurations are used to support the following types of communications requirements:

1. One or more station users at the same location who are physically remote from all customer premises locations housing PBX common equipment
2. One or more station users residing on the customer premises but exceeding the maximum loop length for their desktop telephone equipment
3. Station users physically remote from the customer premises location housing the main PBX common equipment room but supported by local common equipment

The first PBX configuration option supports remote teleworkers. The second and third configurations are based on a distributed common equipment architecture. Some PBX systems have a standard distributed port interface cabinet/carrier design, and others use hardware options designed for remote premises communications requirements.

The first single system PBX networking category supports station users who are remote from all common equipment but require desktop communications support as if they were located at their organization's premises. These type of station users are now referred to as teleworkers. Teleworkers fall into two categories:

1. Fixed teleworker
2. Mobile teleworker

A fixed teleworker has a permanent desktop location relative to the remotely located PBX system. Fixed teleworker solutions are sometimes referred to as small office/home office (SOHO) configurations. Mobile teleworkers are constantly on the move, but need to be linked to their PBX system whenever and wherever they are. Another popular term for a mobile teleworker is *road warrior*. Road warriors are the newest breed of PBX station user and are growing in number at an alarming rate due to the proliferation of mobile computing and communications devices.

There are several available PBX options that support fixed teleworkers who require the same level of communications service and support as station users at the customer premises location. The oldest, and most basic, PBX option supporting off-premises station users is the OPX feature, a solution available for more than 20 years. The OPX feature requires a special local exchange carrier trunk circuit, known as an OPX circuit, to provide PBX system communications and signaling support to a remote analog telephone. The loop length of an OPX circuit connection to the remote station is usually limited to several miles without repeaters. Provisioning of OPX trunk services is also constrained by LATA boundaries. OPX station users have full access to all PBX features and functions but are operationally limited by their analog telephone instruments. No analog telephone supports out-of-band signaling to support multiple line appearances and proprietary PBX display field information. Off-premises and intercom calls to the remote station user are routed over the PBX's OPX trunk. The remote station user can initiate intercom calls to other PBX station users and uses PBX trunk facilities for placing calls outside the PBX system.

SOHO applications requiring a higher performance desktop terminal, similar to the one available at the customer premises location, had limited options until the early 1990s. During the past decade several fixed teleworker solutions have been implemented:

1. Remote ISDN BRI telephones
2. Proprietary and third-party local loop distance extender options
3. Remote IP telephones/softphones

The first SOHO option based on ISDN BRI services was implemented by Intecom in the early 1990s. The PBX system interfaces to the PSTN via a digital trunk circuit by implementing ISDN PRI services, and the remote teleworker uses an ISDN BRI line to support a desktop ISDN BRI telephone that offers performance capabilities comparable to those of proprietary digital telephone instruments. PBX control signaling is carried over the ISDN trunk circuit's D-channel to the remote location. This solution provides remote teleworker access to all PBX system features and functions and can support remote data terminal equipment over the second ISDN BRI B-channel.

A popular SOHO solution has been the use of local loop distance extender equipment to support proprietary digital PBX telephones over the PSTN trunk carrier facilities used to transport voice communications and control signaling between the customer premises and remote location. The first distance extender options were based on hardware equipment at the customer premises location—the proprietary port circuit cards or gateway modules, used to convert the proprietary out-of-band digital desktop control signaling into CAS format for transport over analog trunk circuits to connect PBX and remote teleworker locations. A desktop gateway module at the remote teleworker location supports a proprietary desktop digital telephone. Leading suppliers of local loop extender equipment include MCK Communications and Teltone. The option was developed when residential digital line services, such as ISDN BRI, were not commonly available. The growing availability of digital subscriber line (DSL) services has eliminated the need to convert digital signaling to CAS format. The wideband nature of current digital services available to SOHO locations allows several remote digital telephones to be supported over a single DSL line. Rack-mount carrier gateways at the remote location can support 12, 24, or more digital telephones with one or more digital T1-carrier facilities.

One of the very few early benefits of IP telephony was support of remote IP telephones using dial-up analog or digital lines at a remote location. Almost all IP-PBX systems can remotely support desktop IP telephone instruments or IP softphones via LAN/WAN infrastructures. The remote location configuration may require an analog gateway module or a SOHO router to access the corporate WAN.

PBX Networking

Road warrior options include IP softphone applications running on a notebook/laptop computer, or any DTMF analog telephone or cellular handset, linked to the main PBX system through a proprietary communications/signaling port interface card, gateway module, or teleworker server. The mobile IP softphone option can be implemented by local LAN access to the corporate WAN or a wireless dial-up option wherever the station user happens to be. Several PBX suppliers are currently marketing mobile telephone options behind their communications systems, including Avaya, Nortel Networks, and Siemens. The Avaya Definity and Nortel Networks Meridian 1 options are based on the MCK Communications Mobile EXTender gateway. The EXTender gateway is configured behind the PBX system with the use of standard digital port interface links. The Siemens solution is based on a HiPath Teleworker server linked to the HiPath PBX system. For each road warrior option, the remote desktop telephone or cellular handset appears to the PBX system as an extension of the station user's customer premises desktop voice terminal. Any DTMF telephone can be logged into the PBX to receive and place calls through the centrally located communications system. Road warriors can use intercom; four-digit dialing plans; activate basic call processing features, such as hold, transfer, and conference; and use the private PBX network for long distance calling. Unanswered calls are routed to a call coverage station or VMS mailbox rather than the remote telephone's mailbox.

Distributed PBX Common Equipment Design

Numerous customer configurations require a distributed PBX common equipment design that supports a single system image for all features, functions, and operations. These configurations include a campus environment covering a large geographic area, with loop length runs exceeding supported parameters, and customers with communications requirements across two or more locations with right-of-way cabling options. A distributed common equipment design may consist of any combination of the following design elements:

1. Centralized, dispersed, or distributed call processing
2. Centralized, distributed, or dispersed switch network design
3. Distributed port equipment cabinets/carriers (main and remote equipment rooms)

A distributed call processing design is preferable because local call processing reduces the probability of down time for each port cabinet. Distributed port cabinets linked to a centrally located common control complex (centralized or dispersed design) depend on intercabinet links outside a secure equipment room for all call processing functions, and link failure or problems increase as distances between cabinets increase.

For switching functions a distributed or dispersed design is preferable to a centralized stage because local switching requirements do not depend on intercabinet links for all switched call connections, as the centralized design does. Localized switching reduces intercabinet link communications channel requirements.

The first fully distributed common equipment design was the Intecom IBX S/40, introduced in 1980, which was an immediate success with universities with large campuses. The first IBX installation was at the University of Chicago. Another PBX system based on a distributed cabinet design, which has been popular on university campuses, is the Ericsson MD-110. University systems have been the largest single customer market sector for the Intecom and Ericsson systems during the past 20 years. Intecom PBX systems can support distributed cabinets several miles from the main equipment room housing the central control complex. Since its introduction, the Intecom PBXs have used a fiber optic cabling infrastructure in support of call processing signaling and switch network transmission links. The MD-110 can support more than 200 distributed port cabinets, each with a local common control complex, across virtually unlimited distances over PCM-based transmission links, with the use of copper, broadband fiber, or microwave transmission media.

Several new IP-PBX systems designs using a LAN/WAN infrastructure to support dispersed station users are ideal single system solutions in support of large coverage communications requirements. Systems from recent market entrants such as Cisco Systems and Sphere Communications fall into this design category. The traditional Siemens 300H architecture design is not based on an IP-based LAN/WAN cabling infrastructure, but a recently introduced IP-based remote carrier option is supported over a dark optical fiber cable between the main equipment room and the remote building location. The Alcatel OmniPCX 4400 can use a variety of intercabinet transmission links between distributed cabinet clusters, including PCM, IP, or ATM formats and copper or optical fiber cabling. Avaya, too, can use an ATM LAN/WAN infrastructure to support its EPN equipment cabinets for on-site or off-site requirements. Each of these systems also supports redundant common control capabilities.

Among the new IP-PBX system designs, the Cisco offering stands out. The Cisco AVVID IP telephony system can be configured with up to five active call servers. The call servers can be in the same equipment room, dispersed across the same customer premises, or across multiple customer premises. IP telephones can be supported by one of three servers for redundant call processing control. In case of WAN link failure, Cisco was the first IP-PBX supplier to announce support of basic telephony functions at remote locations with the use of a call processor blade embedded in one of its IP routers. The remote survivable telephony system option is an emergency backup system with limited call processing and feature management capabilities when desktop telephones/softphones do not have access to a centralized call telephony server.

Remote Port Cabinet Options

Customers who want a single PBX system supporting communications requirements across multiple physical locations can use one of many systems offering a remote port cabinet/carrier option, in addition to systems similar to those described in the preceding paragraph. Even though the Ericsson, Alcatel, Cisco, and Sphere systems have a standard single system image distributed architecture capable of supporting station users dispersed across one or more customer premises, some PBXs can support only remote communications requirements with optional hardware. The first PBX system that supported a remote off-premises port cabinet option was the Northern Telecom SL-1. The SL-1 RPE option was introduced in 1982. Since the original SL-1 option most other leading PBX suppliers have introduced their own remote cabinet/carrier options. Current remote options differ greatly across system platforms. Remote port options may include:

1. Multicarrier port cabinets designed to support up to 1,000 ports
2. Single carrier cabinets designed to support about 300 ports
3. Half-shelf carrier cabinets designed to support about 150 ports
4. Single port interface circuit boards designed to support 24 or 36 ports

Customers with significant remote port requirements would benefit from a system that supports a remote equipment option minimizing the number of cabinets/carriers needed at the remote location, instead of configuring 10 or 20 remote interface boards, each with limited port capacities. However, a large remote equipment cabinet would be very expensive if the number of remote ports is limited. Growth requirements at the remote location must be taken into consideration, and the right remote equipment hardware should be selected at the initial installation.

There are several traditional design guideline issues for remote port configurations:

1. Distance between the main and the remote locations
2. Available transmission carrier facilities
3. Traffic engineering requirements
4. Available trunk access at remote location
5. Survivability requirements

Most customer configurations, including a remote port cabinet/carrier, are within a local or regional geographic area. There are few, if any, remote configurations that are transcontinental. PBX remote options using PSTN T1-carrier transmission services can support customer configurations of several hundreds of miles between the main and remote locations. The Avaya Definity ECS remote option over copper T1-carrier is available at distances of up to 3,500 miles, depending on the interexchange carrier. Several PBX systems support remote cabinets with fiber optic cabling, without repeaters, at distances between 6 miles (Nortel Networks Meridian 1, Mitel SX-2000 Light) and 22 miles (Avaya Definity ECS). Fiber optic transmission is not commonly available from exchange carriers and is used mainly for campus settings or when right of way is available.

A benefit of using fiber optic cabling to support remote locations is the high bandwidth capacity of the link. Using a fiber optic link for the Definity or Meridian 1 systems provides the same number of traffic channels between the local TDM bus and the center stage switching complex as the link available for single premises configurations. Using a T1-carrier link limits traffic channels between the locations. Although a T1-carrier circuit using DS1 formatting can support up to 24 voice-grade channels, some of the channels used for the remote option must be reserved for control signaling. Most remote options can be configured with more than one T1-carrier circuit to provide sufficient communications channel capacity. For example, the Definity ECS can support between one and four T1-carrier circuits per remote EPN cabinet, to a maximum of 96 channels. In a maximum T1-carrier implementation, four of the channels are reserved for control signaling, leaving 92 two-way channels for intercabinet voice communications. The dispersed switch network design of the Definity ECS G3r model supports local switching at the remote location, thereby minimizing the number of required center stage switch connections at the main location. The Meridian 1 can support up to three T1-carrier circuits for each remote IPEM cabinet, with 22 available one-way channels per link for all remote location call connections. The centralized switch network design of the intermediate and large Meridian 1 options requires that all remote calls be connected through the center stage switching complex at the main location, thus placing a significant traffic burden on the limited T1-carrier spans. The Siemens Hicom 300H remote communications module (RCM) option is supported with only one or two T1-carrier spans: two channels per span are reserved for control signaling, and the remaining 22 communications channels are one way. Like the Meridian 1, the Hicom 300H centralized switching design requires all remote calls be connected through the main location center stage switch.

The limitations of T1-carrier channel capacity require that traffic engineering support acceptable QoS standards. The number of available communications channels, one- or two-way, will depend on the number of remote location ports. The Definity EPN cabinet can easily support more than 1,000 ports (stations and trunks) at moderate traffic capacity, whereas Avaya usually recommends an 800 station limit. Customers must carefully analyze traffic flow to decide whether one, two, three, or four T1-carrier circuits are needed to support traffic between the remote EPN location and the main or other remote locations. The Nortel and Siemens solutions support fewer remote port capacities per cabinet carrier, but there are far fewer communications channels, and all remote call connections, including direct trunk access at the remote location, require two one-way channels for center stage switch connections.

It is desirable for station users at the remote location to have direct access to local exchange carrier trunk circuits terminated at the remote cabinet/carrier. There are many instances when local area calls placed by station users at the remote customer location would become long distance calls if routed to trunk circuits at the main PBX system. In contrast, it is usually preferable for PSTN long distance or private network calls placed at the remote location to route through the main location, where most trunks are concentrated and trunk access is more highly engineered.

If the control signaling link between the main and remote locations is down or there are transmission errors, the call processing and switching functions at the remote site will be affected, unless there are redundant design elements. The most basic redundant design option is duplication of the T1-carrier or optical fiber links connecting the main and remote locations. Almost all systems with remote port options are available with this option, although some systems have a greater degree of redundancy than others. For example, each of the T1-carrier links supporting the remote Definity EPN cabinets can be fully duplicated; the Meridian 1 design supports two active T1-carrier links and a spare T1-carrier link for emergency backup.

If redundant links are not an option or not configured, another redundant capability is local call control processing and switching capabilities at the remote location. Remote location survivability is an important issue for most customers. For example, there are many customers who support distributed call center operations across locations on a single system platform. Loss of an ACD agent group for a few minutes or hours could translate into significant revenue losses, in addition to reduced customer service levels. If a remote location is strategic to the enterprise operation, available redundant processing must be considered. Intecom, the first company to design a fully distributed PBX cabinet design, also introduced the first survivable remote cabinet option in the mid-1980s. Several other system suppliers have since shipped local call control processing options for their remote port cabinets for emergency situations when the control link to the main common control complex is not available.

Manufacturers of traditional circuit switched PBX systems have also taken advantage of ToIP technology by offering remote carrier options supported by IP-based transmission links. Avaya's R300 Remote Office Communicator and Nortel's i9150 remote options are designed to provide communications support for remote branch offices working behind a larger main location with a centralized PBX system. The R300, based on an Ascend Communications remote access concentrator, at the remote location can support analog, digital, and IP telephones and digital and analog local trunk circuit connections. The R300 is supported over an IP link supported at the main PBX location by Definity IP Media Processing and CLAN circuit boards. The Nortel remote cabinet shelf supports standard Meridian 1 peripherals and has local switching capability. A Meridian 1 supports the IP-connected i9150 using an integrated telephony gateway circuit board. A major benefit of the i9150 is that it has an integrated processor board providing backup call processing should there be link failure to the main location. The Avaya and Nortel solutions are targeted at customers with existing LAN/WAN infrastructures across customer premises locations.

Multiple System Private Networking

PBX private networking has evolved dramatically during the past 25 years. The earliest PBX networking arrangements consisted of two switch nodes linked by a dedicated, private line facility, such as an E&M tie trunk, to save on long distance toll charges. The primary benefit was cost savings. During the late 1970s more complex private tandem network configurations were made available, consisting of a meshed network of private line facilities linking tandem switch PBX nodes, main PBX nodes, and satellite systems. AT&T's first modern private PBX networking option was called Enhanced Private Switched Communications Service (EPSCS). First tariffed in the mid-1970s, it was quickly replaced by the better known and higher-performance ETN offering in the late 1970s. AT&T's innovative PBX private networking option was initially proprietary, but other leading PBX manufacturers at the time, including Northern Telecom (ESN), NEC (EPN), and Rolm (RolmNet), soon offered similar PBX options. After a few years, the competitive PBX offerings became compatible with ETN. By the mid-1980s, a customer could configure a network with a mix of PBX tandem switch nodes from multiple manufacturers.

The original ETN options were based on in-band signaling techniques supporting a network dialing plan and automatic alternate routing between nodes within the network. In addition to cost savings benefits using fixed tariff private line carrier facilities, customers enjoyed greater control over network operation and use. All of this was done initially with narrowband analog trunking facilities. The availability of digital T1-carrier trunk services in the mid-1980s would soon change the rules for PBX networking because in-band signaling would be replaced by out-of-band signaling.

The next step up the evolutionary PBX networking ladder was intelligent network signaling to support transparent feature/function operation between discrete locations served by independent PBX systems. AT&T's DCS was the first intelligent, feature-transparent PBX networking option. The first implementation of DCS required an expensive private data circuit (a 9.6-Kbps DDS line) for the intelligent signaling link between PBX network nodes, but digital voice carrier services using T1-carrier circuits made out-of-band signaling a more economic and feasible solution for implementing intelligent PBX networks. Out-of-band signaling channel PBX systems could communicate with one another at a much higher level than before. The resulting intelligent network configuration would continue to offer customers traditional network transmission cost savings but also provide significant productivity gains and additional cost savings through the use of shared application features/functions. Today's intelligent private networking options may be configured with a variety of digital trunk services based on a variety of transmission and signaling protocols, including ATM and IP.

Multiple PBX system private networks are designed and implemented for a variety of reasons:

1. Reduction of communications expenses (transmission services, maintenance and administrative support)
2. Improved management and control of communications operations
3. Enhanced communications capabilities among station users
4. Shared applications resources

Reduced costs are always a motivating factor for any business venture, and a private network configuration can have a significant effect on monthly telecommunications service bills and personnel outlays. With private lines or a virtual network service offering, PBX networking features can cost-effectively route calls between endpoints over least-cost routes, minimize trunk circuit requirements, monitor and diagnose transmission carrier facilities for efficient operation and problem areas, and minimize support personnel through centralized management centers. PBX traffic carried over private line facilities incurs a fixed monthly cost regardless of the amount of traffic. Private tandem networks are based on concentrated trunk carrier interfaces at select switching node locations; large trunk groups translate into greater traffic handling capacity per individual trunk circuit (see Chapter 4). If a virtual networking service is used, interexchange carrier tariffs reward customers by decreasing billing costs as traffic volume increases, making it an attractive alternative for customers who cannot financially justify dedicated private line facilities.

When the customer communications system infrastructure expands in size and geographic scope, it is vitally important that there is a sense of uniformity across the network. Network heterogeneity requires fewer management tools and reduces personnel and training expenses. Network heterogeneity allows for the dispersion of system administrators across the enterprise, with the capability to provide individual administrators with global responsibilities regardless of their physical locations. Dispersing administrators means fewer administrators because there is no need

PBX Networking

to have local administrators dedicated to a single system. Network administrators can be centralized in a single location for an entire network of communications systems or distributed among several strategic locations, according to the organizational policies of the corporate enterprise.

The fundamental characteristic of a centralized network management center is the support of multiple communications systems and associated administrative/maintenance functions. Integrated network management service tools include fault management, system administration, traffic analysis, directory services, call accounting, support inventory management, and database import/export. Centralization of resources, combined with remote management services, can reduce personnel requirements, the largest single cost item in operating a communications system network.

Fundamentals of PBX Private Networks

The basic elements of a PBX private network are PBX switch nodes and tie trunks. *Tie trunks* are telecommunications channels that directly connect two PBXs. The first analog transmission tie trunks were known as E&M interface signaling trunks. The term E&M originally comes from the works *earth* (earth grounding) and *magnet* (electromagnetic generated tones). The letters E&M have also become known as "ear" and "mouth" or "receive" and "transmit." E&M supervision signaling (on-hook/off-hook signaling) is used for a variety of networking operations, but the most basic function is to pass address signaling (called *party number*) between two endpoint PBXs in the private network. The most basic private network consists of two PBXs and at least one E&M tie trunk.

During the mid-1980s, E&M signaling was incorporated into T1 circuit digital private line services. In the early 1990s, ISDN PRI services were the primary trunking services used for private network tie trunk operations. By the end of the decade, IP-based trunk services could support traditional E&M signaling capabilities for inter-PBX networking requirements.

The first large private networks consisting or three or more PBX systems were known as tandem tie trunk networks (TTTNs). Each PBX was assigned a location code, and each station was assigned a private network extension number. TTTNs were based on a nonhierarchical network of tie trunks. A station user at the originating switch would dial the location code of the destination switch and wait to receive another dial tone signal before dialing the desired extension number. The request for dial tone was carried from switch to switch over E&M tie trunks. The private lines also were used to provide the dial tone signal back to the calling station user.

The first smart ETNs were designed in the late 1970s. An ETN was an enhanced version of a TTTN. It was based on a hierarchy of tie trunks and PBX switching nodes, with multiple call routes between network endpoints. A major innovation was an automatic alternate routing (AAR) program that selected a call route based on the number dialed and the most economic (or preferred) route available at the time the call was placed. The tandem switch node in the early ETNs was equipped with routing tables for determining the best route for on-net calls, had the capability of modifying outpulsed digits (for rerouting and directing calls), and could allow or deny call routing privileges to certain station users. Switch nodes in the ETN one level below the tandem switches are known as *main switches*. Main switches have trunk circuits for local CO switching access, but all network calls must be routed through a directly linked tandem node. Switches working behind the main nodes are known as *satellites* and *tributaries*. These switches are equipped with tie trunks to the main node only. They lack local CO trunk services and attendant operator positions. CO and private network access is through the main switch for all off-premises incoming and outgoing calls.

PBXs with ETN options support the important features described in the next sections.

Uniform Dial Plan (UDP)

A UDP provides for a common multidigit (usually four or five) dial plan that can be shared across a group of networked PBXs for interswitch and intraswitch dialing. The UDP includes a network location code, comparable to a CO code, and a multidigit extension number. The UDP extension number is not necessarily the same number as the station directory number. The network location code is often designated as the RNX; the extension number as XXXX. The network location code determines how the call is routed. Whenever a UDP is used to route a call, the number it outputs is in the form of RNX-XXXX. This must be taken into account so that the correct digit deletion or insertion can be specified within the routing pattern, so that the receiving switch receives digits in the format it expects. The UDP software program automatically translates the RNX-

XXXX number to the directory number associated with the called station for digit analysis and routing by the destination PBX system.

Automatic Alternate Routing (AAR)

AAR provides alternate routing choices for on-net calls carried over the private network. Based on a routing table designed by the customer, the AAR software automatically selects the most preferred (usually the least expensive) route over the hierarchical tie trunk network. If the first route in the program table is not available, another route may be selected if the station user calling privileges warrant a more costly route based on the individual's FRL. Large PBX systems can support several hundred different routing patterns (originating and terminating nodal pairs), with more than a dozen different call routes per routing pattern. The AAR routing patterns are shared with the ARS feature.

AAR also provides for digit modification to allow on-net calls to route over nonprivate PSTN trunk facilities when an on-net call route is not available. Calls rerouted off the network require digit modification to translate an RNX-XXXX number to a direct distance dialing number for national or international calling. On-net calls that are routed off-net are then controlled by the ARS feature of the PBX system.

FRLs and Network Class of Service (NCOS)

FRL and NCOS are features that provide for multiple levels of restriction for users of the AAR and ARS features. FRLs and NCOS allow a certain call type to a specific station user and deny the call to another user. For example, some station users may be allowed to place calls only to private network nodes and not off-net long distance toll calls. A call type that is highly restrictive is international direct distance dialing. FRLs and NCOS are transparent to the station user and are assigned and programmed by the system administrator. Each system facility, station and trunk, is assigned an FRL. Whether or not a call is allowed is based on the originating caller's FRL and the availability of idle trunk circuits within an assigned trunk group required to route the call. If the station user's FRL is equal to or greater than the trunk group FRL, calls may be routed using a trunk circuit in the trunk group. If the station user's FRL is less than the trunk group FRL, the call is denied. Most intermediate/large PBX systems support at least eight FRLs (0 to 7).

The NCOS level of the originating station user determines which call routes can be used to complete the call for a specific routing pattern while the call route is being established across the network. NCOS is also known as *traveling class mark* (TCM) because the assigned restriction level of the originating station user is embedded within the voice communications signal as the call is routed across the network. If trunk facilities at one tandem switch node are busy and an alternate route (more expensive) is available, the NCOS/TCM level determines whether or not the station user is allowed access to a different call route. A call can be blocked or denied anywhere in the network. Another private network feature, advanced routing (*look ahead routing*) can be used to avoid possible blocked calls at tandem switch nodes.

Automatic Circuit Assurance (ACA)

ACA is a feature that helps customers identify trunk errors and problems, particularly for private tie trunk circuits. The PBX system maintains a record of individual trunk performance relative to short and long holding-time calls. Holding time is the period from trunk access to trunk release. A system administrator defines short and long holding-time limits. When a possible trunk circuit failure is detected, the system automatically initiates a referral call to an attendant, station user with a display-based voice terminal, or system manager with a CRT monitor. Based on system measurements of holding times per call, referral calls may be placed to attendant, station, or system manager positions. The display information identifies the call as an ACA call, identifies the trunk group access code and the trunk group member number, and shows the reason for the referral (short or long holding time). The ACA feature provides better telecommunications service through the early detection of faulty trunk circuits and possibly reduces out-of-service time.

Virtual Private Networks (VPNs)

A VPN is based on a custom switched telecommunications service tariffed by an interexchange carrier, which permits a customer to establish a communications path between two stations using a UDP. PBX systems are linked to the carrier's CO facilities using private tie trunks or through dial-up facilities using special access codes. The network facili-

PBX Networking

ties are shared as part of the PSTN. The key benefit of a VPN is a significant reduction in private line services between networked PBX systems. The first voice communications VPN service was AT&T's SDN. SDN was designed to expand the scope of private network solutions to customers who could not justify dedicated private line services in support of their private networking requirements. Other similar services were soon available from AT&T's competitors. VPN telecommunications services, from exchange carriers such as AT&T, MCI Worldcom, and Sprint, simplify private networking applications for business customers because:

1. The backbone private network carrier facilities are maintained and managed by the service provider.
2. The UDP and AAR/ARS databases are centralized in the service provider service control points.
3. Less training and specialized communications equipment is required to design, implement, operate, and maintain a private network.
4. Call accounting records and billing information are provided by the service provider for all on-net and off-net calls.

Most customers with private network requirements use a combination of traditional ETN tie trunk links (analog, mostly digital) and virtual network services for their on-net and off-net calling requirements. Private networks using VPN services can enjoy the following benefits:

- Flexible system and station user port capacity
- Integrated voice/data communications
- System compatibility across different PBX platforms
- Cost effective usage- and distance-sensitive pricing
- Porting of private network numbering plans
- National/international service transparency
- PSTN reliability and QoS standards
- Customized report options
- Centralized network management capabilities
- Secure access via screening provisions and enforced use of authorization codes

There are several options available for PBX access to the carrier network, including local exchange service access and special services access. Local exchange access is usually reserved for very low-traffic volume locations that cannot justify a special access service. Local exchange access is provided through a switched connection from a local exchange carrier's equal access end office. A customer using this access method may presubscribe to the VPN carrier code, or individual station users can dial the carrier code.

Special access arrangements provide direct access to the VPN provider network facilities:

- PSTN analog trunk circuit facilities
- PSTN digital trunk circuit facilities
- Customer-provided access (local bypass)

Digital trunk and customer-provided access may support multiple exchange carrier services, not only VPN service.

The VPN service functions similarly to private line ETNs for on- and off-net calls. Basic call processing features include:

- Seven-digit dialing (national and international calls)
- Advanced numbering plan (flexible multidigit numbering plans)
- Private network interface to ETNs
- Flexible routing when conditions warrant
- Network intercept announcements for call completion procedures
- Network remote access from off-net locations

VPN call management features include:

- **Authorization codes**
- **Originating call screening**—Grouping of callers with same calling privileges
- **Feature screening**—Specifies calling privileges for each screened caller group
- **Access line grouping call screening**—Call restriction by specific off-net telephone numbers
- **Partitioned database management**—Partitioning of VPN locations into multiple autonomous network groups with direct distance dialing and private networks

Exchange carrier VPNs originated as a voice-focused service. During the past few years VPN services have evolved to support voice and data communications networking applications. Most of today's VPN offerings are focused on packet switched services to support customer WAN data communications requirements, but can be used for voice networking needs. As VPN has evolved, PBX networking has evolved with the emergence of ToIP technology and the trend toward VoIP trunking.

Intelligent Feature Transparent Network (IFTN)

AT&T's original ETN offering evolved into DCS in 1982 when the U.S. Navy required a single communications system for its San Diego naval base operations, with a port capacity far greater than the parameter limitations of any single PBX system model available at the time. The AT&T solution was not to design an extremely large PBX system but to intelligently network multiple systems to provide the appearance of one system for most common internal station-to-station user operations. Originally known as the Defense Metropolitan Area Telecommunications System (DMATS), the AT&T Dimension PBX option was renamed the Distributed Communications System (DCS). The AT&T Dimension DCS option became very popular in a short period and forced competitors to develop IFTN offerings of their own. Among the other well-known IFTN brand name options developed almost 20 years ago and still marketed today are Siemens CorNet, NEC CCIS (since enhanced to Fusion CCS), and Fujitsu FIPN.

An IFTN has the property of *transparency* with respect to all on-net calling, and many feature operations. Transparency is the ability of the system, from a station user's perspective, to operate across multiple network nodes in the same way it does at the local node. This allows for a limited digit dialing plan for all on-net calls, thereby eliminating the need for PBX location codes and network extension numbers. All intercom calls are dialed with extension numbers corresponding directly to station user directory numbers.

An IFTN design is based on the ETN model, a hierarchical layer of switching systems interconnected using tie trunks. Direct links between each PBX network node are not required, but there are limitations on the number of transit nodes that can pass intelligent control signals between the originating and terminating nodes. The passing of call handling signals between network nodes is what distinguishes an IFTN from an ETN. Out-of-band common channel signaling techniques are used. Each manufacturer's IFTN offering was based on a proprietary signaling and messaging scheme, thereby limiting the flexibility of the customer network design, although the competing suppliers have worked together during the past decade to develop industry standards for open inter-PBX networking solutions (see the Qsig section later in this chapter).

The idea behind an IFTN is to have the common control complexes of multiple PBX systems communicating with each other to transmit data, customer network information, and command messages for a single system image. Specific details concerning how each PBX system implements its proprietary IFTN service offering are not available. The first implementation of AT&T's DCS offering was based on analog tie trunks and channel-associated private line data circuits for nodal transmission links. When T1-carrier circuit services became available, clear channel signaling techniques were used instead of dedicated data circuits, and analog tie trunks were replaced with voice communications transmission. Twenty-three 64-Kbps channels of the T1-carrier circuit were used as bearer voice communications channels, and one 64-Kbps channel was used for inter-PBX signaling and messaging.

When ISDN PRI services became available in the late 1980s, the B-channels were used for voice communications and the D-channel was used to transport the control information. The most common signaling method used for IFTN networks, based on ISDN PRI service circuit links, is a temporary signaling connection (TSC). A TSC provides a temporary signaling path through ISDN switches for exchanging information between users. There are two TSC types: call associated CA-TSC and non–call associated NCA-TSC.

A CA-TSC refers to a service for exchanging user information messages that are associated with an ISDN B-channel connection by the call reference value of the control data packets. An NCA-TSC is a connection not related to any ISDN B-channel connections. It is an administered virtual connection established for exchanging user information messages on the ISDN D-channel. Once the NCA-TSC has been administered and enabled, it will be active for an extended period. There are two NCA-TSC types: permanent and as needed. A permanent NCA-TSC will remain established while the system is operating. If the connection is lost, an attempt will be made to re-establish it. An as-needed NCA-TSC is established based on user request and the available of TSC facilities. The connection is dropped after a preset period of inactivity.

ISDN PRI transmission services are currently the most commonly used communications and signaling transport links for IFTNs. Some IFTN offerings, such as Siemens CorNet, can be supported only when using ISDN PRI service circuits for circuit switched connections. Most PBX IFTN options also can be implemented with virtual networking services supporting a TSC, such as AT&T's service offering. Other network transmission solutions that support inter-PBX message signaling are ATM trunk carrier services and TCP/IP over a LAN/WAN infrastructure. ATM networking options include T1-carrier CES and TDM/PCM conversion to ATM cell format for transmission over an ATM WAN. An important advantage of the TCP/IP networking option is that dedicated point-to-point signaling links between PBX network nodes are not required because point-to-multipoint signaling is supported by TCP/IP. Tandem switch nodes, the basic network element of circuit switched IFTNs, are not required if IP trunk circuits are used to pass communications and control/message signaling between switch nodes. This eliminates the transit node (hop-though) limitation for control signaling between originating and terminating switch nodes. There are several other advantages to using a LAN/WAN infrastructure as the IFTN network backbone:

1. Reduced PSTN trunk carrier services in support of IFTN networking result in potentially significant cost savings. Using existing IP trunk circuits to carry IFTN voice traffic and signaling between switch nodes eliminates the need for dedicated private line and/or ISDN PRI digital trunk circuits. Fewer physical T1-carrier trunk circuits are needed to carry voice traffic over an IP network because VoIP transport uses voice encoding and compression algorithm standards.
2. PBX system hardware costs decrease because fewer port circuit cards and port cabinet carriers are required. A single IP trunk circuit card can support a greater number of physical IP-based trunk circuit equivalents at a lower cost than traditional TDM/PCM station/trunk port circuit cards.
3. Signaling support is to and from a single LAN-connected applications server across multiple PBX systems, instead of dedicated servers at each location.
4. Network management and maintenance operations are simpler because a single converged voice/data network, instead of dedicated networks, is used to transport voice and data communications. An added benefit is a reduction in human and equipment management support resources.

IFTN Features and Functions

There is no standard level of IFTN feature/function transparency within the PBX industry. Some PBXs support a very high percentage of features and functions across multiple networked systems, up to 90 percent of the total generic software program, and some support less than 50 percent. Almost all PBX system IFTN options support the following basic feature/functions:

- Basic calling with the use of a flexible dialing plan (typically four or five digits)
- Voice terminal display information (calling party/called party name and number, call redirection information)
- Call forwarding services
- Call transfer
- Call conferencing
- Automatic callback
- Bridged call appearance
- Message waiting indication
- Trunk release
- Network-wide attendant services
- Network-wide CDR

An important category of features supported by only a few IFTN solutions is ACD. For example, all 55 of the identifiable NEAX 2400 IPX ACD features are available with Fusion CCS. The NEC ACD Agent Anywhere option is an intelligent network of multiple ACD systems using Fusion CCS links. ACD nodes can communicate with each other and pass and interpret signaling, caller ID, call prompt, and database information across the network. Intelligent interflow routing of callers between nodes improves customer service levels, balances traffic load, and optimizes agent productivity. Fusion CCS also supports centralized management reporting and supervisor workstation data screens. A multiple system ACD network has built-in redundancy to reduce system down time and increase customer satisfaction.

The Agent Anywhere option supports distributed ACD agents behind switch nodes remote from a centralized ACD processor node. ACD agents can be deployed anywhere within a Fusion CCS network, with the only restriction being that the remote switch node be directly linked, (no pass through signaling), to the control switch node. Agent Anywhere can be implemented when using internal or external ACD processing and software options. The Fusion CCS solution supports decentralized agents assigned to the same call split across multiple nodes. Incoming calls to the central control node can be routed to available agents in remote nodes if all agents are busy at the central location. For configurations with local incoming trunking at remote node locations, calls can be queued at the central control node location when no remote agents are available.

Many customers' ACD-based call center systems include a CTI application server. A centralized CTI application server capable of supporting more than one PBX/ACD switch node is less costly and easier to manage and maintain than application servers at each customer location. The NEC Fusion CCS option supports a centralized CTI application server for ACD systems. Another good example of a centralized application server solution used within an IFTN configuration is the Siemens HiPath Allserve 150. A centralized Windows NT application server can support a network of one to four Siemens Hicom 150H systems networked with the Siemens CorNet option implemented over an IP LAN/WAN. The applications, run on a single server for all networked PBXs, include messaging, call center, personal call manager, and call accounting.

Perhaps the most important transparent system operations are management and control from a centralized application server. A central database for all Hicom 150H switch nodes resides in the application server. The centralized server provides one access point to administer and maintain each system. Station move/add/change transactions are implemented as if there were a single system, not multiple switch nodes. A single management or maintenance command can be applied to all switch nodes across the network, instead of inputting individual changes to individual systems. Centralized management system capabilities for all move, add, and change procedures is an important IFTN capability that is not commonly supported by most traditional circuit switched PBX IFTN options but is supported by more of the newer client/server IP-PBX system designs, such as Cisco's AVVID IP telephony system.

Shared applications resources for call center, messaging, and management operations are an important IFTN cost savings benefit. The first IFTN offerings were limited to shared VMS applications. One VMS supported station users across a network of PBX systems. One of the cost savings components is attributable to the lower price for a single, very large messaging system as opposed to the collective cost of several smaller systems with equal voice mailbox and storage capabilities. Another cost savings component is ongoing management and service. Maintaining one messaging system is less costly and more efficient than maintaining several systems. The same cost savings criteria can be attributed to other shared application resources in an IFTN configuration.

Qsig

Qsig is an inter-PBX signaling system designed for multiple PBX system platform networks. The proprietary nature of IFTN solutions restricted customer configuration flexibility to a single supplier's product platform. Qsig in its current form originated during the 1990s as a standardization effort by the IPSN Forum, a group of Western European PBX equipment suppliers, with Siemens and Alcatel at the forefront of the movement. IPSN work efforts were handed off to the ECMA and the International Telecommunications Union (ITU) for the formalization of issuing standards and specifications. Qsig is based on the ITU's Q.93x series of recommendations for basic services and generic features and Q.95x series for supplementary services.

The major benefits for developing Qsig were outlined in the Qsig handbook originally published by the IPNS Forum.

Vendor independence. The nonproprietary nature of Qsig, based on open international standards and supported by all of the leading global PBX suppliers, allows customers to configure an intelligent communications system network when using equipment from more than one supplier.

Guaranteed interoperability. A memorandum of understanding (MoU) signed by the leading global suppliers signifies commitment to Qsig specifications, facilitates interoperability performance tests, and assures customers that they will be able to operate communications networks with a mix of supplier equipment.

Free-form topology. Qsig does not impose the use of a specific network topology, so it can be implemented with any network configuration: meshed, star, main/satellite, etc. Existing networks can, regardless of their topology, be upgraded to Qsig. Newly designed networks can be installed with the most effective and economical topology.

Unlimited number of nodes. There are no nodal limits for a Qsig network. New nodes can be added as needed.

Flexible numbering plan. Qsig does not impose any number plan restrictions for the network, thereby allowing customers to freely adopt customized numbering plans.

Flexible interconnection. Qsig will work over any type transmission network for linking PBX systems, including two- and four-wire analog tie lines, digital leased lines (including ISDN PRI and BRI), radio and satellite links, and VPN services provided by interchange carriers. Associated transmission delays are managed and controlled according to Qsig specifications.

Public ISDN synergy. There is network service compatibility between public and private ISDN transmission facilities. Applications developed for desktop terminals connected directly to a public ISDN network will also be available to desktop terminals provisioned within the Qsig-based customer private network.

Supplementary services for corporate users. Qsig supports private communications features beyond those defined for public ISDN networks, including caller name ID, call intrusion, do not disturb, path replacement, operator services, mobility services, and call completion on no reply.

Feature transparency. Features and functions supported by any network node can be transparently supported across the network to station users configured behind other network nodes. Qsig is structured and organized to adapt to service levels offered by different PBX systems, and it allows each network node to provide only the required level of service. There is an exchange of high-level services between any two nodes, via transit nodes with lower service levels: transit nodes pass communications and control signals between systems.

Innovation. Qsig does not restrict individual PBX manufacturers from developing customized, unique features. A special mechanism within Qsig, generic functional procedures (QSIG GF), provides a standardized method for transporting nonstandard Qsig features. As defined in Qsig GF, the basic rules related to feature transparency allow end-to-end communication through the network, regardless of network structure. Qsig does not prevent the use of innovative, proprietary system features across the customer network and allows for customized new feature development negotiation between PBX suppliers and customers.

Multiapplication domain. Qsig is not restricted to PBX systems and can support applications requiring other peripheral communications equipment, such as VMS, fax servers, data processing equipment, and multipoint conferencing systems.

Evolution. Qsig has an evolutionary path to support communications features, functions, and applications that are developed in the future.

Qsig Architecture

Qsig standards specify a signaling system at the ITU-T ISDN "Q" reference point, which is intended primarily for use on a common channel, although Qsig can be implemented over any suitable inter-PBX connection platform. The "Q" reference point, the logical signaling point between two PBXs, was defined explicitly for the Qsig. The physical connection point to the PBX system is made at the "C" reference (also a new ISDN reference point). There are three sublayer protocols at Layer 3, including the Qsif GF procedures. Qsig GF protocol provides a standardized mechanism to exchange signaling information for the control of supplementary services and additional network features (ANFs) over customer networks.

Qsig basic call (BC) message sequence is an intermediary transit node linking two endpoint PBX systems. Qsig BC is a symmetrical protocol designed for peer-to-peer networks, and it includes transit node capability.

ECMA also has been working on enhancements to its Qsig specifications to support broadband PBX networks. B-Qsig is an extension of Qsig, using many standards as possible available from the ITU-T and ATM forums.

Qsig Supplementary Services and ANFs

The following is a listing of the Qsig supplementary services and ANFs:

- Advice of charge
- Call completion
- Call forwarding and diversion
- Call interception
- Call intrusion
- Call offer
- Call transfer
- Call waiting
- Direct dialing in
- Do not disturb

- Identification services
  - CLID presentation
  - Connected line identification presentation
  - Calling/connected line identification restriction
  - Calling name identification presentation
  - Calling/connected name identification restriction
- Mobile
- Multiple subscriber number
- Operator services
- Path replacement
- Recall
- Subaddressing
- User-to-user signaling

CHAPTER 15

PBX Systems Management and Administration

The well-being of a PBX system is the responsibility of the system administrator. Among the administrator's duties are:

- Administering system port moves, adds, deletions, and changes
- Performing system backup procedures
- Monitoring system performance
- Maintaining system security

After the system is installed and initialized by a vendor, a system administrator can manage and monitor a PBX by using programming tools in the system administration terminal (SAT). SATs have evolved over the years from simple teleprinters, to dumb CRT terminals, to the current PC workstations. System administration access design should have the following elements:

- A color display terminal with a graphic screen format, a menu-driven interface, and online help
- Multiple configurable and active SATs
- Multiple PBX systems support
- Quick and easy access to all station configuration tables
- Formatted screens
- Pull-down menus
- Valid entry choices
- Templates
- Batch processing and transactions scheduling
- Database import/export
- All administration changes written to a core system database
- Fast system response to all administrative programming
- Open system format (SNMP) in support of converged voice and data communications

SATs may be linked directly to the PBX system through an integrated I/O interface, a remote dial-up modem via a RS-232 port connection, or a TCP/IP LAN connection. In addition to a stand-alone SAT, there are two configuration design options. Very large customer installations (single or multiple systems) may support a client/server administration management system, with all design elements—PBXs, management server, and PC clients—linked over a LAN/WAN infrastructure. Another option is Web browser access to system administration programming tools residing in the main PBX. Using Web browser access, any PC with Internet access can be used as a SAT with valid password entry. Table 15-1 summarizes the system administration design options.

PBX Systems Management and Administration

TABLE 15-1
System Administration Design Options

SAT Design Option	PBX Access Method	Administration Software
Direct-link SAT	I/O port, dial-up modem, or LAN	Client based
Client/server system	LAN/WAN	Server based
Web browser access	Intranet/Internet/dial-up	PBX based

System Administration

PBX administration is the process of using system commands to perform a variety of functions to meet customer communications requirements. The basic administration management functions include: system port moves, additions, and changes (MACs); programming of system subscriber and terminal parameters; display configuration parameters; list system subscribers; duplicate customer database; perform system measurements; monitor security conditions and alarms; and configure and monitor system performance.

Recordkeeping also plays an important role in system administration. Records provide a current status of which hardware and features have been installed. For example, the port assignment record provides a view of how a system is initialized and administered. Ports are the physical location on a circuit pack where terminals, trunks, or system adjuncts are connected. Once port numbers are assigned, they become the "address" of the associated equipment or facility in the system. It is necessary that a record be made and kept of all port assignments for system installation and initialization and ongoing administration.

Administration Sequence

After the system is installed, the system administrator must enter the translation data into the system memory via the SAT. Translation data is taken from survey sheets and previous system records and provides a blueprint of what needs to be programmed into the system configuration. When entering the translation data into the system, the system administrator should periodically save the translations on tape. This creates a nonvolatile copy of the translation already entered into the system. If a power outage or system failure occurs, the translation data saved on the tape will not have to be entered again.

PBX system features should be entered into the switch in an ordered manner. The following is the recommended order in which data should be entered into the system:

1. Login and password (change password, if necessary)
2. Dial plan
3. Feature access codes (FACs)
4. System features (class of service and class of restriction)
5. Console parameters
6. Attendant consoles
7. System parameters
8. Voice terminals
9. Data modules
10. Network connection channels
11. Bridged line assignments
12. Group assignment (hunt groups, call coverage, pickup groups, etc.)
13. Trunk groups
14. Paging/code call zone assignments
15. ARS table

Before the customer database data is programmed into the system, the system administrator should review the system hardware configuration to assess the available port circuit interface boards and design layout. Many administration interface screens will require the administrator to input a port or slot identifier. A port or slot is an address that describes the physical location of the installed equipped. Port addresses consist of cabinet, port carrier, card slot, and port circuit card termination identifiers. Each hardware component has a multidigit identifier, and the combination of the hardware component identifiers is the port address.

Dial plan and FACs must be administered before voice terminals, hunt groups, pickup groups, coverage groups, and attendant consoles can be administered. Default values for the dial plan can be changed if they do not meet customer requirements. A standard dialing plan usually supports four- or five-digit extension numbers, but some customers may require more extension digits for system subscribers. For example, a seven-digit dial plan may be required for multisystem intelligent transparent private network configurations. Default FACs can also be changed, but the number of digits assigned to the FAC must agree with that of the dial plan.

The dial plan is used by the system to interpret dialed digits and know how many digits are expected for different call types, such as intercom calls or trunk calls. An important element of the dial plan is the first-digit table. The first digit dialed by a station user may have any one of the following codes: attendant, dial trunk access, extensions, feature access, and miscellaneous (used when more than one code begins with the same digit and requires a second-digit dial table).

Regarding first-digit dialing, North American station users are accustomed to dialing 9 for dial trunk access and 0 to reach an attendant console. In most of Western Europe, the first-digit dial trunk access code is 0, which usually proves confusing to American tourists. Many station users do not understand that access codes are programmable by system administrators, although most systems use the default values programmed by the manufacturer, such as 9 for trunk access in North America.

Miscellaneous codes are usually required when there might be a problem interpreting the first dialed digit. For example, when local 911 emergency services were first introduced, major PBX problems occurred. PBX systems that were programmed to recognize 9 as the first-digit dial trunk access code did not recognize the second dialed digit, 1, as a valid area or exchange code. System administrators were forced to reprogram their first digit tables to interpret a 911 call. Similarly, the revised North American Dialing Plan (NADP) introduced in the mid-1990s forced a reprogramming of the dial plan because trunk calls outside of the local area code required dialing a 1 before an area code for interpretation by local central office switches, and digit restrictions for the second area code digit (previously 0 or 1) were modified. Continuing changes in PSTN dialing requirements require constant updates of PBX dial plans.

Once the dial plan and FACs have been assigned, the system administrator can add voice terminals to the system. A variety of programming commands simplifies the configuration process. For example, the duplicate command can add the same types of voice terminals, instead of repetitive programming of similar information. The terminal extension number, location, type, and user name are entered on the display form with labeled blank fields. For an IP-PBX system, IP voice terminals require similar data entry for the voice station display screen, but also require entry of IP addresses, MAC addresses, and voice codec information. IP addresses are usually assigned by a DHCP server but can be manually administered. QoS programming for IP voice terminals is the responsibility of the LAN administrator, as is performance monitoring of IP telephony metrics, such as call delay and packet loss.

A common misconception is that IP-PBX systems don't require traditional MAC administration because IP voice terminals can be initialized via a DHCP server or physically moved to different LAN connector outlets without administration programming. Despite these capabilities, subscriber and voice terminal parameters must be input for all IP peripherals. With regard to moves, IP-PBX system IP voice terminals require similar data entry for the voice station display screen, but also require the entry of IP addresses, MAC addresses, and voice codec information. IP addresses are usually assigned by a DHCP server but can be manually administered. All PBX systems support a customer station rearrangement feature that allows the movement of digital telephones between telecommunications outlets without administration intervention, exactly like an IP terminal.

Attendant consoles must be added one at a time, and for reliability, attendant consoles should not be assigned to the same circuit pack. Data modules can be assigned after voice terminal administration. Some data modules must be added during voice terminal administration if the voice terminal has a data module. Other data modules can be added separately.

Network connection channels are used to provide switched data access for the following features and functions:

- CDR
- SATs
- Remote SATs
- Property management system (PMS) link
- PMS log printer
- Journal printer
- Recorded announcements
- System printer

Group assignments can be programmed after voice terminals are added. The following groups can be administered:

- Abbreviated dialing (system, group, enhanced)
- Hunt groups
- Call coverage answer groups
- Pickup groups
- Intercom groups
- Terminating extension groups
- Trunk groups

The ARS tables support network access to the PSTN and private networks. Trunk groups must be programmed to ARS. Access to private network facilities include the following network interface types:

- DS1 interface
- TTTN
- Private tandem network
- ISDN PRI
- VPN

Performance Management

PBX performance management records and reports are typically available for the following system measurements.

Trunk Usage and Traffic

Trunk traffic records are kept for all inbound and outbound calls and identify the trunk group and trunk channel, time of call, and duration of call. Individual trunk line counters can measure the number of call attempts, blocked trunk lines, and traffic intensity (Erlangs). Outgoing counters can measure the number of outgoing attempts, successful calls overflowing to another route, calls lost due to blocking, blocked trunks in measurement, and traffic intensity (Erlangs). Incoming trunk route counters can measure the number of incoming call attempts, trunks in the measurement, number of blocked trunks in the measurement, and traffic intensity (Erlangs). Similar statistics are measured for two-way trunk routes.

Attendant Consoles

Attendant counters can measure all attendants in the system or individual attendants positions. Record measurements include number of answered calls, number of calls initiated by attendant, accumulated handling time for all calls, accumulated handling time for recalls, accumulated handling time for calls initiated by attendant, accumulated total delay time for recalls, number of answered recalls, number of abandoned attendant recalls, accumulated waiting time for abandoned calls to an attendant, accumulated waiting time for abandoned recalls, and accumulated response time for all types of calls.

Stations

Station counters can measure individual stations or station group traffic statistics such as number of calls, number of stations in the measurement, number of blocked stations in the measurement, and traffic rating (Erlangs).

Traffic Distribution

Traffic distribution across the internal switching network can be measured for each local TDM bus, traffic over each highway bus, and traffic across the center stage switch by each switch network interface link.

Busy Hour Traffic Analysis

Busy hour traffic analysis measurements for trunks, stations, and the internal switch network can be performed. Busy hour traffic intervals can be programmed for any time of day. Erlang ratings are calculated for individual trunk lines, individual trunk groups, and all trunk groups. CCS ratings are calculated for individual stations or groups of stations.

Processor Occupancy

System call processing performance is measured in terms of BHCs (attempts and completions). The percentage of maximum call processing capacity is reported for programmed intervals. Threshold reports can be generated to monitor system load factors.

Threshold Alarms

For a variety of system hardware devices, it is possible to define a congestion threshold value and measure generated alarms. Alarms are recorded in an alarm record log. The types of devices that can be tracked are tone receivers, DTMF senders and receivers, conference bridges, trunk routes, and modem groups.

Feature Usage

Feature usage counters for selected station features (e.g., call forward, call transfer, add-on conference) and attendant system features (e.g., recall, break-in) can be measured and reported for programmed intervals.

VoIP Gateways

IP-PBX systems collect and store data to track the usage and performance of IP gateway devices, IP phones, and VoIP trunk calls. VoIP information reports include tracking of IP gateway devices and calls that pass through each gateway, gateway congestion, assignment of services or routes to gateways, tracking of phone numbers dialed or originating off-site numbers, and IP gateway addresses.

CDR

CDR data is compiled for all successful incoming and outgoing trunk calls. Call records can be stored in multiple formats (fixed and programmed) per output device. Fixed formats typically conform to standards published by leading call accounting software suppliers, or are proprietary to the PBX system. Programmable formats provide a flexible means to incorporate new data elements in the call record. A variable format allows a record to be defined in terms of its content (from a set of available data elements) and the position of the data elements in the record. This method can be used to construct custom formats.

A system administrator may define programmable CDR formats based on available CRD field data records. Call record fields typically include:

- Date
- Time
- Call duration
- Condition code (categorizes information represented in the call record)
- Trunk access codes

- Dialed number
- Calling number
- Account code
- Authorization code
- FRL for private network calls
- Transit network selection code (ISDN access code to route calls to a specific interexchange carrier)
- ISDN bearer capability class
- Call bandwidth
- Operator system access (ISDN access code to route calls to a specific network operator)
- Time in queue
- Incoming trunk ID
- Incoming ring interval duration
- Outgoing trunk ID

Reports can be generated for any or all of the call record field data.

CDR data is not usually compiled for intraswitch calls (station to station, station to attendant), calls terminated by busy signals, and calls with no answer. When CDR was introduced as a system option in the early 1980s, memory storage was expensive. Any call that did not incur a direct expense was not recorded and stored. Today, PBX systems based on nontraditional designs, such as CTI-based server systems, can collect, store, and process these data records for reporting purposes. Traditional circuit switched PBXs may optionally record a limited number of intraswitch calls for select calling stations or capture data for all calls using optional CTI solutions.

CDR records and call accounting reports are vital to the monitoring and management of the PBX system. It is important to monitor call costs and usage to:

- Bill system subscribers for their communications network use
- Budget and allocate usage charges by department
- Resell telephone services to outside clients
- Monitor PBX effectiveness
- Gather statistical data for performance benchmarking
- Prevent or minimize telephone system abuse and unauthorized access
- Verify monthly service provider bills

There are several optional PBX reports that are useful to system administrators.

Directory

Directory records can include each subscriber's name, with a variety of phone numbers such as primary, published, listed, emergency, and alternate, and authorization code information, job title, employee number, current employment status, and social security number.

Inventory

Inventory records and management is used to administer any kind of inventory product part: PBX common equipment (cabinets, carriers, circuit cards), voice terminals and module options, jacks, and button maps. The reports allow administrators to accurately re-charge items. Inventory can be tracked by data such as user, system (PBX or other networks), jack, serial number, asset tags, trouble calls, recurring and nonrecurring costs, and general ledger codes. The inventory management system also includes records containing the following data: purchase date, purchase order number, depreciation, lease dates, and manufacturer and warranty information.

Cabling

Cabling records keep track of all cable, wire pairs, distribution frames, wiring closets, and all connections (including circuits) down to the position and the pair level. Records include starting and ending locations, description, type, and function. Individual cable lengths are maintained and automatically added, as is the decibel loss for the entire path. Information can be provided on the status of all cable runs and the number of pairs they contain, the status of the pairs, and the type of service they provide.

System Diagnostics and Maintenance

The primary objective of system maintenance is to detect, report, and clear trouble as quickly as possible with minimum disruption of service.

Periodic tests, automatic software diagnostic programs, and fault detection hardware allow most troubles to be traced to an individual assembly in the system. System diagnostic functions include:

- Monitoring of processor status
- Monitoring and testing of all port and service circuit packs
- Monitoring and control of power units, fans, and environmental sensors
- Monitoring of peripherals (voice terminals and trunk circuits)
- Initiating emergency transfer and control to backup systems
- Originatng alarm information and activate alarms

There is a specific maintenance strategy and plan for each of these hardware elements monitored by the system.

The maintenance subsystem software is responsible for initializing and maintaining the system. This software continuously monitors the system and maintains a record of detected errors. The maintenance subsystem also provides a user interface for on-demand testing and contains two general categories: system alarm troubles that are automatically reported to a local maintenance terminal or a remote maintenance center and user-reported troubles resulting from service problems at individual station user terminals.

The major part of maintenance is system-alarmed troubles. PBX system diagnostic circuitry detects and reports most problems automatically. When the trouble is repaired and no longer detected, the alarm is automatically retired. It is not necessary for personnel to retire alarms after a problem is corrected. Dedicated maintenance circuit packs or daughterboards are used in fault detection and repair at many system levels, including the common control complex, expansion cabinets and carriers, and a variety of trunk interface cards, particularly those used for digital trunk connections. Almost all circuit packs have LED indicators to indicate alarm conditions (red) if the system has detected a fault in that circuit pack. A yellow alarm condition indicates the system is running tests on that circuit pack, and a green condition indicates that the circuit pack is operating without problem. In-line error detection circuitry checks for correct operation.

Maintenance tests can be periodic or on demand. Periodic tests run automatically at fixed intervals on a specific schedule. Usually, short tests are run hourly or less; long tests are run every 24 hours. Demand tests are run by the system when it detects a need or by personnel when required during trouble-clearing activities. Demand tests include the periodic tests, and other tests are required only when trouble occurs.

Some nonperiodic tests may be disruptive to system operation. From a terminal, personnel can initiate the same tests the system initiates, and the results are displayed on the terminal screen.

If any part of the system fails any portion of the periodic tests a preset number of times, the system automatically generates an alarm. There are three alarm types common to most systems:

1. **Major alarms**—Failures that cause critical degradation of service and require immediate attention.
2. **Minor alarms**—Failures that cause marginal degradation of service but do not render a crucial portion of the system inoperable. This condition requires action, but its consequences are not immediate. Problems that cause minor alarms might be impaired service in a few trunks or stations or interference with one feature across the entire system.
3. **Warning alarms**—Failures that are localized and cause no noticeable degradation of service. Warning alarms are not reported to the attendant console or a remote location.

The PBX system can usually send an alarm to any customer device such as a light, automatic dialer, a bell, or other equipment. The alarm activation level field on the system parameters maintenance screen must be administered to indicate the alarm level (major, minor, warning, or none) that activates the alarm device.

If the maintenance software detects an error condition related to a specific maintenance element, the system will automatically attempt to repair a problem or operate around it. If a hardware component incurs too many errors, an alarm is generated. Records of each error and alarm are stored. The error log is a record of system errors and can be accessed from a SAT. The error log is useful for analyzing problems that have not caused an alarm or when alarms cannot be retired by replacement of hardware. When errors result in alarms, the alarms are listed in the alarm log. This log can be displayed on a terminal. If several alarms are active, the alarm log can be used to determine the alarms that should be cleared first.

APPENDIX A

Call Processing Feature/Function Glossary and Definitions

This appendix is an abridged glossary of voice calling features that are available on leading PBX systems. This glossary is intended to be representative of the most popularly requested customer features, but it does not include all of the currently available features. Each PBX system feature set is unique, and feature capabilities differ across manufacturers' models. Although the typical PBX station user may commonly use only a small fraction of these features under normal working conditions, each feature has some productivity and/or cost savings value. PBX systems that lack more than 15 percent of these features cannot be said to be competitive in the marketplace.

Station User Features

Add-on Conference
This feature allows a station user to add a third party to an existing two-party conversation.

Automatic Callback
This feature is used when a dialed station is busy. When the feature is activated, the system automatically attempts to call the desired station until the line is free. The calling party is alerted that the called party is available. This saves wasted time dialing when encountering busy signals.

Automatic Intercom

This feature provides a talking path between two voice terminal users. A station user presses a programmed automatic intercom button and lifts the handset, or vice versa. The called user receives a unique intercom alerting signal, and the status lamp associated with the dial or automatic intercom button, if provided, flashes.

Bridged Call Appearance

This feature allows the same line appearance to be programmed to appear on more than one telephone. It is very useful in manager–assistant relationships and call answering position environments. A bridged call appearance can reduce the number of abandoned or lost calls and allows the coverage station to prescreen calls for the called party.

Call-Back Last Internal Caller

This feature allows a station user to automatically consult and call back the last internal caller to the station (unanswered call) by implementing the feature code.

Call Forwarding—All Calls/No Answer/Busy

This feature allows a station user to divert all incoming calls to another programmed station. All Call activation diverts all incoming calls to the station; no answer activation diverts calls after a programmed number of rings; busy activation diverts incoming calls when the station is busy. The features are useful when a station user is away from the desk area, wishes to receive calls at another station, or when there is a desire not to receive calls, but the user wants the call answered. The most common coverage station is a voice messaging port. Call forwarding features decrease abandoned or lost calls and improve call coverage service for the calling party.

Call Forwarding—Follow Me

This feature allows a station user to activate the call forwarding feature from a remote telephone by changing an existing forwarding destination. It provides station users with the capability of changing call forwarding destinations without returning to their desks, and can be used to "follow" station users around the system if they wish to receive calls at different stations.

Call Forwarding—Off Premises

This feature allows a station to forward all calls to an off-premises location outside the system. It allows station users to receive calls at a programmed outside telephone line when they are out of the office.

Call Forwarding—Ringing

This feature allows a station user of a multiple line voice terminal with display to forward an incoming call during the ringing period to another station. The station user can read the display for screening information, such as CLID or calling party name, before activating the feature through a programmed feature button or dial access code. The rerouted call destination is input by the station user after feature activation. The operation is transparent to the calling party.

Call Forwarding—Selective Multiple Line

This feature allows a station user of a multiple line voice terminal to selectively call forward any or all line appearance numbers.

Call Hold

This feature allows a station user to place an existing call in a hold state when there is another incoming call or the station user must leave the desktop area for more than a few seconds. Call hold provides station users with the flexibility of handling multiple concurrent calls without re-establishing the connection after finishing another call.

Call Park

This feature allows a station user to "park" a call at the received station, effectively placing the call in a hold state, retrieve the call at another station, and continue the conversation. A second party can also retrieve the parked call if notified by the first party. The feature provides station users with mobility and eliminates the need to return calls.

Call Pickup

The feature allows a station user to retrieve and answer a call directed to another station (direct), any station in the station user's assigned call pickup group (group), or another call pickup group (designated group). The station user presses a programmed call pickup button or dials the desired feature access code to implement the feature.

Call Transfer
This feature allows a station user to divert an existing call to another station within the system. It eliminates the need for the original calling party to hang up and redial another telephone line to reach the desired or proper called party.

Call Waiting
This feature notifies a station user engaged in conversation that there is another incoming call to the station line number. The notification is usually a special tone or display signal on the telephone. Call waiting reduces lost calls, improves customer service, and reduces the number of calls forwarded to another station or messaging system.

Consecutive Speed Dialing
For speed dialing all station number digits are registered as the speed dial code. This feature allows a common set of partial station number digits to be registered as the speed dial code, and allows the station user to dial the remaining digits of each number to establish the call.

Consultation (Broker) Hold
This feature allows a station user to place one party on hold and confer with a third party on another line. This feature reduces call backs and improves customer service. Some systems allow the station user to toggle back and forth between two lines.

Customer Station Rearrangement
This feature allows station users to physically relocate their multiple line voice terminals internal to the system. When a station user moves between locations, the voice terminal station (number and COS) is logically transferred. This service is activated as follows: dialing the "moving" feature code followed by a personal code; the terminal is out of service; the terminal is disconnected, moved, and reconnected in the new location; the terminal is reinitialized by dialing its extension number, followed by the personal code; the station rings immediately and when answered, the set is validated. The former and latter locations must support the same category of voice terminal.

Discrete Call Observing
This feature allows a supervisor to monitor a conversation between an assistant and a caller on a preselected line. While the supervisor is listening, the voice terminal microphone is off and the assistant is

informed that the feature is activated by a notice on the voice terminal display. During monitoring, the supervisor can take over the call.

Distinctive Ringing
This feature allows the system administrator to define distinct ringing patterns for different call types, such as internal, defined internal line, external, private network, emergency, and private line. Station users can use this feature as a call screening device to decide which calls to answer and which are to be forwarded to a coverage station. The number of distinct patterns differs greatly between different system models.

Dial by Name
This feature allows a station user with a voice terminal equipped with an alphabetical keyboard to call an internal extension or external number by typing in a name using last name, first name, or initials. The directory database can be locally stored in the voice terminal or accessed from a centrally located database in the PBX system or application.

Do Not Disturb
This feature allows a voice terminal user to request that no calls, other than priority calls, terminate at a particular extension number until a specified time. At the specified time, the system automatically deactivates the feature and allows calls to terminate normally at the extension.

Elapsed Call Timer
This feature provides a display of the elapsed time when a multiple line voice terminal is connected to any trunk circuit.

Emergency Access to Attendant
This feature allows emergency calls to be placed to an attendant with special priority status. Calls can be placed automatically when the telephone is in an off-hook state or by dial access. The attendant receives visual and audible feedback when the call is received. The feature is important for situations requiring immediate attendant access.

Executive Busy Override
This feature allows a station user to cut into an internal party's conversation. This feature decreases call backs, saves time reaching the called party, and decreases calls sent to coverage positions.

Executive Calling
This feature allows a station to be assigned VIP class status. The feature allows a VIP station to send a special ringing signal to a called station when idle and automatically send multiple tone bursts to that station when busy.

Executive Access Override
This feature allows a station user to connect a call to an internal extension that is in call forward or do not disturb mode. The service is authorized by COS level.

External Paging with Meet-Me
This feature allows a station user or attendant to dial a local paging equipment access code and connect both parties automatically after the paged party has answered the page and dials a special access code.

Facility Busy Indication
This feature provides multiple line telephone users with a visual indication of the busy or idle status of internal station numbers, trunk groups, hunt groups, or paging zones. Station users can monitor the activity of frequently called numbers with this feature, eliminating encountered busy signals.

Group Listening
This feature allows a station user of a multiple line voice terminal with an integrated speakerphone to place a call using the handset and activate the terminal's built-in speaker, to allow others to listen to the conversation while the station user continues talking through the handset.

Hands-free Dialing
This feature allows a station user of a voice terminal with a built-in speaker to dial and monitor a call without lifting the handset.

Hands-free Intercom
This feature allows a station user of a voice terminal with a built-in speaker to answer a voice call without lifting the handset. The incoming voice call is heard over the speaker.

Help/Information Key
This feature provides a station user of a multiple line voice terminal with display immediate access to help menus for terminal programming

and feature access procedures. Information is displayed by the system in user-friendly way. If a feature access code is changed, it will be displayed automatically in the feature menu. Service consultation shows all the relevant functions and their associated feature codes. This feature allows self-training on the voice terminal and also reduces the need for paper labels on telephone features.

Hot Line
This feature automatically dials calls to preassigned internal stations, off-premises stations, or feature access codes when the handset is lifted. It eliminates the need to dial a number or access code, thereby simplifying and accelerating the process.

Incoming Call Display
This feature, available on telephones with display fields, provides visual notification to the station user of the calling party's station number or incoming trunk group name. The calling party's name may also be displayed with the station number. The feature is a screening device to decide whether to answer or divert the call.

Individual Attendant Access
This feature allows users to access a specific attendant console. Each attendant console can be assigned an individual extension number.

Intercom Dial
This feature allows multiple line voice terminal users to gain quick access to select other voice terminal users within an administered group. Calling voice terminal users lift the handset, press the dial intercom button, and dial the one- or two-digit code assigned to the desired party. The called user receives an alerting tone, and the status lamp associated with the Intercom button, if provided, flashes.

Last Number Redialed
This feature stores the last number dialed by the station user and allows the station user to automatically dial the number by using a programmed feature button or feature access code. It simplifies the calling process, reduces misdialed calls, and saves time.

Line Lockout
This feature removes single line voice terminal extension numbers from service when users fail to hang up after receiving dial tone signals, fol-

lowed by intercept tone signals. The intervals for each tone signal are administrable.

Loudspeaker Paging Access
This feature provides station users or attendants dialing access to voice paging systems. This is useful for paging purposes regardless of the station user's location within the premises environment. It is often used with the call park feature.

Malicious Call Trace
This feature allows a station user to notify a predefined set of station positions that a malicious call is in process. The notified station users can then gather information and data about the call to identify the calling source. The feature is useful when a CLID or ANI is not displayed.

Manual Intercom
This feature allows a station user to call a manual intercom group member by pressing the manual intercom button. All member of a manual intercom group share a common signaling path. When the manual intercom button is pressed, a special tone burst is sent over the voice terminal speakers of all group members. When a group member answers, a speech path is established.

Manual Originating Line Service
This feature connects single line voice terminal users to the attendant automatically when the user lifts the handset. The attendant code is stored in an abbreviated dialing list. When the manual originating line service voice terminal user lifts the handset, the system automatically routes the call to the attendant using the hot line service feature.

Manual Signaling
This feature allows a voice terminal user to signal another voice terminal user. The receiving voice terminal user hears a short burst of tone. The signal is sent each time the button is pressed. If the receiving voice terminal is already being alerted with an incoming call, manual signaling is denied.

Message Waiting
This feature enables multiple line appearance voice terminal users, by pressing a designated button on their own terminals, to light the status lamp associated with the message waiting button at another multiple line appearance voice terminal. Activating the feature causes the lamp

to light on the originating and receiving voice terminals. Either terminal user can cause the lamp to go dark by pressing the button.

Multiparty Conferencing
This feature allows multiappearance voice terminal users to set up multiparty conferences (typically between four to eight station users) without attendant assistance.

Music on Hold
This feature provides music to a party that is on hold, waiting in a queue, parked, or on a trunk call that is being transferred. The music lets the waiting party know that the connection is still in effect. The system provides automatic access to the music source.

Off-hook Alarm
This feature provides a special alerting tone to a station user who does not hang up the handset after receiving a busy signal. The tone signal is sent after a programmed interval.

Padlock
This feature allows a station user to temporarily prevent outgoing calls from the voice terminal. The selection of an external line by feature code, programmable key, or supervision key is controlled. Dialing the appropriate feature code followed by the personal code reactivates direct access.

Paging/Code Call Access
This feature allows voice terminal users, attendants, and tie trunk users to page with coded chime signals. Multiple individual paging zones can be provided.

Personal Speed Dialing
This feature allows a station user to program personal speed dial numbers at the station instrument. The speed dial feature is activated by using a programmed feature button or access code and pressing a one- or two-digit access code. The feature simplifies the dialing process, saves time, and decreases misdialed numbers.

Personalized Ringing
This feature allows users of certain voice terminals to uniquely identify their own calls. Each user can choose one of a number of possible ringing patterns.

Priority Calling
This feature provides a special form of call alerting between internal voice terminal users. The called voice terminal user receives a distinctive, administrable alerting signal.

Private Line
This feature provides a dedicated trunk for direct access to or from the public network for multiple line appearance voice terminal users.

Privacy—Attendant Lockout
This feature prevents an attendant from re-entering a multiple-party connection held on the console unless recalled by a voice terminal user.

Privacy—Manual Exclusion
This feature allows multiple line appearance voice terminal users to keep other users with appearances of the same extension number from bridging onto an existing call.

Recall Signaling
This feature allows an analog station user to place a call on hold and consult with another party or activate a feature. After consulting with that third party, the user can conference the third party with the original party by another recall signal or return to the original party by flashing the switchhook twice.

Ringer Cutoff
This feature allows the user of a multiple line appearance voice terminal to turn certain audible ringing signals on and off. The feature does not affect visual alerting.

Ringing Tone Control
This feature allows station users of multiple line voice terminals to select from a menu of ringing tone melodies and to adjust the volume level of ringing.

Save and Redial
This feature allows a station user to save a specific dialed number and then redial the number at a later time. The station user stores and redials the number by pressing a save and redial feature key.

Call Processing Feature/Function Glossary and Defintions

Secondary Extension Feature Activation
This feature allows a multiple line voice terminal station user to access a line appearance of another extension, and program a limited set of features, such as call forwarding and call pickup, from that extension.

Send All Calls
This feature allows users to temporarily divert all incoming calls to coverage regardless of the assigned call coverage redirection criteria. The feature also allows covering users to temporarily remove their voice terminals from the coverage path.

Step Call
This feature allows a station user or attendant, after dialing a busy station, to dial an idle station by simply dialing an additional digit. The feature can be implemented only if the dialed digits of the first dialed number and the second number are identical, except for the last digit.

Store/Redial
This feature allows a station user of a multiple line voice terminal to store a particular number for later use. A store/redial key is programmed and assigned to this function.

Supervisor/Assistant Calling
This feature allows a station user with a multiple line voice terminal, who is an assistant to a supervisor, to use a call appearance of the supervisor's station to screen calls for the supervisor and announce and/or transfer calls to that extension. The assistant can also dial the supervisor during a busy condition and send a message waiting notification to the supervisor.

Supervisor/Assistant Speed Dial
This feature allows a pair of station users to use a programmed feature key to direct speed calls between a supervisor and an assistant, even if forwarding is validated.

Text Messages
This feature allows station users to leave a short text message for other internal users. Messages are stored in the main system database, and are available for selection via a menu on display-based voice. Calling parties can also receive messages from a voice terminal station that are

preselected by the called party during no answer or busy conditions. There may be three structures of messages: preprogrammed fixed messages fully defined by system management, part programmable messages defined by system management but to be completed by the station user if the voice terminal has an alphanumeric keyboard, and fully programmable messages written entirely by the station user and offered only by sets provided with an alphanumeric keyboard.

Timed Queue
When a multiple line voice terminal station user originates an outgoing trunk call and encounters a no answer or busy condition, the timed queue feature can be implemented. After pressing a programmed feature button or dialing the feature access code, trunk seizure is repeated and the external station number is dialed after a predetermined interval.

Trunk Flash
This feature enables multifunction voice terminals to access CLASS features that are provided by the far-end CO switching system located directly behind the PBX system. CLASS services are accessed by a sequence of flash and dial signals from the station on an active trunk call. The feature can decrease the number of trunk lines connected to the PBX system by performing trunk-to-trunk call transfers at the far-end CO, which eliminates the use of a second trunk line for the duration of the call and frees the original trunk line for the duration of the call. It can also be used to set up a conference call with a second outside call party, which eliminates the need for a second trunk line for the duration of the call.

Trunk-to-Trunk Connection
This feature allows a station user to conference together two outside trunk calls and abandon the connection without dropping the two trunk-to-trunk connections.

Attendant Features

Attendant Auto-Manual Splitting
This feature allows the attendant to announce a call or consult privately with the called party without being heard by the other party on the call.

Call Processing Feature/Function Glossary and Defintions

Attendant Auto-Start/Don't Split
This feature allows an attendant to activate a call by pressing any key on the console dial pad without having to press the start button first. The auto-start feature is deactivated when the don't split button is pressed and cannot be activated until the don't split button is pressed again. The feature decreases the number of feature buttons the attendant needs to press to handle calls.

Attendant Backup Alerting
The attendant backup alerting feature allows other system users to pick up attendant calls when the attendant is unavailable or busy.

Attendant Call Waiting
Attendant call waiting allows an attendant to originate or extend calls to a busy single line voice terminal to wait at the called terminal. The attendant is free to handle other calls.

Attendant Conference
This feature allows an attendant to set up multiple concurrent conference calls that are conferenced together without the assistance of an outside network operator.

Attendant Control of Trunk Group Access
This feature allows the attendant to control trunk groups and prevents voice terminal users from directly accessing a controlled trunk group.

Attendant Delay Announcement
This feature provides an announcement (via an internal announcer card) to external incoming callers who are not answered by the attendant within a predetermined period.

Attendant Direct Station Selection With Busy Lamp Field
This feature allows the attendant to track station status (idle or busy) and to place or extend calls to extension numbers assigned to the system without having to dial the extension number.

Attendant Direct Trunk Group Selection
This feature allows the attendant direct access to an idle outgoing trunk by pressing the button assigned to the desired trunk group.

Attendant Display
This feature shows call-related information that helps the attendant operate the console more efficiently. It also shows personal service and message information. Information is shown on the alphanumeric display on the attendant console.

Attendant Intercept Treatment
This feature allows an attendant to provide information and assistance to callers on internal calls that cannot be completed as dialed. The feature is an alternative to an intercept treatment tone or recorded announcement when a higher level of customer service is required.

Attendant Interposition Call and Transfer
This feature allows an attendant to retrieve an internal call at the conclusion of a conversation and transfer the call to another station. It is used to facilitate internal call transfers.

Attendant Intrusion
This feature enables an attendant to enter an existing call on a multifunction station or analog station and offer a new call or message to the called party.

Attendant Overflow
This feature routes incoming trunk calls to the night service answering position if the attendant does not answer the call within a predetermined period.

Attendant Override of Diversion
This feature allows an attendant to bypass call diversion features invoked by or associated with a dialed extension. A diversion feature is any feature that, when activated, causes the call to alert at a point different from the dialed station. Examples of diversion features are call forwarding—all calls, call coverage, and send all calls.

Attendant Paging System Access
This feature allows the attendant to access and use an internal loudspeaker paging system. It is commonly used by the attendant to alert station users who are at their desks.

Attendant Priority Queue

This feature handles incoming calls to the attendant group or to an individual attendant when the call cannot be immediately terminated to an attendant. These calls are placed in the attendant priority queue in an order based on a priority queue level and timestamp associated with the call. Calls within the same priority level are served on a first come, first served basis. The calling party hears ringback until an attendant answers the call.

Attendant Recall

This feature allows voice terminal users on a two-party call or on an attendant conference call held on the console to recall the attendant for assistance.

Attendant Release Loop Operation

This feature allows the attendant to hold the connection of any call off the console if completion of the call is delayed (such as a call extended to a busy single line voice terminal or to a voice terminal that does not answer). This feature frees the attendant to handle other calls.

Attendant Serial Calling

This feature enables the attendant to transfer trunk calls that return to the same attendant position after the called party hangs up. The returned call may then be transferred to another station within the switch and this can recur. This feature is particularly useful if trunks are scarce and direct inward dialing services are unavailable.

Attendant Through Dialing

This feature allows an attendant to select an outgoing trunk for a station user. The attendant is released from the connection, and the station user can complete the call.

Attendant Trunk-to-Trunk Transfer

This feature allows an attendant to transfer an incoming or outgoing call on one trunk to an outgoing trunk and exit the connection before the called party answers.

Attendant Trunk Group Busy/Warning Indication

This feature provides an attendant with a visual indication that the number of busy trunks in a group has reached a predefined threshold level. It is a feature that is used to warn an attendant that control of trunk group access may be soon be required.

Attendant Trunk Identification
This feature allows an attendant to identify the specific trunk circuit that is being used on a call. It is useful whenever it is necessary to identify trunk group use or faulty trunk circuits.

Busy Verification of Terminals and Trunks
This feature allows attendants and specified multiple line voice terminal users to make test calls to trunks, voice terminals, and hunt direct department calling and UCD groups. These test calls check the status of an apparently busy resource.

Straightforward Outward Completion
This feature allows an attendant to complete an outgoing trunk call for a voice terminal user without requiring the voice terminal user to hang up.

System Features

Account Codes
This feature allows a station user to input a multiple digit code for certain types of outgoing trunk calls. Each code is associated with a unique file record, usually used for account billing purposes.

Answer Detection
This feature detects the state of outgoing trunk calls that do not receive network answer supervision to improve the accuracy of the call duration field in CDRs.

Authorization Codes
This feature allows a station user to input a personal identification code as a means for extending the control of system users' calling privileges and security for remote access callers. Authorization codes may be used for any or all of the following reasons: allow a calling user to override the FRL assigned to the originating station or trunk, restrict individual incoming tie trunks and remote access trunks from accessing an outgoing trunk, identify certain calls on CDRs for cost-allocation purposes, and provide additional security control for the system.

Call Processing Feature/Function Glossary and Defintions

Automated Attendant
This feature allows the system to answer incoming trunk calls with no intervention of an attendant position. The system will provide the caller with a message or dial tone and allow the caller to directly dial an internal extension number.

Automatic Alternate Routing
This feature provides alternative routing choices for private on-network calls. When implemented, the system automatically selects the most desirable (normally the least expensive) route among multiple trunking facilities for private network calls. AAR also provides digit modification to allow on-network calls to route through the public network when an on-network route is not available.

Automatic Call Distribution
This feature provides the automatic connection of incoming calls to specific splits (hunt groups) of station users (agents). Calls to a specific split are automatically distributed among the agents assigned to that split. If agents are not available, the call can queue to the split to wait for an agent to become available.

Automatic Camp-on
When a DID call has been terminated at a busy station, the call is "camped-on" to the called station. When busy station becomes idle, it is automatically connected to the camped-on incoming trunk call.

Automatic Circuit Assurance
This feature assists users in identifying possible trunk malfunctions. The system maintains a record of the performance of individual trunks relative to short and long holding-time calls. The system automatically initiates a referral call to an attendant or display-equipped voice terminal user when a possible failure is detected.

Automatic Number Identification
This feature allows the system to receive an incoming caller's local telephone company trunk billing number and display the number on a station user's voice terminal with a display. The ANI is transmitted over an incoming digital trunk circuit with the use of in-band or out-of-band signaling techniques.

Automatic Recall
This feature alerts a voice terminal after a fixed interval that a call it transferred has been placed on hold, is camped-on, or continues ringing with no answer.

Automatic Route Selection
This feature routes calls over the public network based on the preferred (normally the least expensive) route available at the time the call is placed. ARS provides a choice of routes for any given public network call. The following types of trunk groups can be accessed by ARS: local CO, FX, WATS, tie trunk, T1/E1, ISDN PRI, and IP WAN. The system selects the most preferred (normally least expensive) route for the call. Interexchange carrier code dialing is not required on routes selected by the system. Interexchange carrier codes are assigned in translations to best benefit the customer on any given call. These codes are inserted as needed to guarantee automatic carrier selection.

Automatic Transmission Measurement System
This feature provides for trunk facilities to be measured for satisfactory transmission performance. The performance of the trunks are evaluated according to measurements produced by a series of analog tests and are compared against user-defined threshold values.

Call Coverage (Multiple Call Forwarding, Split Call Forwarding)
This feature provides automatic redirection of calls that meet specified criteria to alternate answering positions in a call coverage path. Lead coverage paths can be administered to apply to all calls all the time, internal or external calls, or to apply to a specific day of the week or a specific time of the day. Different coverage paths are administered based on incoming call origination, type, or time.

Call-by-Call Service Selection
This feature allows a single ISDN-PRI trunk group to carry calls to many services or facilities or to carry calls using different interexchange carriers. The feature typically uses the same routing tables and routing preferences that are used by AAR and ARS. The service or facility used on an outgoing CBCSS call is determined by information assigned in the AAR/ARS routing patterns. Without CBCSS each trunk group must be dedicated to a specific service or facility. CBCSS eliminates this requirement by allowing a variety of services to use a single trunk group. These services are specified on a call-by-call basis.

Call Processing Feature/Function Glossary and Defintions

Call Detail Recording
This feature records detailed call information on all incoming and outgoing calls on specified trunk groups and extensions administered for intraswitch recording and sends this information to a CDR output device. The CDR output device provides a detailed printout that can be used by the system administrator to compute call costs, allocate charges, analyze calling patterns, detect unauthorized calls, and keep track of unnecessary calls.

Call Log
This feature stores dialed station numbers and incoming identification numbers (internal CLID, CLASS CLID, ANI) at a multiple line voice terminal with a display. The numbers that are stored are those of the most recently dialed and incoming calls. There is a limited amount of stored and displayed numbers that varies by system. Pressing a call log button brings up the display. Calls to numbers appearing in the call log display field can be dialed automatically through menu control keys.

Centralized Attendant Service
This feature allows services performed by attendants in a private network of switching systems to be concentrated at a central, or main, location. Although all incoming calls to the network are routed to a main PBX system, each branch in the centralized attendant service configuration has its own listed directory number or other type of access from the public network. Incoming trunk calls to the branch and attendant-seeking voice terminal calls are routed to the centralized attendants over a release link trunk. The centralized attendants are located at the main location.

Class of Restriction
This feature defines different classes of call origination and termination privileges. Systems may have one restriction class, one with no restrictions, or as many restriction classes as necessary to effect the desired restrictions.

Class of Service
This feature determines whether or not voice terminal users can access any or all station or system features and functions. There are many COS levels that can be programmed by the system administrator; each level is associated with a defined feature/function set that may contain one or many features and functions. Each COS level allows or denies

access to the defined feature/function set. Every system user is assigned a COS level by the system administrator.

Controlled Private Calls
This feature allows the system operator to charge station users for personal outgoing calls. The following items define station users: extension number (virtual or real), personal identification number, and call restriction table for private calls. The user can make a private call according to the following rules: only from his own set, from every authorized set in the subnetwork, or from only a few selected sets.

Delayed Ringing
This feature allows trunks and station lines to ring immediately at the dialed destination station and, after a programmed interval, at a secondary station that shares the same line appearance as the original destination station.

Dial Plan
The dial plan is the system's guide to digit translation. When a digit is dialed, the system must know what to expect based on that digit. The dial plan, or first-digit and second-digit tables, established during administration for each system provides information to the switch on what to do with dialed digits. The tables define the intended use of a code beginning with a specific first digit or specific pair of digits. These digits tell the system how many digits to collect before processing the full digit string.

Dialed Number Identification Service
This feature provides a display of the listed directory number of an incoming trunk call to the attendant position. The display can be the actual digits of the number, or an alphanumeric name or identifier. This screening feature allows an attendant to better handle the incoming call and provides a higher level of customer service.

Direct Department Calling
This feature allows direct inward access to an answering group other than the attendant even if the system does not use the DID feature. A direct department answering group can consist of voice terminals and individual attendants. One extension number is assigned to all voice terminals and individual attendants. Incoming calls to a direct department group can be internal or external. With this feature, an incoming call

rings the first available voice terminal or individual attendant in the administered sequence. If the first group member in the sequence is active on a call (busy) or has had calls temporarily redirected, the call routes to the next group member, and so on. Incoming calls always try to complete at the first group member in the administered sequence. Calls are not evenly distributed among the group members.

Direct Inward Dialing
DID connects calls from the public network directly to a dialed extension number without attendant assistance. Specialized DID trunk circuits are required to implement this feature. DID reduces attendant workload and facilitates connections between an external calling and an internal called party.

DID Call Waiting
This feature allows an incoming call on a DID trunk circuit to be automatically camped-on to the destination station if the destination station is busy.

Direct Inward System Access
DISA allows system users who are off-premise to dial into the system, input a special access code, and use the system facilities even though the caller is not using an internal voice terminal. It allows access to the system's optimally priced trunk network facilities and other cost savings features.

Direct Inward Termination
This feature automatically routes incoming network exchange calls to a preselected station without attendant assistance. The called party can process the call in a manner similar to any normal trunk call.

Direct Outward Dialing
This feature allows voice terminal users to access the public network without attendant assistance. Station users dial a defined trunk access code (such as 9) for public network connection and dialing.

Extended Trunk Access
This software feature provides a means for routing calls that are not defined in the first- or second-digit tables or the feature/trunk access code tables. This feature makes use of an extended trunk access routing pattern or node number for determining how to route an unidentified call.

Facility Restriction Levels

FRLs provide multiple levels of restriction for users of the AAR or ARS features. FRLs provide a method of allowing certain calls to specific users and denying the same calls to other users. For example, certain users may be allowed to use CO trunks to other corporate locations, whereas other users may be restricted to the less expensive private network lines. The FRLs are defined and programmed into the system by the administrator and are transparent to the station user. Regular dialing procedures are unaffected.

Facility Test Calls

This feature allows a voice terminal user to make test calls to access specific trunks, DTMF receivers, time slots, and system tones. The test call makes sure the facility is operating properly. The feature is implemented by dialing an access code.

Forced Account Code

This feature forces a station user to enter an account code for all outgoing trunk calls. The account code must be entered before dialing the outgoing number. Calls are processed only after the account code is entered and verified. Some systems allow calls to be classified into multiple groups, such as: call with a controlled project number, call with an uncontrolled project number, and call without a project number. The choice of call depends on a data system configuration based on two parameters: with/without a project number or with/without a controlled project number.

Hoteling

This feature allows users whose stations are translated to their own preferences and permissions to associate those preferences and permissions with any compatible terminal. These include the definitions of terminal buttons, abbreviated dial lists, and COS and restriction class permissions assigned to the user's station.

House Phone

This feature allows station users to use selected voice terminals to reach an attendant by simply going off-hook.

Hunting

This feature routes calls to a station within a predefined ordered group after checking for station idle or busy status. Calls are routed to another group when all stations are busy. Hunting is accomplished through the

ACD, direct department calling, and UCD features. The order of hunting is defined under each feature. Under direct department calling, call distribution is not uniform across hunt group members.

Integrated Directory
This feature allows internal system users with display-equipped terminals to access the system database, use the touch-tone buttons to key in a name, and retrieve an extension number from the system directory. The directory contains an alphanumeric listing of the names and extension numbers assigned to all voice terminals administered in the system.

Modem Pooling
This feature allows switched connections between digital data endpoints (data modules) and analog data endpoints and data modems. The analog data endpoint can be a trunk or a line circuit.

Multiple Listed Directory Numbers
This feature allows a publicly published number for each incoming and two-way (incoming side) FX and local CO trunk group assigned to the system. This feature also allows DID numbers to be treated as listed directory numbers.

Music on Hold
This feature plays music for a caller who is on hold, waiting in a queue, or on a trunk call in the process of being transferred. The feature provides a means to let callers know they are still connected to the system.

Night Service
This feature directs all calls for the primary and daytime attendant consoles to a night console. It is typically activated when an attendant presses the night button on the principal attendant console and deactivated by pressing the night button again. Night service also can be activated and deactivated from one station in the system by use of a night service button assigned to that station.

Off-Hook Alarm
This feature allows a station user to call an attendant or any preselected programmed station by simply staying off-hook for a preprogrammed period. The calling number is automatically displayed at the attendant console or the preselected station.

Off-Premises Station
The feature allows a voice terminal outside the switching system location to be connected to the system via specialized CO trunk circuits. The voice terminal must be analog and must be registered with the Federal Communications Commission.

Open System Speed Dial
A station user can select a system speed number by dialing a system speed code or name and then dialing the relevant final sequence of numbers to select the external party.

Power Failure Transfer
This feature provides service to and from the local telephony company to a designated station during power failures affecting the PBX system. During PFT mode, no other system features can be activated.

Property Management System Interface
PMS interface provides a communications link between the system and a customer-owned PMS. The PMS allows a customer to control certain features used in hospital-type and hotel/motel-type environments. The communications link allows the PMS to interrogate the system and allows information to be passed between the system and the PMS.

Recent Change History
This feature allows system administrators to view or print a history report of the most recent administration and maintenance changes. The history report also lists each time a user logs in or off the system. This report may be used for diagnostic, information, or security purposes.

Restricted Incoming Station
When a station is configured to receive only incoming calls, the station user receives a busy signal as soon as the handset is picked up.

Restriction—Controlled, Inward/Outward, Toll/Code, Trunk, Voice Terminal
A series of features that allows an administrator or attendant to activate or deactivate defined trunk access and I/O calling privileges for a station or group of stations.

Call Processing Feature/Function Glossary and Defintions

Route Advance
This feature automatically routes outgoing trunk calls over alternate facilities when the first choice trunk group is busy. This feature is implemented only if a station user selects the first choice trunk group with a dial access code. The system advances through a series of alternate trunk groups only if the first-choice trunk group is busy.

Shared Tenant Service
A system manager can partition the PBX to provide telecommunications services to multiple tenant groups. The tenant groups can have independent dial plans, CDR, ARS and call routing tables, attendant groups, and COS/class restriction levels. Each group is logically partitioned from the others for all premises telecommunications services. The number of partitioned tenants depends on the system.

System Speed Dial
A station user can call a number by dialing system speed codes or names. The list of system speed codes can be common to all system users or split into different lists. With splits, the users can access different lists according to their COSs. Each outside system speed code corresponds to the access feature code of a trunk group or external trunk (public or private).

Timed Reminder
This feature allows the system to be programmed to automatically call stations at specified times. When the called party answers, the station is connected to a recorded announcement or music source.

Trunk Answer Any Station
This feature allows any station to answer an incoming call trunk when the system is in night service mode. A common alert signal is sent to all stations, and any station can answer the call. The answering station can extend the call to any other station by using call transfer.

Trunk Callback Queuing
This feature places outgoing calls in an ordered queue (first in, first out) when all trunks are busy. The voice terminal user is automatically called back when a trunk becomes available. The voice terminal receives a distinctive three-burst alerting signal when called back.

Uniform Call Distribution

This feature allows direct inward access to an answering group other than the attendant. A UCD answering group can consist of voice terminals and individual attendants. One extension number is assigned to all voice terminals and individual attendants. Incoming calls to a UCD group can be internal or external. With UCD, an incoming call rings the member of the group that has not received a UCD group call for the longest period (the most idle member). Incoming calls to a UCD group extension number are distributed evenly across the group members.

Uniform Dial Plan

A UDP may be established during administration as part of the dial plan. This plan provides a common extension number plan that can be shared across a group of switches. If a UDP is to be established, all extension numbers (in the UDP numbering plan) must be the same length.

Virtual Extensions

This feature permits the assignment of circuits that do not physically exist, to be used for secondary extensions on multiple line voice terminals.

Voice Message System Interface

This feature provides a signaling interface between a PBX and an external VMS. The interface allows the VMS to activate message waiting indicators on PBX voice terminals.

APPENDIX B

PBX/IP-PBX Cost and Pricing Issues

PBX system pricing is more of an art than a science. Each PBX system model's unique design and proprietary hardware and software elements make it very difficult to develop definitive pricing statements that cover all PBX systems. It is easiest to analyze PBX system pricing by breaking a system down into its basic components: main control cabinet and software, expansion port cabinets, port circuit cards, telephone instruments, cabling, and installation. The information in this appendix can be used as pricing guidelines for customers developing budgets for new system purchases.

Common Control Cabinet and Software

The most basic PBX system consists of a single control cabinet/carrier equipped with the minimum number of system and service circuit cards. The basic assembly typically includes switch network elements, such as center stage switch and/or local TDM bus, and a power supply unit. This equipment is sometimes referred to as the basic system assembly. Starter system pricing for a basic system assembly must also include the cost of the generic software program. Common control cabinet pricing directly depends on system port capacity. The greater the maximum system port capacity of the system, the greater the basic assembly prices. Table B-1 presents the likely basic assembly price ranges for several PBX system configurations based on port capacity ranges.

TABLE B-1

Price Ranges for PBX Systems Based on Port Capacity

Basic Assembly Starter Kit	Price Range
Small PBX system (<80 stations)	$3,000–$5,000
Intermediate PBX system (80–400 stations)	$5,000–$20,000
Large PBX system (400–1,000 stations)	$20,000–$35,000
Very large PBX system (>1,000 stations)	$50,000–$100,000

The basic assembly unit typically includes a software right-to-use (RTU) fee for a minimum of 100 station users. Software RTU fees are usually based on the number of required active ports. RTU fees may be based on the number of ports, regardless of size, or on a sliding scale as the system port capacity increases. Beyond the RTU fee bundled into the basic assembly price, the typical RTU fee is about $50 per port. For example, an RTU fee for 100 ports is about $5,000. Some manufacturers may bundle their software fees into the price of port circuit cards but not disclose this information in their price lists.

Optional software features are priced based on feature complexity. Most station and attendant features are bundled into the generic software program price and RTU fee. Advanced networking applications, such as private networking and IFTN, and advanced ACD features are always priced as software options. Private networking software is typically priced by system model size. Pricing ranges from $5,000 for a small PBX system to more than $20,000 for a very large system. Intelligent networking options usually require private networking software as a prerequisite and are priced comparably to the private networking software option price range. ACD software options are priced based on feature complexity on a per-agent basis (based on agent group size). ACD pricing ranges from $100 to $1,000 per agent. MIS reporting systems to support ACD operations range from about $200 to $1,000 per agent, independent of requisite processing and database management hardware.

Basis PBX systems management software is usually bundled into the basic assembly price. Advanced systems management software options supporting sophisticated GUI and programming tools are priced at about $5,000 to $10,000 per user client. A server-based management system would be priced far less per client user but at a higher fixed price per system, exclusive of server and client equipment.

Redundant common pricing is also based on system size. A few manufacturers include with their standard system assembly redundant com-

PBX/IP-PBX Cost and Pricing Issues

mon control elements, including main system processor and memory (software generic, customer database), at no additional cost to the customer. This is reflected in the basic assembly price ranges. Most manufacturers offer redundancy as an option, if available for a particular system model. Table B-2 lists the likely optional common control (processing and/or memory, if available) redundancy price ranges for several PBX system configurations based on port capacity ranges.

TABLE B-2

Optional Common Control Redundancy Price Ranges Based on Port Capacity

Common Control Redundancy	Price Range
Small PBX system (<80 stations)	$1,000–$2,500
Intermediate PBX system (80–400 stations)	$2,500–$10,000
Large PBX system (400–1,000 stations)	$5,000–$10,000
Very large PBX system (>1,000 stations)	$25,000–$50,000

A redundant power supply module and a UPS are system options associated with the common control cabinet. No small PBX system models are available with redundant power supplies. The option price for redundant power supplies in intermediate to very large PBX system models will typically range from $10,000 to $50,000; the exact price correlates with the system port capacity. UPS option prices vary by system port capacity and reserve power time. UPS pricing for a small PBX system can start at about $5,000 and go up to $75,000 for very large system models.

Expansion Port Cabinets/Carriers

Stackable port carriers are priced by carrier function. An expansion port carrier that is required to support local processing and/or highway bus switching functions will be priced higher than a basic port carrier with no local processing or expansion interface capabilities. The price range for expansion port carriers includes the price of local processor and interface elements, if required. Multicarrier port cabinet pricing is based on maximum equipped port carrier shelves. The price ranges shown in Table B-3 include local processing and interface elements, if required.

TABLE B-3
Price Ranges for Port Cabinets/Carriers

Port Cabinets/Carriers	Price Range
Basic port carrier	$3,000–$5,000
Expansion port carrier	$5,000–$10,000
Multicarrier cabinet (4–6 carriers)	$35,000–$50,000

Port Circuit Cards

Port circuit card prices are based on port circuit terminations. Table B-4 lists likely price ranges based on the number of port circuit terminations. For example, a 24-port circuit card priced at $4,800 would be priced at $200 per port.

TABLE B-4
Price Ranges for Port Circuit Cards

Port Circuit Card	Price Range per Port
Analog station line	$175–$250
Digital station line	$175–$250
ISDN station line	$200–$300
IP station line (control signaling only)	$1,000–$3,000 (per card)
IP station line (gateway only)	$150–$300 (per gateway)
IP station line (control signaling and gateway)	$150–$300 (per gateway)
CO GS/LS trunk	$200–$250
DID trunk	$225–$300
E&M tie trunk	$250–$300
Digital trunk (T1/E1)	$2,500–$5,000 (per card)
IP trunk	$250–$500 (per gateway)

Voice Terminals

Voice terminals represent the largest cost variable in a PBX system because there is a great range of telephone instruments and options.

PBX/IP-PBX Cost and Pricing Issues

Table B-5 shows the likely price ranges for the current portfolio of PBX voice terminals and common options.

TABLE B-5
Price Ranges for Voice Terminals

Voice Terminal Type	Price Range
Basic analog telephone	$25–$50
Analog featurephone/SP	$50–$125
Desktop audioconferencing unit	$300–$1000
Single line digital telephone	$75–$100
Basic multiline digital telephone	$100–$200
Basic multiline digital telephone/SP	$200–$300
Advanced multiline digital telephone/display and SP	$300–$600
ACD digital telephone	$750–$900
Entry IP telephone	$100–$250
Basic multiline IP telephone/display and SP	$250–$500
Advanced multiline IP telephone/browser display and SP	$500–$1,000
IP audioconferencing unit	$800–$1200
Options	
Analog or digital adapter	$200–$250
IP adapter	$250–$350
Add-on line module (10–16 buttons)	$150–$200
Data/API module+A33	$200–$300

Systems Management

Most PBXs bundle basic systems management software into the basic assembly price. Customers must provide their own terminal equipment. Advanced systems management software options that provide sophisticated GUIs and multiple system support and require customer-provided PC clients are priced starting at about $5,000. Multiple user license fees are available at discount. The price for server-based systems management option starts at about $25,000. The new generation of client/server IP-PBXs support Web-based systems management access and may or may not be bundled in the price of a customized server cabinet with generic operating software.

Fault management and traffic reporting software are usually priced as an option that is priced based on system port capacity limits. Starting prices begin at about $3,000 for basic reporting and $5,000 for advanced reporting.

Third-party call accounting software is based on system subscribers. Basic call accounting software packages can start as low as $1,000 for up to 100 station users. Software packages for intermediate and large line size systems start at about $5,000, or about $10 to $25 per station user, depending on the number and types of accounting reports.

Installation and Cabling

These hardware prices are exclusive of installation charges added by the distributor to configure, initialize, and physically install the PBX system. Installation pricing is typically based on unit cost per hardware component and ranges from about 5 to 10 percent of total hardware costs.

PBX cabling costs are based on equipped port capacity and typically range from about $75 to $125 per installed station based on the complexity of the system configuration. This cost estimate includes all equipment room, distribution frame, and wiring closet costs and assumes a single location installation, not a multiple location configuration with intramachine trunk requirements (private and/or PSTN). Cabling costs can be significantly reduced if existing frames and wiring are reused.

IP-PBX Pricing

IP-PBX system pricing differs from that of traditional PBXs. Table B-6 lists estimated costs to upgrade an installed circuit switched PBX system to a converged IP-enabled IP-PBX platform.

A client/server IP-PBX design does not use traditional common equipment but still includes many of the same cost elements of a circuit switched PBX. Table B-7 shows estimated client/server IP-PBX price ranges.

PBX/IP-PBX Cost and Pricing Issues

TABLE B-6
IP-PBX Price Ranges

System Price Element	Price Range
Common control upgrade	$0–$25,000
Software generic upgrade	$0–$50/port
IP port interface	$50–$200/port
IP telephone (multiline)	$250–$700
Audio codec user license	$0–$200/port
Desktop power	$25–$100/port
LAN/WAN upgrade	$0–$100/port
Typical upgrade cost/station	$500–$1,000/port
Typical upgrade cost/trunk	$100–$500/port

TABLE B-7
Client/Server IP-PBX Price Ranges

System Price Element	Price Range
Call telephony server with software	$10,000–$15,000
IP telephony, entry	$100–$250
IP telephony, advanced	$300–$500
IP telephony, Web browser	$600–$800
PSTN gateway, T1 interface	$3,000–$5,000
Analog station gateway	$200–$400
In-line power/station	$50–$100
LAN/WAN upgrade/station	$50–$100
Price range/station (200 stations)	$600–$1,000

These price estimates include audio codec license fees, desktop AC power transformer, and systems management software, but not a PC client management terminal. Installation fees are about 10 percent of system hardware costs. Pricing also assumes an existing LAN/WAN infrastructure. If it is a green field installation, the LAN/WAN infrastructure costs would be about $100 to $150 per station and cabling would be about $50 to $100 per station.

Appendix B

PBX Price Curve

PBX system prices are based on fixed and variable cost elements. The major fixed cost element of a circuit switched PBX system price is the basic system assembly, including common control. Port carrier, port circuit card, terminal costs, cabling, and installation costs depend on equipped stations and trunks. The system price also depends on the model port capacity and percentage of equipped ports. If fixed cost elements can be allocated to a larger number of equipped ports, the system price per port is slightly higher than the incremental port cost. PBX model price curves decline exponentially before reaching a price floor at about 75 percent of maximum port capacity.

Pricing for each PBX model is very high at the start of the price curve and declines as equipped port size increases to maximum capacity. Although a customer can minimize system pricing by installing the model with the lowest pricing at a fixed port capacity, port growth requirements and upgrade costs to the next larger model must be considered. Most customers typically require a minimum of 100 percent port expansion of a new PBX system and will select a system that is not optimally priced at initial installation.

APPENDIX C

Client/Server IP-PBX RFP Example

A request for proposal (RFP) is a customer document used in the PBX system purchase process. The document states the bidder ground rules for proposing a PBX system that satisfies the customer's communications needs and wants.

A PBX RFP includes sections on overall system requirements, port land traffic capacities, design issues, generic software features, terminals, applications support, and systems management. More than a listing of system design attributes and features, it also includes system installation and cutover milestones, terms and conditions for bidding and contract negotiations, and legal clauses to protect the customer.

The sample RFP was prepared by Siemens ICN, the leading global supplier of PBX systems. Although exhibiting some slight bias toward Siemens product capabilities, the document can be used as a framework to evaluate any supplier's enterprise voice communications systems.

Communications System Request for Proposal for [Customer Name]

Proposal Due Date: [ResponseDate]

Customer Requirements

Background Information

This Request for Proposal (RFP) is intended to provide a standard base from which to evaluate alternatives for communications systems and to allow the bidder flexibility in providing the most appropriate and cost-effective solution. The acceptance of a Response to Request for Proposal does not obligate [CustomerName] ([Customer]) to purchase any system.

Customer Contacts

For questions about the Request for Proposal or [Customer]'s system requirements, contact:

Contact Name: _____
Phone Number: _____
Fax number: _____
Email Address: _____

Proposal Filing Date

Responses should received by (_____) p.m., on (Day), (Date). Bids received after that time may not be considered. Deliver bids to:

Name: _____
Company: _____
Address: _____
Telephone: _____
# of hard copies _____
# of soft copies _____

Schedule of Events

Event Date _____
Request for Proposal Issued _____
Bidders Conference _____
Site Visit _____
Last Day for Questions/Clarifications _____
Response to Bidder Questions _____
Response to Request for Proposal Due _____
Bidder Presentation _____
Purchasing Decision _____
Contract Signing _____
Solution Implementation Date _____

Proposal Format

The Response to the RFP must be organized in the following manner:

- Table of Contents
- Section 1 Executive Overview—This section summarizes the response, structured so that anyone reading only this section has a clear understanding of the proposed solution. Bidders must clearly identify the benefits afforded [Customer] through the implementation of the proposed solution.
- Section 2 Statement of Compliance—Statement of compliance or noncompliance with RFP requirements, pricing, and terms and conditions. If an alternative solution provides required functionality, highlight any operational variances and additional costs at the end of the Statement of Compliance table.
- Section 3 Bidder Profile
- Proposed Requirements—This section provides the [Customer]'s proposal evaluation team sufficient information to assess compliance with the requirements. Describe optional features only if included in the proposed solution.
  Section 4—Communication Platform
  Section 5—Workpoint Clients
  Section 6—Messaging Applications
  Section 7—Customer Interaction Applications

Section 8—Mobility Solutions
Section 9—Management
Section 10—Availability
- Section 11 Configuration, Pricing and Terms and Conditions
- Appendices
  - Customer Requirements
  - References
  - Products and Services Literature

Customer Environment and Requirements

[Customer]'s current environment is as follows: (customer insert information regarding current environment).

The proposed solution must address the following objectives: (customer insert information regarding relevant business objectives).

[Customer] will base their decisions regarding responses to this Request for Proposal on the following criteria. (Sample criteria appear below. Replace with Customer's Decision Criteria).

Key Decision Criteria
- Global platform, using the same communications hardware and software nationally and internationally, adjusted as necessary for local regulations and operating conditions.
- Global availability, covering remote serviceability and technical support of the entire proposed solution, including applications. Availability infrastructure must provide a single point of contact, domestically as well as worldwide.
- Asset protection, providing a platform that accommodates future technologies and allows a smooth, cost effective migration path. The solution must
  - be based on an open architecture
  - adhere to industry standards
  - ensure modular, cost-effective growth in the base system
  - permit addition of application capacities and processing power as needed
  - interoperate with IP-PBXs
- Efficient call processing through distributed architecture
- Increased efficiencies at the desktop, helping individual employees communicate better and work more productively at the desktop or other workpoint

Client/Server IP-PBX RFP Example

- Management, providing maximum flexibility for rapid, efficient and cost-effective configuration changes affecting personnel and associated equipment, and supporting management control of call abuse through reporting features and security precautions

Within the specified criteria, each bidder is invited to design a solution meeting [Customer]'s objectives. Preference will be given to the bidder providing a comprehensive, cost-effective, single-vendor solution for current specifications, future capacity requirements, and ongoing service and support.

The following sections detail the hardware, software, and wiring requirements necessary to configure the proposed solution. Bidder is responsible for clarifying any items in accordance with the Schedule of Events.

Required Equipment

Communications Platform Switching Equipment	Equipped[1]	Wired[2]
Digital voice ports		
Analog voice/fax/data ports		
Analog CO trunks		
Analog tie trunks		
Digital trunks T1 trunks		
Mobility users		
ATM backboning		
Redundancy option		
Attendant position		
Switching to host data connections/modem connections		
System failure by-pass lines		
Message announcer		
Paging adapter		

---

[1] Equipped means the specified quantities available at implementation.

[2] Wired means the specified quantities can be accommodated by the proposed solution with only the addition of deskset devices/licenses and line/trunk cards.

Appendix C

Server Application Requirements	Equipped	Wired
Private networking		
Integration with virtual private networks		
Integration with QSIG-compliant third-party switches		
Integration with narrowband/wideband ATM switches		
Voice mail/voice messaging users		
Integrated email capability		
Integrated fax capability		
Mobility users		
CIC (customer interaction center) agent software licenses æ basic group call control		
CIC agent software licenses æ virtual group skills-based call control		
CIC agent soft client licenses		
CIC supervisor soft client licenses		
CIC simulation licenses		
CIC reporting interface licenses		
CIC call-by-call reporting feature		
CIC system status monitoring application		
CIC information collection application		
CIC self-service application		
CIC email application		
CIC automated callback application		
CIC web-based application		
CIC outbound call application		
CIC CTI API		
CIC message board application		
CIC real-time messaging to the desktop		
Remote worker application		
Open seating application		
Campus wireless system		

Client/Server IP-PBX RFP Example

Digital Station Equipment	Up to # Lines/Keys	Speaker	Display	Add-on Module Bay	Quantity
Single-line digital phone	1/4				
Multi-line digital phone	4/8	✔			
Multi-line digital phone	4/8	✔		✔	
Multi-line digital phone	12	✔	✔		
Multi-line digital phone	12	Two-way	✔		
Multi-line digital phone	12	Two-way	✔	✔	
Multi-line digital phone	12	Full duplex	✔	✔	
Multi-line digital phone with line capacity expansion option	29/76	Two-way	✔	✔	
Multi-line digital phone with line capacity expansion option	29/76	Full duplex	✔	✔	
Campus wireless system phone	1/4		✔		

Appendix C

Add-on Station Equipment	Quantity

Modular analog adapter—supports any EIA/TIA-464 analog device

Modular phone adapter—allows one phone line to support two telephones

Modular control adapter—provides TAPI-compliant

CTI application connectivity Modular data adapter—supports industry-standard data device transmission

Modular ISDN adapter—provides a BRI S- or T-interface

Modular headset adapter—supports amplified and non-amplified Electret-type headsets

Modular headset/recorder adapter—supports amplified carbon-type headsets and 600-ohm recording devices

Modular privacy adapter—supports public-key encryption and decryption

Modular distance adapter—allows digital telephone operation up to 3.4 miles from switching platform across standard, 22-gauge wire

KEM (key expansion module)—adds up to 16 feature/line keys with associated LEDs/KEM

Other Equipment	Quantity

Workpoint Soft Client software

Remote worker adapters

CSU/DSU

MDF

Recorders/announcers

Analog phones for by-pass and fax machine

Wall mount jacks

Headsets

System administration terminals

System administration printers

Security devices

Call accounting software

Quality assurance system

Workforce management system

Predictive dialing system

CRM software application

Client/Server IP-PBX RFP Example

Messaging[3]	Quantity
Average message length (in seconds)	
Messages per voice mail mailbox	
Messages per user/hour	
Peak hour traffic hours in total daily traffic	
Calls per peak hour	
Maximum number of users proposed system to support	
Minimum hours/megabytes message storage capacity	
Maximum hours/megabytes additional message storage capacity	
Maximum number of channels per server	
Automated attendant channels (if channels are not universal)	
Automated attendant/call processing levels	
Voice forms applications Information services applications	
Number of multi-lingual automated attendants, voice forms and information services	
Number of outcalling destinations per subscriber	
Number of subscribers licensed for text-to-speech application	
Number of voice-only subscribers	
Number of email user licenses	
Number of fax user licenses	

Additional Equipment

[Customer] to complete the following:
Provide the following additional hardware, software, or services. Bidder is responsible for including all required products and services necessary to provide functionality defined in this RFP in the purchase price.

Section 1 Executive Overview

Provide an overview of the proposed solution.

---

[3]Only single vendor solutions will be considered. The voice messaging system must assure integrated compatibility and serviceability. The switching to voice messaging interface should be sized to provide a P.01 grade of service.

Appendix C

Section 2 Statement of Compliance

Provide a statement of compliance/noncompliance with requirements. If an alternative solution provides required functionality, highlight operational variances and additional costs at the end of this table.

Requirement	Comply	Comply with Alternative	Does Not Comply
Section 3 Bidder Profile			
3.1 Bidder History and Experience			
3.2 Strategic Direction Investment Protection Support for IP Telephony			
3.3 Research and Development			
3.4 Customer Participation			
3.5 Quality—ISO 9000 Certification			
3.6 Other Products and Services			
Section 4 Communication Platform			

4.1 Proposed Solution
  Server Based
  Digital Switching
  Voice and Data over Single, Twisted-Pair
  Self-Diagnostic Routines
  Modular Design
  Interoperate with IP-PBXs
  Wired for Configuration			

4.2 Support for Open Standards			
4.3 System Architecture			

4.4 Redundancy
  Power Supply
  No Single Point of Failure in Power System
  Optional Redundant CPU, Memory, and Switching			

4.5 Distance Limitations			
4.6 Maximum Capacities & Busy Hour Traffic Handling			
4.7 Software Features			
4.8 Optional Features Proposed			
4.9 Interface to Networks Like System Integration			

Client/Server IP-PBX RFP Example

Requirement	Comply	Comply with Alternative	Does Not Comply
Non-bidder System Integration			
Public Switching Integration			
4.10 Private Network Interfaces			
4.11 Public Network Interfaces			
4.12 Trunk Signaling Protocols			
4.13 Automatic System Restart			
4.14 Ability to Support CTI and Third-Party Products			
4.15 IP Telephone Integration			
4.16 Facility Requirements			
Section 5 Workpoint Clients			
5.1 Telephone Features			
Digital			
Single, Twisted-pair Wiring			
Optional Display			
Speaker—handsfree dialing, 2-way & full-duplex			
5.2 Proposed Models			
5.3 Context-Sensitive Help & User-Customizable Help			
5.4 Digital Cordless Phone			
5.5 User-Installed Modules			
5.6 Support Phones Wired Up to ? Mile from Switch			
5.7 Privacy			
5.8 Line Powered Phone			
5.9 Soft Client			
5.10 Attendant Position			
5.11 Attendant Console Features			
Section 6 Messaging Applications			
6.1 Fully Integrated Messaging Solution			
6.2 Support for Open Standards			
6.3 System Architecture			
6.4 Expansion Capabilities			
6.5 Digital Networking			
6.6 Features			
6.7 Optional Features Proposed			

Requirement	Comply	Comply with Alternative	Does Not Comply
6.8 Security			
6.9 User Interface			
6.10 Receiving Messages			
6.11 Sending Messages			
6.12 Mobile Users			
6.13 System Management			
6.14 Implementation			
6.15 Facility Requirements			
6.16 Multiple Language Support			
Section 7 Customer Interaction Applications			
7.1 Customer Interaction Solution			
7.2 Support for Open Standards			
7.3 System Architecture			
7.4 System Capacity and Expansion Capabilities			

7.5 Basic Call Control Capabilities
    ANI/DNIS
    Call Volume
    Performance Criteria
    Priority Queuing			

7.6 Advanced Call Control Capabilities
    Agent Skills
    Customer Preference
    Inbound and Outbound Call Levels
    Multimedia			

7.7 Caller Notification of Wait Time			
7.8 Transfer to Voice Messaging Application			

7.9 GUI Administration Tool
    Real-time Reconfiguration
    Change Priority of Multiple Calls Simultaneously
    View Details of Orphaned Calls
    Retain Customized Settings
    Graphical Editor
    What-if Modeling			

7.10 Soft Client
    Online Help
    Ability to Reserve Calls
    Change Call Priority			

7.11 Real-time Call Handling and Performance Status
    User Customizable Display
    Logged-on Agent
    Agent Status			

Client/Server IP-PBX RFP Example

Requirement	Comply	Comply with Alternative	Does Not Comply
Queue Status			
Thresholds			
7.12 Statistical and Configuration Reporting			
Available on Terminal Display			
Available in Hardcopy			
Downloadable to PC			
Open Storage			
7.13 Graphical Reporting			
7.14 Call-by-Call Reporting			
7.15 Self Service			
7.16 Script Development			
7.17 Workforce Management System			
7.18 Integrated Email Call Control			
7.19 Auomated Callbacks			
7.20 Web-based Interaction			
7.21 Outbound Dialing			
7.22 Quality Assurance System			
7.23 Host-Based CTI Call Control			
7.24 Multiple Access Levels			
7.25 Multi-Site Support			
7.26 Professional Services			
7.27 CRM Applications			
Section 8 Mobility Solutions			
8.1 Mobile Business			
8.2 Support for Open Standards			
8.3 Remote Workers			
8.3.1 Remote Worker Architecture			
8.3.2 Remote Worker Capacity			
8.3.3 Remote Worker Features			
8.3.4 System Administration Tools			
8.4 Open Seating Solution			
8.5 Campus Wireless Solution			
8.5.1 Campus Wireless Architecture			
8.5.2 Campus Wireless Telephone			
8.5.3 Campus Wireless Features			
8.5.4 Campus Wireless Specifications			
8.5.5 Wireless Transmission Security			

Requirement	Comply	Comply with Alternative	Does Not Comply
8.6 Single Number Wireless Solution			

Section 9 Management

9.1 Proposed System Administration Tool			
9.2 Support for Open Standards			
9.3 System Administration Architecture			
9.4 Single Point of Access			
9.5 System Administration Features			
9.6 Security			
9.7 Training Requirements			
9.8 Optional System Management Tools			

Section 10 Availability

10.1 Bidder Implementation Responsibilities			
10.2 Customer Implementation Responsibilities			
10.3 Implementation Team and Master Project Schedule			
10.4 Training Requirements			
10.5 Ongoing Education			
10.6 Warranty Coverage			
10.7 Maintenance Coverage			
10.8 Price Protection Options			
10.9 Maintenance Support			
10.10 Internal Diagnostics			
10.11 Remote System Monitoring			
10.12 Spare Parts Availability			
10.13 Availability Options			

Section 11 Configuration, Pricing and Terms, and Conditions

11.1 Configuration/Pricing			
11.2 Payment Options			
11.3 Standard Terms and Conditions			

Client/Server IP-PBX RFP Example

Section 3 Bidder Profile

Complete the following table:

Corporate Profile:
Corporate Name:
Corporate Address:
Number of Years Doing Business:

Sales Team Profile:
Sales Person/Account Executive
e-mail address
Telephone Number
Fax Number
Address
Technical Advisor
Telephone Number
Address
Sales Manager
Telephone Number
Address

24-Hour Support Center

If the proposing bidder does not manufacture the recommended solution, please provide manufacturer's name and bidder/manufacturer relationship, including future support for proposed solution. If bidder does not provide installation, warranty, or maintenance services, explain manufacturer/subcontractor responsibilities.

3.1 Bidder History and Experience

Bidder must have over 10 years' experience in providing communications technology and provide 24-hour, 7-day a week support. Provide a history of the bidder's company and experience in the manufacturing, implementation, and support of the proposed solution.

3.2 Strategic Direction

Provide a description of future development plans, including investment protection and support for IP Telephony within the proposed solution.

3.3 Research and Development

Bidder should describe their commitment to ongoing research and development.

3.4 Customer Participation

Describe how installed-base customers participate in defining product and services requirements.

3.5 Quality

Bidder must be ISO 9000 certified. Describe the bidder's quality programs in the development and manufacture of communication and information networks.

3.6 Other Products and Services

Briefly describe products and services available but not included in the proposed solution (e.g., directory services, professional services, packet-based switching IP telephony solution, e-Business, etc.). Do not include product literature/brochures as response to this section.

Section 4 Communication Platform

4.1 Proposed Solution

The proposed communications platform requirements include server-based, digital switching capable of integrated voice and data communications over standard, single twisted-pair telephone wire. It must have stored program control, self-diagnostic routines, modular design, and optional duplication of critical subsystems. The proposed solution must be able to interoperate with packet-based switches, telco switches, and third-party switching solutions.

The bidder must guarantee that expansion from the "equipped for" configuration to the "wired for" configuration requires only the addition of workpoint clients and printed circuit cards. All cabinets, carriers, power supplies, etc., for the "wired for" system must be included in this Response to RFP.

Describe the proposed solution, including capability to interface with existing information and communication infrastructure. Identify advantages that differentiate proposed solution.

4.2 Support for Open Standards

Describe communication platform support for open standards.

4.3 System Architecture

Describe the software and hardware architecture of the proposed solution.

4.4 Redundancy

The proposed solution must offer multiple speech paths ensuring there is no single point of failure down to the shelf level.

It must offer, as an option, a redundant CPU and memory, redundant switching unit, redundant switching unit to remote shelf links, and redundant power supplies, ensuring there is no single point of failure in the power system. Describe the level of redundancy of the proposed solution and identify which components are not redundant.

4.5 Distance Limitations

Describe how far users on a remote node can be from the main communication platform location. Identify any redundancy issues or additional operating costs required to support remote location.

4.6 Maximum Capacities

Complete the following table with the maximum capacities of the proposed solution.

Cabinet Limits

Number of cabinets
Shelves per cabinet
Line and trunk shelves
Total ports
Total lines
Fiber-linked extension shelves
Remote modules
Remote modules—ports per system

System Capacities

Trunks
Trunk groups
T1 spans
Logical lines
Number of voice/data highways
Bandwidth
Busy hour call completions
Attendant consoles
Voice messaging channels
Call forwarding hops
CTI links
Class of service
Digit screening
LCR (least cost routing) outbound dialing rules
Pickup group members
Private network extension digits
Private network extensions in database
Maximum number of conference call parties on non-ISDN trunks

The proposed solution must be nonblocking[4] and include sufficient time slots, software, and hardware capacity to provide the following minimum busy-hour traffic handling capability for the "wired for" configuration:

- 12 CCS voice
- 36 CCS data
- 32 CCS (minimum) per CIC agent

Explain how the addition of station users, trunks, and/or data communications affects the traffic handling capability of the proposed solution.

4.7 Software Features

Briefly describe basic and optional applications. Include a complete list of standard and optional software features in the Appendices.

4.8 Optional Features Proposed

Describe optional features included in the proposed solution.

4.9 Interface to Networks

The proposed solution must support private and public networks, allowing integration with other like systems, non-bidder systems, and public switching systems. Describe private and public network capabilities.

4.10 Private Network Interfaces

Describe feature transparency provided by private network interfaces.

---

[4]A communication platform is nonblocking if sufficient switching network bandwidth exists for all ports to be simultaneously involved in two-way conversation. The term nonblocking does not imply that sufficient resources, other than bandwidth, are available for all stations to make calls simultaneously.

4.11 Public Network Interfaces

Specify public network trunk interfaces and services that are supported.

4.12 Trunk Signaling Protocols

Specify signaling protocols supported.

4.13 Automatic System Restart

In the event of total system failure, all stored program control devices must, upon systems recovery and without outside intervention, immediately initialize all system software instructions and customer-specific data. Bidders must describe the time involved and the steps required to accomplish the restart procedure.

4.14 Ability to Support CTI and Third-Party Products

Describe ability to support host CTI and third party products, such as fax and video. Specify standards supported.

4.15 IP Telephone Integration

Describe capability of proposed solution to communicate with packet-based switch and, if optional, what additional equipment is required.

4.16 Facility Requirements

Summarize space, physical access, weight, temperature, heat dissipation, humidity, and electrical requirements for the proposed solution. Include a complete list of environmental specifications in the Appendices.

Section 5 Workpoint Clients

5.1 Telephone Features

The proposed solution must support a variety of digital telephone instruments, requiring no more than single twisted-pair wiring. These phones must optionally provide displays, two-way speakerphones, duplex speaker, and user-installed modules to provide functionality only when required.

Describe telephone set features and models available.

5.2 Proposed Models

Describe telephone models proposed and include pictures.

5.3 Context-Sensitive Help

Describe real-time, context-sensitive help feature, and user-customizable help files.

5.4 User-Installed Modules

The proposed telephones must accommodate user-installed telephone modules. Briefly describe capabilities available through the user installable modules. Include the tools and technical support required to add functionality to the telephone.

5.5 Distance Adapter

The proposed solution must support distances beyond 4,500 feet and up to 18,000 feet from the switching equipment via a straight copper connection. Describe.

5.6 Encryption

The proposed solution must support an optional Public Key encryption algorithm, providing a secure, scrambled voice conversation between a local and remote user. Users must be able to deactivate this feature for a standard unencrypted call. Describe.

5.7 Local Power Supply Requirements

Phones must be line powered through the switching equipment. Describe any local power requirements.

5.8 Soft Client

Describe soft client capabilities.

5.9 Attendant Position

Describe the proposed attendant console.

5.10 Attendant Console Features

For the features listed below, the bidder must indicate if the requirement is Standard (S), Optional (O), or Not Available (NA). Clearly explain any exception immediately following the table.

Required Features	S	O	NA
Access to Special Devices			
CIC Participation			
Alarm Indicator			
Alphanumeric Display			
Answer an External Call			
Answer an Internal Call			
Answering Priority			
Busy Override			

Client/Server IP-PBX RFP Example

Required Features	S	O	NA
Call External Numbers			
Call Internal Numbers			
Call Waiting			
Console Self-Test			
Control Of System and Station Features			
Direct Trunk Select			
Display Called Party Name and Number			
Display Called Trunk Number			
Display Calling Party Name and Number			
Display Calling Party Status			
Display Calling Trunk Number			
Display Time and Date			
Extension of Calls			
Hold			
Inter-Console Calling and Transfer			
Intercept			
Message Waiting Activation			
Multiple Console Operation			
Multiple Language Support			
Notebook			
Overflow Answer Position			
Programmability			
Queuing For Busy Station			
Queuing for Busy Trunk			
Remote Attendant			
Repertory (Repdial) Keys			
Ring Tone Select			
Ring Tone Volume Control			
Saved Number Redial			
Serial Calling			
Switched Loop Operation			
System Hold			
System Speed			
Trunk-to-Trunk Connection			
Two-Way Splitting			
Uniform Call Distribution			
Verification of Busy Extensions			
Verification of Busy Trunks			

Section 6 Messaging Applications

6.1 Integrated Messaging Solution

The ([Customer] specify whether a unified messaging or voice messaging solution is required) unified voice messaging solution must be fully integrated with the proposed communication platform. Describe messaging solution capabilities and capacity, with emphasis on how [Customer]'s operations will benefit.

6.2 Support for Open Standards

Describe messaging application support for open standards. List the clients that can be used with your proposed solution. For proprietary clients, detail minimum hardware and software requirements.

6.3 System Architecture

Describe the software and hardware architecture of the proposed messaging solution, including media supported. If a unified messaging solution is proposed, describe requirements for e-mail and fax integration.

6.4 Expansion Capabilities

Describe process for adding users when needed.

6.5 Digital Networking

Describe digital networking capabilities.

6.6 Features

Describe basic messaging, call processing, and outcalling functionality. Briefly describe any optional features available from the proposed messaging solution. Include a complete list of standard and optional software features in the Appendices.

6.7 Optional Features Proposed

Describe optional features included in the proposed messaging solution.

6.8 Security

Describe security features of the proposed messaging solution.

6.9 User Interface

Describe user interfaces.

6.10 Receiving Messages

Describe how the user receives messages.

6.11 Sending Messages

Describe how the user sends messages.

6.12 Mobile Users

Describe features supporting mobile users.

6.13 System Management

Describe messaging solution management and administration tools.

6.14 Implementation

Describe implementation and training process.

6.15 Facility Requirements

Briefly describe space, physical access, weight, temperature, heat dissipation, humidity, and electrical requirements for the proposed messaging solution.

6.16 Multiple Language Support

Describe foreign languages support.

Section 7 Customer Interaction Applications

7.1 Customer Interaction Solution

[Customer]'s customer interaction center requires effective administration, call control, reporting, and customer interaction center management information. The bidder must demonstrate expertise in design, implementation, and support.

The proposed solution must efficiently and economically route, queue, and distribute calls in response to the following customer interaction center characteristics ([Customer] modify sample environment characteristics shown below):

Requirement	Fill In or Place Check Mark in Appropriate Box		
Call Control	Multiple group for homogeneous agent skills	Skills-based for specific requirements	Custom routing
	Supervisor ability to assign calls to specific agents	Agent ability to reserve calls	
Desktop	Supervisor softclient	Agent softclient	Real-time messaging
Reporting	Historical	Graphical	Call-by-call
Integration	Self service	Workforce mgmt	Email
	Callback	Internet	Outbound dialing
	Quality assurance	Host-based CTI	eCRM

Client/Server IP-PBX RFP Example

Requirement		Fill In or Place Check Mark in Appropriate Box	
Capacity	Number of agents	Number of supervisors	
Environment	Single location	Networked locations	Multi-vendor PBX integration

The proposed customer interaction solution must be fully integrated with the proposed communication platform. Describe solution capabilities, with emphasis on how [Customer]'s operations will benefit.

7.2 Support for Open Standards

Describe customer interaction application support for open standards.

7.3 System Architecture

Describe the software and hardware architecture of the proposed customer interaction solution. Identify hardware and software required for the customer interaction application to integrate fully with the communication platform.

7.4 System Capacity and Expansion Capabilities

Describe proposed solution capacities and whether agents can be added at any time.

7.5 Basic Call Control Capabilities

At a minimum, the proposed solution must be able to provide call control based on:

- ANI/DNIS
- Call volumes
- Performance criteria
- Priority queuing

Describe call control methodology.

7.6 Advanced Call Control Capabilities

As an option, the proposed solution must be able to provide call control based on:

- Agent skills
- Customer preference
- Inbound and outbound call levels
- Multimedia

Describe call control methodology.

7.7 Caller Notification of Wait Time

The proposed solution must be able to notify callers of expected wait times and "place" in queue and support information collection (such as an automated attendant feature) using "internal" hardware and software. Describe how the application calculates wait time and any optional hardware or software required.

What impact does announcement of wait time have on a caller's state in queue?

7.8 Transfer to Voice Messaging Application

After a configurable time, the caller should be able to transfer to a voice messaging system to leave a callback message. If the caller chooses to continue waiting rather than hanging up after leaving a message, describe how the call is placed back in queue.

7.9 GUI Administration Tool

Supervisors must be able to reconfigure call control and assignments in real time, change priority of multiple calls simultaneously, view details of orphaned calls, and retain customized settings regardless of log-on location. The solution must use a GUI administration tool and provide a graphical editor and what-if modeling as standard. Describe.

7.10 Soft Client

Describe the soft clients available for agent and supervisor use. Soft client must provide on-line help, ability to reserve calls, or change call priority. For proprietary clients, detail minimum hardware and software requirements.

If standard telephones can support agents and supervisors, describe specific features provided.

7.11 Real-time Call Handling and Performance Status

Supervisor terminals must show, in real time, all logged-on agents, the status of each agent, caller queue information, and thresholds and alarms. Users must be able to customize displays. Describe the proposed solution's real time display capabilities for assisting supervisors with managing the customer interaction center.

Describe real-time display information provided to agents.

7.12 Statistical and Configuration Reporting

[Customer] requires sophisticated reporting to track and further enhance its CIC operations. Reports must be available on terminal display and paper printout and be able to be downloaded to a PC. The proposed solution must provide open storage capability.

Describe the number of and type of information standard statistical, configuration, and audit reports provide.

7.13 Graphical Reporting

The proposed solution must provide graphical reports as a standard feature. Describe.

7.14 Call-by-Call Reporting

The proposed solution must provide call-by-call reporting as an optional feature. Describe.

7.15 Self Service

The proposed solution must support self service (e.g., IVR) integration as an option. Can callers retain their place in queue while using IVR features? Describe ability to support inbound calling, call control services, messaging for agents, speech recognition, text-to-speech, TDD and CTI, and integration with customer interaction application.

7.16 Script Development

Describe the design tools and environment for IVR script development, the method used to test applications and changes prior to putting them into production, and the method of putting changes into production.

7.17 Workforce Management System

The proposed solution must provide forecasting and scheduling capabilities as an option. Describe.

7.18 Integrated Email Call Control

The proposed solution must integrate customer email messages as an option. Describe.

7.19 Callbacks

The proposed solution must support customer callback requests via IVR, Web, and voice requests as an option. Describe

7.20 Web-Based Interaction

The proposed solution must support customer-initiated contact through the Internet as an option. Describe.

7.21 Outbound Dialing

The proposed solution must support automated outbound dialing as an option. Describe.

7.22 Quality Assurance System

The proposed solution must support quality monitoring as an option. Describe.

7.23 Host-Based CTI Call Control

The proposed solution must support host-based CTI applications as an option. Describe the capabilities of the proposed solution to simultaneously route a call and data screen populated with the caller's identity, location, or reason for calling.

7.24 Multiple Access Levels

The proposed solution must provide multiple access levels. Describe.

7.25 Multi-Site Support

The proposed solution must support multi-site networking as an option. Describe.

7.26 Professional Services

Describe professional services available.

7.27 CRM Applications

Describe customer resource management (CRM) applications available.

Section 8 Mobility Solutions

8.1 Mobile Business

Mobility solutions must support individual workers at remote sites (teleworkers), open seating at corporate locations, and campus wireless capabilities. Describe mobility solutions available.

8.2 Support for Open Standards

Describe mobility solutions support for open standards.

8.3 Remote Workers

Describe individual remote worker solution.

8.3.1 Remote Worker Architecture. Describe the hardware and software required for individual remote workers, including client licensing options.

8.3.2 Remote Worker Capacity. Describe remote worker capacity.

8.3.3 Remote Worker Features. Describe remote worker features.

8.3.4 System Administration Tools. Describe remote worker administration tools.

8.4 Open Seating Solution

Describe corporate hoteling solution.

8.5 Campus Wireless Solution

Describe campus wireless solution.

Delete the following questions if proposed solution does not include campus wireless:

8.5.1 Campus Wireless Architecture. Describe campus wireless system hardware and software architecture.

8.5.2 Campus Wireless Telephone. Describe campus wireless telephone and optional accessories. The wireless telephone must be able to be used either as a stand-alone wireless telephone or as a wireless extension of the wired desktop telephone.

8.5.3 Campus Wireless Features. The proposed wireless system must be fully integrated with the communication platform. Full feature transparency is required. Describe campus wireless features.

8.5.4 Campus Wireless Specifications. List campus wireless solution and wireless telephone specifications.

8.5.5 Wireless Transmission Security. Describe the system's ability to provide secure transmission during conversations.

8.6 Single-Number Wireless Solution

What single-number (corporate wireless) solution is available?

Section 9 Management

9.1 Proposed System Administration Tool

System administration tools must provide maximum flexibility for rapid, efficient, and cost-effective configuration changes affecting personnel and associated telephone equipment. Describe system administration tools for the proposed solution.

9.2 Support for Open Standards

Describe system management tool support for open standards, LDAP, and SNMP.

9.3 System Administration Architecture

Describe hardware and software architecture for the proposed system administration tool.

9.4 Single Point of Access

Describe extent to which proposed solution can provide single point of access for system-wide and networked environments.

9.5 System Administration Features

Describe how the proposed system administration tool contributes to [Customer]'s operating efficiency.

9.6 Security

Please define the security features that are incorporated within the proposed solution.

9.7 Training Requirements

Describe training requirements for proposed system administration tool.

9.8 Optional System Management Tools

Describe optional system management capabilities.

Section 10 Availability

10.1 Bidder Implementation Responsibilities

The selected bidder is responsible for the complete engineering of the proposed solution. Charges for all of services should be included in the total implementation price of the proposed solution. Briefly describe bidder implementation services.

10.2 Customer Implementation Responsibilities

[Customer] will provide cabling to the MDF. Describe other customer implementation responsibilities.

10.3 Implementation Team and Master Project Schedule

Describe the key team members responsible for the implementation of the proposed solution.

Upon award of contract, bidder will provide a master project schedule, identifying the tasks the bidder and [Customer] will perform. Include a sample milestone.

10.4 Training Requirements

The successful bidder will conduct end-user training on [Customer] premises, tailored specifically to [Customer]'s particular requirements (e.g., telephone user, soft client user, attendant position, administrator, CIC agent, CIC supervisor, administration support, executive, as appropriate). Describe on-site training. If required training is performed off site, please describe.

10.5 On-going Education

Describe additional education available.

10.6 Warranty Coverage

Please state the standard warranty provided for the bidder's entire proposed solution. Describe any optional or extended warranty coverage available. Bidder should include a copy of their warranty terms.

10.7 Maintenance Coverage

Briefly describe the various maintenance programs and options available to [Customer] following the warranty period.

10.8 Price Protection Options

Describe options that allow additional discounts or price guarantees for future service.

10.9 Maintenance Support Resources

All maintenance during the warranty period and under any maintenance agreements shall be performed by the successful bidding organization using personnel employed by the bidder and at no additional cost to [Customer] other than those charges identified in the applicable warranty/maintenance agreement.

Bidder should identify the address of the bidder's local service center and the number of service personnel trained on the proposed solution. Include in this section any other support levels available to [Customer] in supporting the proposed solution.

10.10 Internal Diagnostics

Please describe the diagnostic capability and alarms inherent within the communication platform.

10.11 Remote System Monitoring

The successful bidder will provide routine system monitoring, remote diagnostics and remote repair to assure the continued operation of all system components covered by bidder warranty or maintenance contracts. Describe remote system monitoring and any on-site monitoring available.

10.12 Spare Parts Availability

Describe the availability of spare parts maintained for critical hardware.

10.13 Availability Options

Describe optional customer support services available.

Section 11 Configuration, Pricing, and Terms and Conditions

11.1 Configuration/Pricing

Describe the pricing methodology and itemize all charges for individually identifiable components of the proposed solution, including all associated installation, programming, and cabling costs. Pricing must include charges for all components required to connect all applications, all design charges, Telco interface charges, and training charges (workpoint user, attendant position, system management, CIC agent, CIC supervisor training, etc.). The purchase price excludes estimated freight and sales taxes.

11.2 Payment Options

Bidder shall offer cash and leasing payment options. Describe.

11.3 Standard Terms and Conditions

The bidder shall include a copy of standard terms and conditions as part of the Response to RFP.

Appendices

The bidder shall include such other materials as required to supplement information contained in other section of this Response to RFP.

APPENDIX D

PBX/IP-PBX System Feature and Function Matrices

Appendix D

Terminals

	3Com	3Com	3Com
SUPPLIER			
MODEL NAME	NBX 1102	NBX 2102	NBX 2102-1R
INDICATE TYPE (ONLY ONE TYPE PER MODEL)			
DIGITAL			
PROPRIETARY CORDLESS			
ISDN BRI			
IP TELEPHONE (IF YES COMPLETE LINES 199-202)	NO	YES	YES
G.711 and/or G.729 ENCODING	g.711 only	YES	YES
AUTO SELF DISCOVERY/DHCP	YES	YES	YES
INTEGRATED ETHERNET HUB	YES	YES	YES
INTEGRATED ETHERNET PORT(S) (INDICATE #)	2	2	2
INTEGRATED WEB BROWSER	NO	YES	YES
PC-BASED SOFTPHONE	YES	YES	YES
LOOP LENGTH (24 AWG) (# FEET)	ntwk dependent	ntwk dependent	ntwk dependent
# WIRING PAIRS	2	2	
MAX # LINE/CALL APPEARANCE KEYS	12	12	12
TOTAL # FIXED FEATURE BUTTONS	16	16	16
TOTAL # PROGRAMMABLE LINE/FEATURE BUTTONS	16	16	16
SOFTKEY BUTTONS			
# DISPLAY-ASSOCIATED CONTEXT SENSITIVE KEYS			
# SELECT/CURSOR CONTROL KEYS			
SHIFT KEY OPERATION FOR LINE/FEATURE ACCESS	NO	NO	NO
ACD SPECIFIC DESIGN & FEATURE BUTTONS	YES	YES	YES
VOLUME CONTROL (SPECIFY BUTTON, SLIDE, ETC)	button	button	button
DEDICATED MESSAGE WAITING INDICATOR	YES	YES	YES
ON-HOOK DIALING	YES	YES	YES
HANDSFREE INTERCOM	YES	YES	YES
TWO-WAY SPEAKERPHONE	YES	YES	YES
SIMPLEX OPERATION	YES	YES	YES
FULL DUPLEX OPERATION	simulated	simulated	simulated
DISPLAY FIELD (LINES x CHARACTERS)	2x16	2x16	2x24
STORED DATA CAPABILITIES			
LAST INCOMING CALLS (INDICATE #)	NO	NO	NO
LAST NUMBERS DIALED (INDICATE #)	YES	YES	YES
FULLY INTEGRATED SYSTEM DIRECTORY ACCESS	YES	YES	YES
FULLY INTEGRATED PERSONAL DIRECTORY ACCESS	NO	NO	NO
DIRECTORY ACCESS DEPENDENT ON LDAP-TYPE SERVER	NO	NO	NO
HEADSET INTERFACE (INDICATE #)	1	1	1
MODULE ADAPTER BAYS/INTERFACES (INDICATE #)	NO	NO	1
MODULE OPTIONS			
RS-232 DATA MODULE	NO	NO	Infared
APPLICATIONS PROGRAMMING INTERFACE	YES	YES	YES
ANALOG ADAPTER	YES	YES	YES
PROPRIETARY DIGITAL ADAPTER	NO	NO	NO
DISTANCE ADAPTER	NO	YES	YES
SECURITY MODULE	NO	NO	NO
BRI ADAPTER	NO	NO	YES
IP ADAPTER	NO	YES	YES
ADD-ON BUTTON MODULE OPTION			
# MODULES	3	3	3
# BUTTONS/MODULE	100	100	100
SHIFT KEY OPERATION	YES	YES	YES
LINE POWERED (STANDARD OPERATION)	option	option	option
LOCAL AC POWER (OPTION-DEPENDENT)	std	std	std

PBX/IP-PBX System Feature and Function Matrices

SUPPLIER	Alcatel	Alcatel	Alcatel	Alcatel	Alcatel	Alcatel
MODEL NAME	4004	4010	4020	4035	4022	4037
INDICATE TYPE (ONLY ONE TYPE PER MODEL)						
DIGITAL		Y	Y	Y		
PROPRIETARY CORDLESS						
ISDN BRI						
IP TELEPHONE (IF YES COMPLETE LINES 199-202)	Y				Y	Y
G.711 and/or G.729 ENCODING	OPT	OPT	OPT	OPT		
AUTO SELF DISCOVERY/DHCP	Y	Y	Y	Y	Y	Y
INTEGRATED ETHERNET HUB	Y	Y	Y	Y	Y	Y
INTEGRATED ETHERNET PORT(S) (INDICATE #)	Three Port Switch	Three Port Switch	Three Port Switch	Three Port Switch	Three Port Switch	Three Port Switch
INTEGRATED WEB BROWSER	1 (AF)	1 (AF)	1 (AF)	1 (AF)	1 (AF)	1 (AF)
PC-BASED SOFTPHONE	N	N	N	N	N	N
LOOP LENGTH (24 AWG) (# FEET)	4000	4000	4000	4000	4000	4000
# WIRING PAIRS	1	1	1	1	2	2
MAX # LINE/CALL APPEARANCE KEYS	0	8	12	24	12	24
TOTAL # FIXED FEATURE BUTTONS	0	5	10	7	10	7
TOTAL # PROGRAMMABLE LINE/FEATURE BUTTONS	8	8	12	24	12	24
SOFTKEY BUTTONS						
# DISPLAY-ASSOCIATED CONTEXT SENSITIVE KEYS	0	0	12	24	12	24
# SELECT/CURSOR CONTROL KEYS	NA	NA	NA	NA	NA	NA
SHIFT KEY OPERATION FOR LINE/FEATURE ACCESS	N	N	N	N	N	N
ACD SPECIFIC DESIGN & FEATURE BUTTONS	N	N	N	N	N	N
VOLUME CONTROL (SPECIFY BUTTON, SLIDE, ETC)	Button	Button	Button	Button	Button	Button
DEDICATED MESSAGE WAITING INDICATOR	Y	Y	Y	Y	Y	Y
ON-HOOK DIALING	N	Y	Y	Y	Y	Y
HANDSFREE INTERCOM	N	N	Y	Y	Y	Y
TWO-WAY SPEAKERPHONE						
SIMPLEX OPERATION	N	N	N	N	N	N
FULL DUPLEX OPERATION	N	N	N	Y	Y	Y
DISPLAY FIELD (LINES x CHARACTERS)	NA	1 x 20	1 x 20	2 x 40	1 x 20	2 x 40
STORED DATA CAPABILITIES						
LAST INCOMING CALLS (INDICATE #)	N	16	16	16	16	16
LAST NUMBERS DIALED (INDICATE #)	1	1	1	1	1	1
FULLY INTEGRATED SYSTEM DIRECTORY ACCESS	N	Y	Y	Y	Y	Y
FULLY INTEGRATED PERSONAL DIRECTORY ACCESS	Y	Y	Y	Y	Y	Y
DIRECTORY ACCESS DEPENDENT ON LDAP-TYPE SERVER	Y	Y	Y	Y	Y	Y
HEADSET INTERFACE (INDICATE #)	N	N	N	N	N	N
MODULE ADAPTER BAYS/INTERFACES (INDICATE #)	1	1	1	1	1	1
MODULE OPTIONS						
RS-232 DATA MODULE	Y	Y	Y	Y	Y	Y
APPLICATIONS PROGRAMMING INTERFACE	Y	Y	Y	Y	Y	Y
ANALOG ADAPTER	Y	Y	Y	Y	Y	Y
PROPRIETARY DIGITAL ADAPTER	N	N	N	N	N	N
DISTANCE ADAPTER	N	N	N	N	N	N
SECURITY MODULE						
BRI ADAPTER	Y	Y	Y	Y	Y	Y
IP ADAPTER	N	N	N	N	N	N
ADD-ON BUTTON MODULE OPTION	O	O	2	2	2	2
# MODULES	NA	NA	20/40	20/40	20/40	20/40
# BUTTONS/MODULE		NA	NA	NA	NA	NA
SHIFT KEY OPERATION	Y	Y	Y	Y	Y	Y
LINE POWERED (STANDARD OPERATION)						
LOCAL AC POWER (OPTION-DEPENDENT)					OPT	OPT

Appendix D

SUPPLIER	AVAYA	AVAYA	AVAYA	AVAYA	AVAYA	AVAYA	AVAYA
MODEL NAME	5402	6402D	6408D+	6416D+M	4606	4612	4624	IP Softphone
INDICATE TYPE (ONLY ONE TYPE PER MODEL)								
DIGITAL	Yes	Yes	Yes	Yes				
PROPRIETARY CORDLESS								
ISDN BRI								
IP TELEPHONE					Yes	Yes	Yes	Yes
G.711 and/or G.729 ENCODING					G.711/G.729 A&B/H.323V2	G.711 /G.729 A&B/H323V2	G.711 /G.729 A&B/H323V2	G.711 /G.729 A&B/H323V2
AUTO SELF DISCOVERY/DHCP					Yes	Yes	Yes	Yes
INTEGRATED ETHERNET HUB					Yes	Yes	Yes	Sftw
INTEGRATED ETHERNET PORT(S) (INDICATE #)					2	2	2	Sftw
INTEGRATED WEB BROWSER					NA	NA	NA	24
PC-BASED SOFTPHONE								24
LOOP LENGTH (24 AWG) (IF FEET)	3500	3500	3500	3500				
# WIRING PAIRS	2	2	1	1	308	308	308	Variable
MAX # LINE/CALL APPEARANCE KEYS	0	0	8	16	2	2	2	Variable
TOTAL # FIXED FEATURE BUTTONS	0	0	8	12	0	10	10	ICON
TOTAL # PROGRAMMABLE LINE/FEATURE BUTTONS				16	0	12	24	NA
SOFTKEY BUTTONS								
# DISPLAY-ASSOCIATED CONTEXT SENSITIVE KEYS	NA	12	12	12	NA	4	4	GUI
# SELECTION/SOS CONTROL KEYS	NA	NA	4	4	0	4	4	Yes
SHIFT KEY OPERATION FOR LINE/FEATURE BUTTONS	NA	NA	NA	NA	Yes	Yes	Yes	Optional
ACD SPECIFIC DESIGN & FEATURE BUTTONS	NA	NA	NA	NA	Yes	Yes	Yes	Optional
VOLUME CONTROL (SPECIFY BUTTON, SLIDE, ETC)	Button	Button	Button	Button	Button	Button	Button	NA
DEDICATED MESSAGE WAITING INDICATOR	Yes	Yes	Yes	Yes	Yes	Yes	Yes	2x24
ON-HOOK DIALING	Yes	Yes	Yes	Yes	Yes	Yes	Yes	
HANDSFREE INTERCOM	No	No	Yes	Yes	Yes	Yes	Yes	Yes
TWO-WAY SPEAKERPHONE	No	No	Yes	Yes	NA	NA	NA	No
SIMPLEX OPERATION	No	No	No	No	Yes	Yes	Yes	No
FULL DUPLEX OPERATION	No	No	No	No	Yes	Yes	Yes	No
DISPLAY FIELD (LINES x CHARACTERS)	NA	2 x 16	2 x 24	2 x 24	2 x 16	2 x 24	2 x 24	No
STORED DATA CAPABILITIES								
LAST INCOMING CALLS (INDICATE #)	0	0	0	0	0	0	0	Yes
LAST NUMBERS DIALED (INDICATE #)	1	1	1	1	1	1	1	No
FULLY INTEGRATED SYSTEM DIRECTORY ACCESS (speed dial)	Yes	Yes	Yes	Yes	Yes	Yes	Yes	No
FULLY INTEGRATED PERSONAL DIRECTORY ACCESS	Yes	Yes	Yes	Yes	Yes	Yes	Yes	NA
DIRECTORY ACCESS DEPENDENT ON LDAP-TYPE SERVER	Optional	Optional	Optional	Optional	No	No	No	NA
HEADSET INTERFACE (INDICATE #)	1	1	1	1	1	1	1	NA
MODULE ADAPTER BAYS/INTERFACES (INDICATE #)	NA	NA	NA	NA	NA	NA	NA	NA
MODULE OPTIONS								
RS-232 DATA MODULE	No	No	No	NA	NA	No	No	NA
APPLICATIONS PROGRAMMING INTERFACE	No	No	No	NA	NA	No	No	NA
ANALOG ADAPTER	No	No	No	NA	NA	No	No	NA
PROPRIETARY DIGITAL ADAPTER	No	No	No	NA	NA	No	No	NA
DISTANCE ADAPTER	No	No	No	NA	NA	No	No	NA
SECURITY MODULE security in software	No	No	No	NA	NA	No	No	NA
IR ADAPTER	No	No	No	NA	NA	No	No	NA
ADD-ON BUTTON MODULE OPTION								
# MODULES	NA	NA	Yes	Yes	NA	NA	NA	NA
# BUTTONS/MODULE	NA	NA	24	24	NA	NA	NA	NA
SHIFT KEY OPERATION	NA	NA	No	No	NA	Yes	Yes	NA
LINE POWERED (STANDARD OPERATION)	NA	Yes	Yes	Option	Option	Option	Option	NA
LOCAL AC POWER (OPTION-DEPENDENT)	NA	NA	NA	STD	STD	Yes	Yes	NA

PBX/IP-PBX System Feature and Function Matrices

	Cisco Systems 7910	Cisco Systems 7910-SW	Cisco Systems 7940	Cisco Systems 7960	Cisco Systems Softphone	Cisco Systems 7935
SUPPLIER						
MODEL NAME						
INDICATE TYPE (ONLY ONE TYPE PER MODEL)						
DIGITAL						
PROPRIETARY CORDLESS						
ISDN BRI						
IP TELEPHONE (IF YES COMPLETE LINES 199-202)	YES	YES	YES	YES		YES
G.711 and/or G.729 ENCODING	YES (Both)	YES (Both)	YES (Both)	YES (Both)	YES (Both)	YES (Both)
AUTO SELF DISCOVERY/DHCP	YES	YES	YES	YES		YES
INTEGRATED ETHERNET HUB	NO	YES	NO	NO		NO
INTEGRATED ETHERNET PORT(S) (INDICATE #)	YES (1)	YES (2)	YES (2)	YES (2)		YES (1)
INTEGRATED WEB BROWSER	NO	NO	YES	YES		NO
PC-BASED SOFTPHONE					YES	
LOOP LENGTH (24 AWG) (# FEET)	300	300	300	300	300	300
# WIRING PAIRS	2	2	2	2	2	2
MAX # LINE/CALL APPEARANCE KEYS	1/2	1/2	2/4	6/12/28/40	6/12	1/2
TOTAL # FIXED FEATURE BUTTONS	7	7	9	9	9	7
TOTAL # PROGRAMMABLE LINE/FEATURE BUTTONS	6	6	2	34	4	3
SOFTKEY BUTTONS						
# DISPLAY-ASSOCIATED CONTEXT SENSITIVE KEYS	0	0	4	4	4	3
# SELECT/CURSOR CONTROL KEYS	0	0	1	1	1	3
SHIFT KEY OPERATION FOR LINE/FEATURE ACCESS	NO	NO	YES	YES	YES	YES
ACD SPECIFIC DESIGN & FEATURE BUTTONS	YES (Button)	YES (Button)	YES (Button)	YES (Button)	YES (Slide)	YES (Button)
VOLUME CONTROL (SPECIFY BUTTON, SLIDE, ETC)	YES	YES	YES	YES	YES	YES
DEDICATED MESSAGE WAITING INDICATOR	YES	YES	YES	YES	YES	YES
ON-HOOK DIALING	N/A	N/A	N/A	N/A	N/A	N/A
HANDSFREE INTERCOM	YES	YES	YES	YES	YES	YES
TWO-WAY SPEAKERPHONE	YES	YES	YES	YES	YES	YES
SIMPLEX OPERATION						
FULL DUPLEX OPERATION			YES	YES		YES
DISPLAY FIELD (LINES x CHARACTERS)	2 X 24	2 X 24	100 X 145 Pixels	100 X 145 Pixels	PC Display	Pixel-based
STORED DATA CAPABILITIES						
LAST INCOMING CALLS (INDICATE #)						
LAST NUMBERS DIALED (INDICATE #)	1	1	15	15	15	1
FULLY INTEGRATED SYSTEM DIRECTORY ACCESS						
FULLY INTEGRATED PERSONAL DIRECTORY ACCESS						
DIRECTORY ACCESS DEPENDENT ON LDAP-TYPE SERVER						
HEADSET INTERFACE (INDICATE #)	0	0	1 (RS-232)	1 (RS-232)	1	0
MODULE ADAPTER BAYS/INTERFACES (INDICATE #)	0	0	N/A	N/A	N/A	0
MODULE OPTIONS	N/A	N/A				N/A
RS-232 DATA MODULE						
APPLICATIONS PROGRAMMING INTERFACE						
ANALOG ADAPTER						
PROPRIETARY DIGITAL ADAPTER						
DISTANCE ADAPTER						
SECURITY MODULE						
BRI ADAPTER						
IP ADAPTER						
ADD-ON BUTTON MODULE OPTION				2		
# MODULES				14		
# BUTTONS/MODULE						
SHIFT KEY OPERATION						
LINE POWERED (STANDARD OPERATION)	STD	STD	STD	14	N/A	N/A
LOCAL AC POWER (OPTION-DEPENDENT)	OPT	OPT	OPT	OPT	STD	STD

Appendix D

	ERICSSON DBC210	ERICSSON DBC211	ERICSSON DBC212	ERICSSON DBC213	ERICSSON DBC413	ERICSSON CORDLESS
SUPPLIER						
MODEL NAME						
INDICATE TYPE (ONLY ONE TYPE PER MODEL)						
DIGITAL	YES	YES	YES	YES	YES	
PROPRIETARY CORDLESS						YES
ISDN BRI						
IP TELEPHONE (IF YES COMPLETE LINES 199-202)					YES	
G.711 and/or G.729 ENCODING					YES	
AUTO SELF DISCOVERY/DHCP					YES	
INTEGRATED ETHERNET HUB					NO	
INTEGRATED ETHERNET PORT(S) (INDICATE #)					1	
INTEGRATED WEB BROWSER					NO	
PC-BASED SOFTPHONE						
LOOP LENGTH (24 AWG) (# FEET)	3300	3300	3300	3300		
# WIRING PAIRS	1	1	1	1		NA
MAX # LINE/CALL APPEARANCE KEYS	3	3	3		2	NA
TOTAL # FIXED FEATURE BUTTONS	5	5	5	5	3	1
TOTAL # PROGRAMMABLE LINE/FEATURE BUTTONS	6	10	10	20	5	NA
SOFTKEY BUTTONS					20	NA
# DISPLAY-ASSOCIATED CONTEXT SENSITIVE KEYS	0	0	0	4	4	NA
# SELECT/CURSOR CONTROL KEYS	NA	NA	NA	NA	NA	NA
SHIFT KEY OPERATION FOR LINE/FEATURE BUTTONS	NA	NA	NA	NA	NA	NA
ACD SPECIFIC DESIGN & FEATURE BUTTONS						
VOLUME CONTROL (SPECIFY BUTTON, SLIDE, ETC)	BUTTON	BUTTON	BUTTON	BUTTON	BUTTON	BUTTON
DEDICATED MESSAGE WAITING INDICATOR	YES	YES	YES	YES	YES	YES
ON-HOOK DIALING	YES	YES	YES	YES	YES	YES
HANDSFREE INTERCOM	NO	YES	YES	YES	YES	NO
TWO-WAY SPEAKERPHONE	YES	YES	YES	YES	YES	NA
SIMPLEX OPERATION	NO	NO	NO	NO	NO	NA
FULL DUPLEX OPERATION	NO	NO	NO	NO	NO	NA
DISPLAY FIELD (LINES x CHARACTERS)	NA	NA	2 X 20	3 X 40	3 X 40	2 X 12
STORED DATA CAPABILITIES						
LAST INCOMING CALLS (INDICATE #)	NA	NA	NA	16	16	3
LAST NUMBERS DIALED (INDICATE #)	1	1	1	1	1	
FULLY INTEGRATED SYSTEM DIRECTORY ACCESS	NO	NO	NO	YES	NO	NO
FULLY INTEGRATED PERSONAL DIRECTORY ACCESS	NO	NO	NO	NO	NO	YES
DIRECTORY ACCESS DEPENDENT ON LDAP-TYPE SERVER						
HEADSET INTERFACE (INDICATE #)	1	1	1	1	1	YES
MODULE ADAPTER BAYS/INTERFACES (INDICATE #)	1	1	1	1	1	NO
MODULE OPTIONS						
RS-232 DATA MODULE	YES	YES	YES	YES	NO	
APPLICATIONS PROGRAMMING INTERFACE	NA	NA	NA	NA	NO	
ANALOG ADAPTER	NO	NO	NO	NO	NO	
PROPRIETARY DIGITAL ADAPTER	NO	NO	NO	NO	NO	
DISTANCE ADAPTER	YES	YES	YES	YES	NA	
SECURITY MODULE	NO	NO	NO	NO	NA	
BRI ADAPTER	YES	YES	YES	YES	NA	
IP ADAPTER	NO	NO	YES	NO	NA	
ADD-ON BUTTON MODULE OPTION	NO	NO	NO	YES	NO	
# MODULES				2		
# BUTTONS/MODULE				17		
SHIFT KEY OPERATION						
LINE POWERED (STANDARD OPERATION)	YES	YES	YES	NO	YES	
LOCAL AC POWER (OPTION-DEPENDENT)	NA	NA	NA	YES	NA	

PBX/IP-PBX System Feature and Function Matrices

	Mitel Networks SS4001	Mitel Networks SS4015	Mitel Networks SS4025	Mitel Networks SS4125	Mitel Networks SS4150	Mitel Networks 5010 IP Phone	Mitel Networks 5020 IP Phone	Mitel Networks 5140 IP Appliance
SUPPLIER								
MODEL NAME								
INDICATE TYPE (ONLY ONE TYPE PER MODEL)								
DIGITAL	YES	YES	YES	YES	YES	No	No	No
PROPRIETARY CORDLESS						No	No	No
ISDN BRI						No	No	No
IP TELEPHONE (IF YES COMPLETE LINES 199-202)						Yes	Yes	Yes
G.711 and/or G.729 ENCODING						Yes	Yes	Yes
AUTO SELF DISCOVERY/DHCP						Yes	Yes	Yes
INTEGRATED ETHERNET HUB						2	2	2
INTEGRATED ETHERNET PORT(S) (INDICATE #)						No	No	Yes
INTEGRATED WEB BROWSER								
PC-BASED SOFTPHONE								
LOOP LENGTH (24 AWG) (# FEET)	3300	3300	3300	3300	3300	325 feet/100 meters	325 feet/100 meters	325 feet/100 meters
# WIRING PAIRS	1	1	1	1	1	4	4	4
MAX # LINE/CALL APPEARANCE KEYS	NO	7	14	14	14	14	7	9
TOTAL # FIXED FEATURE BUTTONS	5	10	10	10	10	8	6	8
TOTAL # PROGRAMMABLE LINE/FEATURE BUTTONS	NO	7	14	14	14	14	7	8
SOFTKEY BUTTONS	0	0	10	10	10	3		6
# DISPLAY-ASSOCIATED CONTEXT SENSITIVE KEYS						2	2	
# SELECT/CURSOR CONTROL KEYS						Yes	Yes	Directional Pad
SHIFT KEY OPERATION FOR LINE/FEATURE ACCESS	BUTTON	BUTTON	BUTTON	BUTTON	BUTTON	Button	Button	Button
ACD SPECIFIC DESIGN & FEATURE BUTTONS	YES	YES	YES	YES	YES	Yes	Yes	Yes
DEDICATED MESSAGE WAITING INDICATOR	NO	YES	YES	YES	YES	Yes	Yes	Yes
VOLUME CONTROL (SPECIFY BUTTON, SLIDE, ETC)	NO	NO	YES	YES	YES	Yes	Yes	Yes
ON-HOOK DIALING	NO	YES	YES	YES	YES	Yes	Yes	Yes
HANDSFREE INTERCOM						No	No	No
TWO-WAY SPEAKERPHONE						2X20	2X20	320X240
FULL DUPLEX OPERATION	NO	2X20	2X20	2X20	40 CHARACTERS	Yes	Yes	Yes
SIMPLEX OPERATION								INA
DISPLAY FIELD (LINES x CHARACTERS)						10 different numbers	10 different numbers	INA
STORED DATA CAPABILITIES						10	10	No
LAST INCOMING CALLS (INDICATE #)						Yes	Yes	No
LAST NUMBERS DIALED (INDICATE #)						Yes	Yes	No
FULLY INTEGRATED SYSTEM DIRECTORY ACCESS	YES	YES	YES	YES	YES	1	1	No
FULLY INTEGRATED PERSONAL DIRECTORY ACCESS						No	No	No
DIRECTORY ACCESS DEPENDENT ON LDAP-TYPE SERVER						No	No	No
HEADSET INTERFACE (INDICATE #)						No	No	No
MODULE ADAPTER BAYS/INTERFACES (INDICATE #)						No	No	No
MODULE OPTIONS						PKM		No
RS-232 DATA MODULE	NA	NA	NA	NA	YES	1	No	No
APPLICATIONS PROGRAMMING INTERFACE						12 or 48 button PKM		No
ANALOG ADAPTER						No	No	No
PROPRIETARY DIGITAL ADAPTER						Yes	Yes	Yes
DISTANCE ADAPTER						Yes	Yes	Yes
SECURITY MODULE								
BRI ADAPTER								
IP ADAPTER								
ADD-ON BUTTON MODULE OPTION								
# MODULES								
# BUTTONS/MODULE								
SHIFT KEY OPERATION								
LINE POWERED (STANDARD OPERATION)								
LOCAL AC POWER (OPTION-DEPENDENT)								

Appendix D

SUPPLIER	NEC	NEC	NEC
MODEL NAME	DTP-8D	DTP-16D	DTP-32D
INDICATE TYPE (ONLY ONE TYPE PER MODEL)			
DIGITAL	YES	YES	YES
PROPRIETARY CORDLESS	NA	NA	NA
ISDN BRI	NA	NA	NA
IP TELEPHONE	Optional snap on base	Optional snap on base	Optional snap on base
G.711 and/or G.729 ENCODING	OPT	OPT	OPT
AUTO SELF DISCOVERY/DHCP	OPT	OPT	OPT
INTEGRATED ETHERNET HUB	OPT	OPT	OPT
INTEGRATED ETHERNET PORT(S) (INDICATE #)	1	1	1
INTEGRATED WEB BROWSER	AF	AF	AF
PC-BASED SOFTPHONE	OPT/SP10	OPT/SP10	OPT/SP10
LOOP LENGTH (24 AWG) (# FEET)	4000	4000	4000
# WIRING PAIRS	1	1	1
MAX # LINE/CALL APPEARANCE KEYS	8	16	32
TOTAL # FIXED FEATURE BUTTONS	8	8	8
TOTAL # PROGRAMMABLE LINE/FEATURE BUTTONS	8	16	32
SOFTKEY BUTTONS	YES	YES	YES
# DISPLAY-ASSOCIATED CONTEXT SENSITIVE KEYS	YES	YES	YES
# SELECT/CURSOR CONTROL KEYS	YES	YES	YES
SHIFT KEY OPERATION FOR LINE/FEATURE ACCESS	YES	YES	YES
ACD SPECIFIC DESIGN & FEATURE BUTTONS	YES	YES	YES
VOLUME CONTROL (SPECIFY BUTTON, SLIDE, ETC)	Up-down Toggle	Up-down Toggle	Up-down Toggle
DEDICATED MESSAGE WAITING INDICATOR	YES	YES	YES
ON-HOOK DIALING	YES	YES	YES
HANDSFREE INTERCOM	YES	YES	YES
TWO-WAY SPEAKERPHONE	YES	YES	YES
SIMPLEX OPERATION	STD	STD	STD
FULL DUPLEX OPERATION	OPT	OPT	OPT
DISPLAY FIELD (LINES x CHARACTERS)	3 X 24	3 X 24	3 X 24
STORED DATA CAPABILITIES	YES	YES	YES
LAST INCOMING CALLS (INDICATE #)	25	25	25
LAST NUMBERS DIALED (INDICATE #)	5	5	5
FULLY INTEGRATED SYSTEM DIRECTORY ACCESS (speed dial)	YES	YES	YES
FULLY INTEGRATED PERSONAL DIRECTORY ACCESS	YES	YES	YES
DIRECTORY ACCESS DEPENDENT ON LDAP-TYPE SERVER	YES	YES	YES
HEADSET INTERFACE (INDICATE #)	OPT	OPT	OPT
MODULE ADAPTER BAYS/INTERFACES (INDICATE #)	1 TO 3	1 TO 3	1 TO 3
MODULE OPTIONS	YES	YES	YES
RS-232 DATA MODULE	NA	NA	NA
APPLICATIONS PROGRAMMING INTERFACE	YES	YES	YES
ANALOG ADAPTER	YES	YES	YES
PROPRIETARY DIGITAL ADAPTER	YES	YES	YES
DISTANCE ADAPTER	YES	YES	YES
SECURITY MODULE security in software	YES	YES	YES
BRI ADAPTER	NA	NA	NA
IP ADAPTER	YES	YES	YES
ADD-ON BUTTON MODULE OPTION	YES	YES	YES
# MODULES	8	8	8
# BUTTONS/MODULE	60	60	60
SHIFT KEY OPERATION	YES	YES	YES
LINE POWERED (STANDARD OPERATION)	STD	STD	STD
LOCAL AC POWER (OPTION-DEPENDENT)	OPT	OPT	OPT

PBX/IP-PBX System Feature and Function Matrices

	Nortel Networks M3901	Nortel Networks M3902	Nortel Networks M3903	Nortel Networks M3904	Nortel Networks M3905	Nortel Networks i2004	Nortel Networks i2050
SUPPLIER							
MODEL NAME							
INDICATE TYPE (ONLY ONE TYPE PER MODEL)							
DIGITAL	YES	YES	YES	YES	YES		
PROPRIETARY CORDLESS							
ISDN BRI							
IP TELEPHONE (IF YES COMPLETE LINES 199-202)						YES	YES
G.711 and/or G.729 ENCODING						YES	
AUTO SELF DISCOVERY/DHCP						YES	
INTEGRATED ETHERNET HUB						Switch Cartridge	
INTEGRATED ETHERNET PORT(S) (INDICATE #)						1	
INTEGRATED WEB BROWSER							
PC-BASED SOFTPHONE							YES
LOOP LENGTH (24 AWG) (# FEET)	3500'	3500'	3500'	3500'	3500'	30'	
# WIRING PAIRS	1 PAIR	1 PAIR	1 PAIR	1 PAIR	1 PAIR		
MAX # LINE/CALL APPEARANCE KEYS	4	1	4	12	8	6	6
TOTAL # FIXED FEATURE BUTTONS	4	8	14	14	13	13	4
TOTAL # PROGRAMMABLE LINE/FEATURE BUTTONS	6	4	8	16	12	16	6
SOFTKEY BUTTONS							
# DISPLAY-ASSOCIATED CONTEXT SENSITIVE KEYS	0	3	4	4	4	4	0
# SELECT/CURSOR CONTROL KEYS	4	4	4	4	4	4	0
SHIFT KEY OPERATION FOR LINE/FEATURE ACCESS			YES	YES		YES	Quick Access Tray
ACD SPECIFIC DESIGN & FEATURE BUTTONS					YES		
VOLUME CONTROL (SPECIFY BUTTON, SLIDE, ETC)	ROCKER	ROCKER	ROCKER	ROCKER	ROCKER	ROCKER	GUI SLIDE
DEDICATED MESSAGE WAITING INDICATOR	YES	YES	YES	YES	YES	YES	YES
ON-HOOK DIALING	YES	YES	YES	YES	NO HOOK	YES	NO HOOK
HANDSFREE INTERCOM	NO				NO		YES
TWO-WAY SPEAKERPHONE		YES	YES	YES	YES	YES	
SIMPLEX OPERATION							YES
FULL DUPLEX OPERATION							
DISPLAY FIELD (LINES x CHARACTERS)	NO	2x24	3x24	5x24	4x24	5x24	2x24
STORED DATA CAPABILITIES							
LAST INCOMING CALLS (INDICATE #)			10	100			
LAST NUMBERS DIALED (INDICATE #)			5	20			
FULLY INTEGRATED SYSTEM DIRECTORY ACCESS			YES	YES		YES	YES
FULLY INTEGRATED PERSONAL DIRECTORY ACCESS			YES	YES			
DIRECTORY ACCESS DEPENDENT ON LDAP-TYPE SERVER							
HEADSET INTERFACE (INDICATE #)	1	1	1	1	2	1	1
MODULE ADAPTER BAYS/INTERFACES (INDICATE #)		1	2	2	2		
MODULE OPTIONS							
RS-232 DATA MODULE	NO	NO	NO	NO	NO	NO	
APPLICATIONS PROGRAMMING INTERFACE	NO	NO	NO	NO	NO	YES	
ANALOG ADAPTER	NO	YES	NO	NO	YES	NO	
PROPRIETARY DIGITAL ADAPTER	3rd Party	3rd Party	3rd Party	3rd Party	3rd Party	NO	
DISTANCE ADAPTER	NO	NO	NO	NO	NO		
SECURITY MODULE	NO	NO	NO	NO	NO		
BRI ADAPTER	YES	YES	YES	YES	YES		
IP ADAPTER	NO	NO	NO	NO	NO		
ADD-ON BUTTON MODULE OPTION				YES	YES		
# MODULES				2	2		
# BUTTONS/MODULE				22	22		
SHIFT KEY OPERATION			YES	YES			
LINE POWERED (STANDARD OPERATION)	STD	STD	STD	STD	STD		
LOCAL AC POWER (OPTION-DEPENDENT)	N/A	OPT	OPT	OPT	OPT	OPT	YES

Appendix D

	Siemens optiPoint 500 ENTRY	Siemens optiPoint 500 BASIC	Siemens optiPoint 500 Standard SL	Siemens optiPoint 500 Standard	Siemens optiPoint 500 Advance	Siemens optiPoint 400 Standard	Siemens optiClient 330	Siemens optiClient 350	Siemens optiClient 360
SUPPLIER									
MODEL NAME									
INDICATE TYPE (ONLY ONE TYPE PER MODEL)									
DIGITAL	YES	YES	YES	YES	YES	N/A	N/A	N/A	N/A
PROPRIETARY CORDLESS	NO	NO	NO	NO	NO	N/A	N/A	N/A	N/A
ISDN BRI	NO	NO	NO	NO	NO	N/A	N/A	N/A	N/A
IP TELEPHONE	VIA IP ADAPTER	VIA IP ADAPTER	VIA IP ADAPTER	YES	YES	YES (10/100 SWITCHED)	YES	YES	YES
G.711 and/or G.729 ENCODING						2	YES	YES	YES
AUTO SELF DISCOVERY/DHCP						NO	NO	NO	NO
INTEGRATED ETHERNET HUB						Ethernet compliant	Ethernet compliant	Ethernet compliant	Ethernet compliant
INTEGRATED ETHERNET PORT(S) (INDICATE #)						2 plus 2 for power	PC-client	PC-Client	PC-Client
PC-BASED SOFTPHONE							10	3	3
LOOP LENGTH (24 AWG) (# FEET)	3000	3000	3000	3000	3000				
MAX # LINE/CALL APPEARANCE KEYS	1	1	1	1	1				
# WIRING PAIRS	4	2	2	2	2	5 (3 dialog and 2 control)	Soft feature buttons	Soft feature buttons	Soft feature buttons
TOTAL # FIXED FEATURE BUTTONS	8	3	12	12	19	24 (12 two-level)	Soft feature buttons	Soft feature buttons	Soft feature buttons
TOTAL # PROGRAMMABLE LINE/FEATURE BUTTONS	0	3	3	3	3				
SOFTKEY BUTTONS	0	3	3	3	3	YES	YES	YES	YES
# SELECT/CURSOR CONTROL KEYS	YES	YES	YES	YES	YES	YES	YES	YES	YES
SHIFT KEY OPERATION FOR LINE/FEATURE ACCESS	0	0	0	0	0	NO	N/A	N/A	N/A
ACD SPECIFIC DESIGN & FEATURE BUTTONS	2 buttons	2 buttons	2 buttons	2 buttons	2 buttons	BUTTON	YES	YES	YES
VOLUME CONTROL (SPECIFY BUTTON, SLIDE, ETC)	NO	NO	NO	NO	NO	YES	YES	YES	YES
DIOLULS MESSAGE WAITING INDICATOR	YES	YES	YES	YES	YES	NO	NO	NO	NO
ON-HOOK DIALING	YES	YES	YES	YES	YES				
HANDSFREE INTERCOM	YES	YES	YES	YES	YES	YES	YES	YES	YES
TWO-WAY SPEAKERPHONE							PC Based or optional	PC Based or optional	PC Based or optional
DUPLEX OPERATION	NO	NO	YES	YES	YES	YES	YES	YES	YES
FULL DUPLEX OPERATION	NO	2x24	2x24	2x24	2x24	GUI	GUI	GUI	GUI
DISPLAY FIELD (LINES & CHARACTERS)	NO	NO	NO	NO	NO	20	N/A	N/A	N/A
STORED DATA CAPABILITIES	NO	YES	YES	YES	YES	20	20	20	20
LAST INCOMING CALLS (INDICATE #)	YES	YES	YES	YES	YES	YES	YES	YES	YES
LAST NUMBERS DIALED (INDICATE #)	YES	YES	YES	YES	YES	YES	YES	YES	YES
FULLY INTEGRATED SYSTEM DIRECTORY ACCESS (speed dial)	YES	YES	YES	YES	YES		NO	NO	NO
FULLY INTEGRATED PERSONAL DIRECTORY ACCESS	NO	NO	NO	NO		YES			
DIRECTORY ACCESS DEPENDENT ON LDAP-TYPE SERVER	NO	NO	YES	YES	YES	Optional	Optional	Optional	Optional
HEADSET INTERFACE (INDICATE #)	NA	ONE	ONE	ONE	two	NO	NO	NO	NO
MODULE ADAPTER BAYS/INTERFACES (INDICATE #)	NO	USB	USB	USB	USB				
MODULE OPTIONS									
RS-232 DATA MODULE	NO	USB	USB	USB	USB	NO	NO	NO	NO
APPLICATIONS PROGRAMMING INTERFACE	NO	YES	YES	YES	YES	YES	N/A	N/A	N/A
ANALOG ADAPTER	NO	YES	YES	YES	YES	YES	N/A	N/A	N/A
PROPRIETARY DIGITAL ADAPTER	YES	YES	YES	YES	YES	20	N/A	N/A	N/A
DISTANCE ADAPTER	NO	YES	YES	YES	YES		N/A	N/A	N/A
SECURITY MODULE security in software	NO	YES	YES	YES	YES				
BRI ADAPTER	YES	YES	YES	YES	YES	N/A	N/A	N/A	N/A
IP ADAPTER	0	2	2	2	30	N/A	N/A	N/A	N/A
ADD-ON BUTTON MODULE OPTION	0	30	30	30	STD	N/A	N/A	N/A	N/A
# MODULES	0	YES	YES	YES	YES	N/A	N/A	N/A	N/A
# BUTTONS/MODULE	NO	STD	STD	STD	STD	YES	N/A	N/A	N/A
SHIFT KEY OPERATION	STD	YES	YES	YES	YES	YES	N/A	N/A	N/A
LINE POWERED (STANDARD OPERATION)	NO								
LOCAL AC POWER (OPTION-DEPENDENT)									

PBX/IP-PBX System Feature and Function Matrices

Suppliers

SUPPLIER	**Alcatel**
SYSTEM ARCHITECTURE DESIGN/TECHNOLOGY (Model)	**OmniPCX 4400**
	(Distributed Model)
CAPACITIES	
MAX # PORTS	50000+2000 per site
MAX # STATIONS	50000
MAX # ATTENDANT CONSOLES	80
MAX # TRUNKS	2000 per site
MAX # TRUNK GROUPS	2000
MAX # TRUNKS/GROUP	400
SWITCHING SYSTEM/DESIGN	
TOPOLOGY	Distributed
CENTER STAGE SWITCH DESIGN	TDM (ATM, optional)
LEVEL OF DUPLICATION (CHECK ALL THAT APPLY)	
CENTER STAGE SWITCH	Yes
LOCAL SWITCH/BUS	Yes
INTERCABINET SWITCH LINKS	Yes
TRAFFIC CAPACITIES	
CCS RATING/STATION @MAXIMUM CAPACITY	
# TALK SLOTS/SYSTEM	90720
# TALK SLOTS/LOCAL TDM BUS	120 channels X 28 Links
# SIMULTANEOUS 2-PARTY CONVERSATIONS/SYSTEM	non blocking
MAXIMUM # 3-PARTY CONFERENCES	no limit
PROCESSING SYSTEM DESIGN	
TOPOLOGY	Distributed
MAIN CPU (INDICATE MAKE/MODEL)	Intel Pentium
MAIN CPU OPERATING SYSTEM (INDICATE PLATFORM)	Chorus Mix
EXTERNAL APPLICATION SERVERS (INDICATED STD OR OPT)	
PROCESSOR TYPE(S) (INDICATE MAKE/MODEL)	Intel Pentium
OPERATING SYSTEM PLATFORM	Windows NT
FULLY DUPLICATED ELEMENTS (CHECK ALL THAT APPLY)	
MAIN CPU (STD OR OPT)	Optional
LOCAL PROCESSOR (STD OR OPT)	NA
INTERCABINET PROCESSOR LINKS (STD OR OPT)	Optional
MEMORY ELEMENTS	
MAIN MEMORY PLATFORM (INDICATE RAM, EPROM, ET AL)	RAM, EPROM
MAIN SYSTEM MEMORY (# MB)	64
DUPLICATED MAIN MEMORY (STD OR OPT)	OPT
CUSTOMER DATABASE MEMORY (INDICATE DISK, DRIVE, ET AL)	DISK
CUSTOMER DATABASE MEMORY (#MB)	2000
DUPLICATED CUSTOMER DB MEMORY (STD OR OPT)	OPT
SYSTEM RELOAD TIME (SPECIFY #MINUTES)	3
BATTERY BACKUP (SPECIFY TIME)	external
CALL PROCESSING RATING (BUSY HOUR CALL COMPLETIONS)	
MAX BHCC RATING	150,000
LOADED BHCC RATING (ACD/CTI CONFIGURATION)	
PORT CABINET DESIGN	
CONTROL/PORT CABINET DESIGN	
WALL MOUNTABLE	Y
FLOOR-BASED	Y
NO PORT CABINET REQUIRED	NA
CABINET TYPE (CHECK ALL THAT APPLY PER MODEL)	
STACKABLE CARRIERS	Y
MULTICARRIER FRAME	Y
SERVER CABINET W/PCI CARD SLOTS	NA
# PORT CARD SLOTS	
STACKABLE CARRIER SHELF	8
MULTICARRIER CABINET SHELF	28
SERVER CABINET W/PCI CARD SLOTS	NA
# PORT CARRIERS PER MULTICARRIER CABINET /CABINET STACK	2
UNIVERSAL PORT CARD SLOTS	Y
SYSTEM POWER REQUIREMENTS (CHECK ALL OPTIONS)	
AC	
DC	Y
CIRCUIT CARD DENSITY (INDICATE # PORT TERMINATIONS)	
ANALOG STATION	12/24
DIGITAL STATION	16/32
ISDN BRI STATION	8
IP STATION (# Gateways)	8/16/30/60 DSP
SOFTPHONE CTI-BASED STATION	2000/server
ATTENDANT CONSOLE	16/32
DATA-ONLY STATION	NA
WIDEBAND (N x DS0) STATION	NA
GS/LS CO TRUNK	8
DID TRUNK	8
E&M TIE LINE TRUNK	4
IP TRUNK	8/16/30/60
DIGITAL TRUNK (VOICE GRADE T1-CARRIER)	2
DIGITAL TRUNK (ISDN PRI)	2
AUXILIARY TRUNK	distributed on each coupler
MODEM POOL CARD	42
DTMF (# TONE DETECTORS)	distributed on each coupler
SERIAL DATA INTERFACE (# RS-232C CONNECTIONS)	16

Appendix D

SUPPLIER	Definity	Definity	Definity	Definity
SYSTEM ARCHITECTURE DESIGN/TECHNOLOGY (Model)	ONE	ProLogix	G3si	G3r
CAPACITIES				
MAX # PORTS	618	600	2,800	29,000
MAX # STATIONS	240	500	2,400	25,000
MAX # ATTENDANT CONSOLES	15	15	15	27
MAX # TRUNKS	300	400	400	4,000
MAX # TRUNK GROUPS	99	99	99	666
MAX # TRUNKS/GROUP	99	99	99	255
SWITCHING SYSTEM/DESIGN				
TOPOLOGY	Centralized	Centralized	Distributed	Dispersed
CENTER STAGE SWITCH DESIGN	TDM Bus	TDM Bus	N/A	TDM Bus (ATM, option)
LEVEL OF DUPLICATION (CHECK ALL THAT APPLY)				
CENTER STAGE SWITCH				
LOCAL SWITCH/BUS				Optional
INTERCABINET SWITCH LINKS			Optional	Optional
TRAFFIC CAPACITIES				
CCS RATING/STATION @MAXIMUM CAPACITY	36	36	Variable	Variable
# TALK SLOTS/SYSTEM	483	483	1449	21,252
# TALK SLOTS/LOCAL TDM BUS	483	483	483	483
# SIMULTANEOUS 2-PARTY CONVERSATIONS/SYSTEM	241	241	723	10,604
MAXIMUM # 3-PARTY CONFERENCES	161	161	483	7,098
PROCESSING SYSTEM DESIGN				
TOPOLOGY	Centralized	Centralized	Centralized	Centralized
MAIN CPU (INDICATE MAKE/MODEL)	Pentium	Pentium	MIPS 2000	MIPS 3000
MAIN CPU OPERATING SYSTEM (INDICATE PLATFORM)	Windows 2000	Windows 2000	Oryx-Pecos (UNIX)	Oryx-Pecos (UNIX)
EXTERNAL APPLICATION SERVERS (INDICATED STD OR OPT)	Optional	Optional	Optional	Optional
PROCESSOR TYPE(S) (INDICATE MAKE/MODEL)	Variable	Variable	Variable	Variable
OPERATING SYSTEM PLATFORM	Variable	Variable	Variable	Variable
FULLY DUPLICATED ELEMENTS (CHECK ALL THAT APPLY)				
MAIN CPU (STD OR OPT)			Optional	Optional
LOCAL PROCESSOR (STD OR OPT)	NA	NA	NA	NA
INTERCABINET PROCESSOR LINKS (STD OR OPT)			Optional	Optional
MAIN MEMORY PLATFORM (INDICATE RAM, EPROM, ET AL)				
MAIN SYSTEM MEMORY (# MB)	128	32	32	128
DUPLICATED MAIN MEMORY (STD OR OPT)	NA	NA	Optional	Optional
CUSTOMER DATABASE MEMORY (INDICATE DISK, DRIVE, ET AL)	Flash ROM	Flash ROM	Flash ROM	Magneto-optical drive
CUSTOMER DATABASE MEMORY (#MB)	16	16	16	96
DUPLICATED CUSTOMER DB MEMORY (STD OR OPT)	NA	NA	Optional	Optional
SYSTEM RELOAD TIME (SPECIFY #MINUTES)	< 5 min	< 5 min	< 5 min	< 5 min
BATTERY BACKUP (SPECIFY TIME)	NA UPS	NA UPS	10 min std, 2-8 hrs optional	10 min std, 2-8 hrs optional
CALL PROCESSING RATING (BUSY HOUR CALL COMPLETIONS)				
MAX BHCC RATING	5,000	10,000	20,000	135,000
LOADED BHCC RATING (ACD/CTI CONFIGURATION)	2,000	10,000	20,000	70,000
PORT CABINET DESIGN				
CONTROL/PORT CABINET DESIGN				
WALL MOUNTABLE	Yes	Yes	NA	NA
FLOOR-BASED	NA	Optional	Yes	Yes
NO PORT CABINET REQUIRED	NA	NA	NA	NA
CABINET TYPE (CHECK ALL THAT APPLY PER MODEL)				
STACKABLE CARRIERS	Yes	Yes	Yes	Yes
MULTICARRIER FRAME	NA	NA	Yes	Yes
SERVER CABINET W/PCI CARD SLOTS	NA	NA	NA	NA
# PORT CARD SLOTS				
STACKABLE CARRIER SHELF	10	10	18	18
MULTICARRIER CABINET SHELF	NA	NA	20	20
SERVER CABINET W/PCI CARD SLOTS	NA	NA	NA	NA
# PORT CARRIERS PER MULTICARRIER CABINET /CABINET STACK	3	3	4/5	4/5
UNIVERSAL PORT CARD SLOTS	Yes	Yes	Yes	Yes
SYSTEM POWER REQUIREMENTS (CHECK ALL OPTIONS)				
AC	Yes	Yes	Yes	Yes
DC	NA	NA	Yes	Yes
CIRCUIT CARD DENSITY (INDICATE # PORT TERMINATIONS)				
ANALOG STATION	24	24	24	24
DIGITAL STATION	24	24	24	24
ISDN BRI STATION	24	24	24	24
IP STATION (GATEWAYS)	64	64	64	64
ATTENDANT CONSOLE	24	24	24	24
DATA-ONLY STATION	24	24	24	24
WIDEBAND (N x DS0) STATION	24	24	24	24
GS/LS CO TRUNK	8	8	8	8
DID TRUNK	8	8	8	8
E&M TIE LINE TRUNK	4	4	4	4
IP TRUNK	8	8	8	8
DIGITAL TRUNK (VOICE GRADE T1-CARRIER)	1 T1/E1/PRI	1 T1/E1/PRI	1 T1/E1/PRI	1 T1/E1/PRI
DIGITAL TRUNK (ISDN PRI)	1 T1/E1/PRI	1 T1/E1/PRI	1 T1/E1/PRI	1 T1/E1/PRI
AUXILIARY TRUNK	4	4	4	4
MODEM POOL CARD	2	2	2	2
DTMF (# TONE DETECTORS)	8	8	8	8
SERIAL DATA INTERFACE (# RS-232C CONNECTIONS)	8	8	8	8

PBX/IP-PBX System Feature and Function Matrices

SUPPLIER	Ericsson
SYSTEM ARCHITECTURE DESIGN/TECHNOLOGY	MD110
CAPACITIES	
MAX # PORTS	40,000
MAX # STATIONS	26,000
MAX # ATTENDANT CONSOLES	250
MAX # TRUNKS	5000
MAX # TRUNK GROUPS	250
MAX # TRUNKS/GROUP	99
SWITCHING SYSTEM/DESIGN	
TOPOLOGY	Dispersed
CENTER STAGE SWITCH DESIGN	Crosspoint Switch
LEVEL OF DUPLICATION (CHECK ALL THAT APPLY)	
CENTER STAGE SWITCH	NA
LOCAL SWITCH/BUS	Optional
INTERCABINET SWITCH LINKS	YES
TRAFFIC CAPACITIES	
CCS RATING/STATION @MAXIMUM CAPACITY	36
# TALK SLOTS/SYSTEM	100,000
# TALK SLOTS/LOCAL TDM BUS	256
# SIMULTANEOUS 2-PARTY CONVERSATIONS/SYSTEM	15,000
MAXIMUM # 3-PARTY CONFERENCES	2000
PROCESSING SYSTEM DESIGN	
TOPOLOGY	Distributed
MAIN CPU (INDICATE MAKE/MODEL)	Motorola 68030
MAIN CPU OPERATING SYSTEM (INDICATE PLATFORM)	
EXTERNAL APPLICATION SERVERS (INDICATED STD OR OPT)	
PROCESSOR TYPE(S) (INDICATE MAKE/MODEL)	Intel Pentium
OPERATING SYSTEM PLATFORM	Windows NT
FULLY DUPLICATED ELEMENTS (CHECK ALL THAT APPLY)	
MAIN CPU (STD OR OPT)	Optional
LOCAL PROCESSOR (STD OR OPT)	NA
INTERCABINET PROCESSOR LINKS (STD OR OPT)	Standard
MEMORY ELEMENTS	
MAIN MEMORY PLATFORM (INDICATE RAM, EPROM, ET AL)	RAM
MAIN SYSTEM MEMORY (# MB)	64MB
DUPLICATED MAIN MEMORY (STD OR OPT)	NA
CUSTOMER DATABASE MEMORY (INDICATE DISK, DRIVE, ET AL)	RAM
CUSTOMER DATABASE MEMORY (#MB)	32MB
DUPLICATED CUSTOMER DB MEMORY (STD OR OPT)	OPT
SYSTEM RELOAD TIME (SPECIFY #MINUTES)	2.5 - 9 MINUTES
BATTERY BACKUP (SPECIFY TIME)	STD 1 - 4 HOURS
CALL PROCESSING RATING (BUSY HOUR CALL COMPLETIONS)	
MAX BHCC RATING	535,000
LOADED BHCC RATING (ACD/CTI CONFIGURATION)	144,000
PORT CABINET DESIGN	
CONTROL/PORT CABINET DESIGN	
WALL MOUNTABLE	OPT
FLOOR-BASED	STD
NO PORT CABINET REQUIRED	NA
CABINET TYPE (CHECK ALL THAT APPLY PER MODEL)	
STACKABLE CARRIERS	STD
MULTICARRIER FRAME	
SERVER CABINET W/PCI CARD SLOTS	NO
# PORT CARD SLOTS	
STACKABLE CARRIER SHELF	15
MULTICARRIER CABINET SHELF	NO
SERVER CABINET W/PCI CARD SLOTS	NO
# PORT CARRIERS PER MULTICARRIER CABINET /CABINET STACK	3 OR 4
UNIVERSAL PORT CARD SLOTS	YES
SYSTEM POWER REQUIREMENTS (CHECK ALL OPTIONS)	
AC	YES
DC	YES
CIRCUIT CARD DENSITY (INDICATE # PORT TERMINATIONS)	
ANALOG STATION	16
DIGITAL STATION	16
ISDN BRI STATION	4
IP STATION (# Gateways)	16
SOFTPHONE CTI-BASED STATION	16
ATTENDANT CONSOLE	10
DATA-ONLY STATION	16
WIDEBAND (N x DS0) STATION	na
GS/LS CO TRUNK	8
DID TRUNK	8
E&M TIE LINE TRUNK	4
IP TRUNK	na
DIGITAL TRUNK (VOICE GRADE T1-CARRIER)	24
DIGITAL TRUNK (ISDN PRI)	30
AUXILIARY TRUNK	8
MODEM POOL CARD	NA
DTMF (# TONE DETECTORS)	8
SERIAL DATA INTERFACE (# RS-232C CONNECTIONS)	16

Appendix D

SUPPLIER	Mitel Networks	Mitel Networks	Mitel Networks
SYSTEM ARCHITECTURE DESIGN/TECHNOLOGY (Model)	SX-2000 Light	SX-200 ML	SX-200 EL
CAPACITIES			
MAX # PORTS	4224	192	768
MAX # STATIONS	3950	192	650
MAX # ATTENDANT CONSOLES	16	11	11
MAX # TRUNKS	1300	200	200
MAX # TRUNK GROUPS	212	50	50
MAX # TRUNKS/GROUP	175	50	50
SWITCHING SYSTEM/DESIGN			
TOPOLOGY			
CENTER STAGE SWITCH DESIGN	Centralized Crosspoint Switch Matrix	Centralized Crosspoint Switch Matrix	Centralized Crosspoint Switch Matrix
LEVEL OF DUPLICATION (CHECK ALL THAT APPLY)			
CENTER STAGE SWITCH	Yes	Yes	Yes
LOCAL SWITCH/BUS			
INTERCABINET SWITCH LINKS			
TRAFFIC CAPACITIES			
CCS RATING/STATION @MAXIMUM CAPACITY	26 CCS	33 CCS	33 CCS
# TALK SLOTS/SYSTEM	1536	248	248
# TALK SLOTS/LOCAL TDM BUS	NA	NA	NA
# SIMULTANEOUS 2-PARTY CONVERSATIONS/SYSTEM	768	248	248
MAXIMUM # 3-PARTY CONFERENCES	64	5	5
PROCESSING SYSTEM DESIGN			
TOPOLOGY	Dispersed	CENTRALIZED	CENTRALIZED
MAIN CPU (INDICATE MAKE/MODEL)	Motorola 68020	Motorola 68020	Motorola 68020
MAIN CPU OPERATING SYSTEM (INDICATE PLATFORM)			
EXTERNAL APPLICATION SERVERS (INDICATED STD OR OPT)			
PROCESSOR TYPE(S) (INDICATE MAKE/MODEL)	Pentium	Pentium	Pentium
OPERATING SYSTEM PLATFORM	Windows NT	Windows NT	Windows NT
FULLY DUPLICATED ELEMENTS (CHECK ALL THAT APPLY)			
MAIN CPU (STD OR OPT)	Optional		NI
LOCAL PROCESSOR (STD OR OPT)			
INTERCABINET PROCESSOR LINKS (STD OR OPT)	YES		
MEMORY ELEMENTS			
MAIN MEMORY PLATFORM (INDICATE RAM, EPROM, ET AL)		RAM	RAM
MAIN SYSTEM MEMORY (# MB)	32 MB	4 MB	4 MB
DUPLICATED MAIN MEMORY (STD OR OPT)	YES OPTIONAL		
CUSTOMER DATABASE MEMORY (INDICATE DISK, DRIVE, ET AL)	HARD DISK	NVRAM	NVRAM
CUSTOMER DATABASE MEMORY (#MB)	128 MB	1 MB	1 MB
DUPLICATED CUSTOMER DB MEMORY (STD OR OPT)	YES OPTIONAL	STD OFF BOARD PC	STD OFF BOARD PC
SYSTEM RELOAD TIME (SPECIFY #MINUTES)	4 MIN	2 minutes	2 minutes
BATTERY BACKUP (SPECIFY TIME)	OPT	68 hours	68 hours
CALL PROCESSING RATING (BUSY HOUR CALL COMPLETIONS)			
MAX BHCC RATING	43346 BHC		
LOADED BHCC RATING (ACD/CTI CONFIGURATION)	43346 BHC		
PORT CABINET DESIGN			
CONTROL/PORT CABINET DESIGN			
WALL MOUNTABLE	NO	YES	YES
FLOOR-BASED	YES	YES	YES
NO PORT CABINET REQUIRED			
CABINET TYPE (CHECK ALL THAT APPLY PER MODEL)			
STACKABLE CARRIERS	YES	YES	YES
MULTICARRIER FRAME	YES		
SERVER CABINET W/PCI CARD SLOTS			
# PORT CARD SLOTS		8	8
STACKABLE CARRIER SHELF	12	8	8
MULTICARRIER CABINET SHELF	23		
SERVER CABINET W/PCI CARD SLOTS			
# PORT CARRIERS PER MULTICARRIER CABINET /CABINET STACK			
UNIVERSAL PORT CARD SLOTS	YES	YES	YES
SYSTEM POWER REQUIREMENTS (CHECK ALL OPTIONS)			
AC	YES	YES	YES
DC	YES		
CIRCUIT CARD DENSITY (INDICATE # PORT TERMINATIONS)			
ANALOG STATION	16	12	12
DIGITAL STATION	16	12	12
ISDN BRI STATION	15	12	12
IP STATION (# Gateways)			
SOFTPHONE CTI-BASED STATION	16		
ATTENDANT CONSOLE	16	12 (Digital Card)	12 (Digital Card)
DATA-ONLY STATION	16	12	12
WIDEBAND (N x DS0) STATION	APPLICATION DEPENDANT		
GS/LS CO TRUNK	8	6	6
DID TRUNK	8	6	6
E&M TIE LINE TRUNK	4	4	4
IP TRUNK	Via Gateway (60 Channels)		
DIGITAL TRUNK (VOICE GRADE T1-CARRIER)	48	2	2
DIGITAL TRUNK (ISDN PRI)	48	2	2
AUXILIARY TRUNK	16	12	12
MODEM POOL CARD	16	12	12
DTMF (# TONE DETECTORS)	16	4/16	4/16
SERIAL DATA INTERFACE (# RS-232C CONNECTIONS)	16	12	12

PBX/IP-PBX System Feature and Function Matrices

SUPPLIER	NEC	NEC
SYSTEM ARCHITECTURE DESIGN/TECHNOLOGY (Model)	NEAX2000 IPS	NEAX2400 IPS
CAPACITIES		
MAX # PORTS	752	24576
MAX # STATIONS	512	24576
MAX # ATTENDANT CONSOLES	8	60
MAX # TRUNKS	496	24576
MAX # TRUNK GROUPS	64	899
MAX # TRUNKS/GROUP	256	255
SWITCHING SYSTEM/DESIGN		
TOPOLOGY	Centralized	Dispersed
CENTER STAGE SWITCH DESIGN	TDM Bus	TDM Bus
LEVEL OF DUPLICATION (CHECK ALL THAT APPLY)		
CENTER STAGE SWITCH	Optional	Optional
LOCAL SWITCH/BUS	NA	Optional
INTERCABINET SWITCH LINKS	NA	Optional
TRAFFIC CAPACITIES		
CCS RATING/STATION @MAXIMUM CAPACITY	36	36
# TALK SLOTS/SYSTEM	1024	32768
# TALK SLOTS/LOCAL TDM BUS	1024	2048
# SIMULTANEOUS 2-PARTY CONVERSATIONS/SYSTEM	512	12288
MAXIMUM # 3-PARTY CONFERENCES	24	512
PROCESSING SYSTEM DESIGN		
TOPOLOGY	Centralized	Dispersed
MAIN CPU (INDICATE MAKE/MODEL)	AMD /FLAN	Pentiuim 133
MAIN CPU OPERATING SYSTEM (INDICATE PLATFORM)	Proprietary	Proprietary
EXTERNAL APPLICATION SERVERS (INDICATED STD OR OPT)		
PROCESSOR TYPE(S) (INDICATE MAKE/MODEL)		NEC V50
OPERATING SYSTEM PLATFORM		Proprietary
FULLY DUPLICATED ELEMENTS (CHECK ALL THAT APPLY)		
MAIN CPU (STD OR OPT)	Optional	Optional
LOCAL PROCESSOR (STD OR OPT)		Optional
INTERCABINET PROCESSOR LINKS (STD OR OPT)		Optional
MEMORY ELEMENTS		
MAIN MEMORY PLATFORM (INDICATE RAM, EPROM, ET AL)	FLSH ROM	256MB RAM, 2MB FLASH ROM
MAIN SYSTEM MEMORY (# MB)	6	256MB
DUPLICATED MAIN MEMORY (STD OR OPT)	OPT	OPT
CUSTOMER DATABASE MEMORY (INDICATE DISK, DRIVE, ET AL)	NA	ON-LINE RAM (HDD BACKUP)
CUSTOMER DATABASE MEMORY (#MB)	4-Jun	VARIABLE
DUPLICATED CUSTOMER DB MEMORY (STD OR OPT)	STD/OPT	OPT
SYSTEM RELOAD TIME (SPECIFY #MINUTES)	IMMEDIATE	5 MINUTES
BATTERY BACKUP (SPECIFY TIME)	8 HOUR	OPT
CALL PROCESSING RATING (BUSY HOUR CALL COMPLETIONS)		
MAX BHCC RATING	8000	46000
LOADED BHCC RATING (ACD/CTI CONFIGURATION)	5500	46000
PORT CABINET DESIGN		
CONTROL/PORT CABINET DESIGN		
WALL MOUNTABLE	YES	
FLOOR-BASED	YES	STD
NO PORT CABINET REQUIRED	NA	
CABINET TYPE (CHECK ALL THAT APPLY PER MODEL)	NA	
STACKABLE CARRIERS	YES	STD
MULTICARRIER FRAME	NA	
SERVER CABINET W/PCI CARD SLOTS	NA	
# PORT CARD SLOTS		
STACKABLE CARRIER SHELF	8	18
MULTICARRIER CABINET SHELF		
SERVER CABINET W/PCI CARD SLOTS		
# PORT CARRIERS PER MULTICARRIER CABINET /CABINET STACK		
UNIVERSAL PORT CARD SLOTS	YES	Ltd
SYSTEM POWER REQUIREMENTS (CHECK ALL OPTIONS)		
AC	STD	STD
DC		
CIRCUIT CARD DENSITY (INDICATE # PORT TERMINATIONS)		
ANALOG STATION	8	16/24/30
DIGITAL STATION	8	16/23/30
ISDN BRI STATION	2	4,8
IP STATION (# Gateways)	448	32
SOFTPHONE CTI-BASED STATION	8	32
ATTENDANT CONSOLE	8	2
DATA-ONLY STATION	2	8
WIDEBAND (N x DS0) STATION	2	
GS/LS CO TRUNK	8	16
DID TRUNK	4	8
E&M TIE LINE TRUNK	4	8
IP TRUNK	16	30
DIGITAL TRUNK (VOICE GRADE T1-CARRIER)	24/31	24
DIGITAL TRUNK (ISDN PRI)	24	24
AUXILIARY TRUNK	4	4
MODEM POOL CARD		
DTMF (# TONE DETECTORS)	8	16
SERIAL DATA INTERFACE (# RS-232C CONNECTIONS)	4	4

Appendix D

SUPPLIER SYSTEM ARCHITECTURE DESIGN/TECHNOLOGY (Model)	NORTEL NETWORKS OPTION 11C	NORTEL NETWORKS OPT 11C MINI	NORTEL NETWORKS OPTION 61C	NORTEL NETWORKS OPTION 81C
CAPACITIES				
MAX # PORTS	800	800	2000	16,000
MAX # STATIONS	800	800	2000	16,000
MAX # ATTENDANT CONSOLES	63	63	63	63
MAX # TRUNKS	1080	360	2000	16,000
MAX # TRUNK GROUPS	512	512	512	512
MAX # TRUNKS/GROUP	254	254	254	254
SWITCHING SYSTEM/DESIGN				
TOPOLOGY	Centralized	Centralized	Centralized	Centralized
CENTER STAGE SWITCH DESIGN	TDM bus	TDM bus	Switch Matrix	Fiber Loop Switch Matrices
LEVEL OF DUPLICATION (CHECK ALL THAT APPLY)				
CENTER STAGE SWITCH	NO	NO	YES	YES
LOCAL SWITCH/BUS	NA	NA	NO	NO
INTERCABINET SWITCH LINKS	NA	NA	NO	NO
TRAFFIC CAPACITIES				
CCS RATING/STATION @MAXIMUM CAPACITY	VARIABLE	VARIABLE	VARIABLE	VARIABLE
# TALK SLOTS/SYSTEM	1500	1500	960	7680
# TALK SLOTS/LOCAL TDM BUS	300	300	120	120
# SIMULTANEOUS 2-PARTY CONVERSATIONS/SYSTEM	400	400	480	3840
MAXIMUM # 3-PARTY CONFERENCES	32	5	20	160
PROCESSING SYSTEM DESIGN				
TOPOLOGY	Centralized	Centralized	Dispersed	Dispersed
MAIN CPU (INDICATE MAKE/MODEL)	M68040	M68040	M68040	Pentium II
MAIN CPU OPERATING SYSTEM (INDICATE PLATFORM)	VXWorks	VXWorks	VXWorks	VXWorks
EXTERNAL APPLICATION SERVERS (INDICATED STD OR OPT)				
PROCESSOR TYPE(S) (INDICATE MAKE/MODEL)	Pentium	Pentium	Pentium	Pentium
OPERATING SYSTEM PLATFORM	Windows NT	Windows NT	Windows NT	Windows NT
FULLY DUPLICATED ELEMENTS (CHECK ALL THAT APPLY)				
MAIN CPU (STD OR OPT)	NO	NO	YES	YES
LOCAL PROCESSOR (STD OR OPT)	NA	NA	NO	NO
INTERCABINET PROCESSOR LINKS (STD OR OPT)	NA	NA	NO	NO
MEMORY ELEMENTS				
MAIN MEMORY PLATFORM (INDICATE RAM, EPROM, ET AL)				
MAIN SYSTEM MEMORY (# MB)	48M	48M	48-160M	48-256M
DUPLICATED MAIN MEMORY (STD OR OPT)			STD	STD
CUSTOMER DATABASE MEMORY (INDICATE DISK, DRIVE, ET AL)				
CUSTOMER DATABASE MEMORY (#MB)	16M	16M	32-80M	32-80M
DUPLICATED CUSTOMER DB MEMORY (STD OR OPT)				
SYSTEM RELOAD TIME (SPECIFY #MINUTES)	1-2	1-2	1-2	1-2
BATTERY BACKUP (SPECIFY TIME)	VARIABLE	VARIABLE	VARIABLE	VARIABLE
CALL PROCESSING RATING (BUSY HOUR CALL COMPLETIONS)				
MAX BHCC RATING	58K	58K	135K	320K
LOADED BHCC RATING (ACD/CTI CONFIGURATION)	CONFIG. DEP.	CONFIG. DEP.	CONFIG. DEP.	CONFIG. DEP.
PORT CABINET DESIGN				
CONTROL/PORT CABINET DESIGN				
WALL MOUNTABLE	YES	YES		
FLOOR-BASED	YES	YES	YES	YES
NO PORT CABINET REQUIRED				
CABINET TYPE (CHECK ALL THAT APPLY PER MODEL)				
STACKABLE CARRIERS	YES	YES	YES	YES
MULTICARRIER FRAME	YES	YES		
SERVER CABINET W/PCI CARD SLOTS				
# PORT CARD SLOTS				
STACKABLE CARRIER SHELF	10	4	16	16
MULTICARRIER CABINET SHELF	10	4		
SERVER CABINET W/PCI CARD SLOTS				
# PORT CARRIERS PER MULTICARRIER CABINET /CABINET STACK	5	10	4	4
UNIVERSAL PORT CARD SLOTS	YES	YES	YES	YES
SYSTEM POWER REQUIREMENTS (CHECK ALL OPTIONS)				
AC	YES	YES	YES	YES
DC	YES	NO	YES	YES
CIRCUIT CARD DENSITY (INDICATE # PORT TERMINATIONS)				
ANALOG STATION	16	16	16	16
DIGITAL STATION	16	4/8/16	16	16
ISDN BRI STATION	8	8	8	8
IP STATION (# Gateways)	30	30	30	30
SOFTPHONE CTI-BASED STATION	24	24	24	24
ATTENDANT CONSOLE	16	16	16	16
DATA-ONLY STATION	8	8	8	8
WIDEBAND (N x DS0) STATION				
GS/LS CO TRUNK	8	8	8	8
DID TRUNK	8	8	8	8
E&M TIE LINE TRUNK	4	4	4	4
IP TRUNK	24	24	24	24
DIGITAL TRUNK (VOICE GRADE T1-CARRIER)	2	2	2	2
DIGITAL TRUNK (ISDN PRI)	2	2	2	2
AUXILIARY TRUNK	8	8	8	8
MODEM POOL CARD	6	6	6	6
DTMF (# TONE DETECTORS)	8	8	8	8
SERIAL DATA INTERFACE (# RS-232C CONNECTIONS)	4	4	4	4

PBX/IP-PBX System Feature and Function Matrices

SUPPLIER	**SHORELINE**
SYSTEM ARCHITECTURE DESIGN/TECHNOLOGY (Model)	Shoreline3
CAPACITIES	
MAX # PORTS	10,000
MAX # STATIONS	10,000
MAX # ATTENDANT CONSOLES	200
MAX # TRUNKS	10,000
MAX # TRUNK GROUPS	100
MAX # TRUNKS/GROUP	10,000
SWITCHING SYSTEM/DESIGN	
TOPOLOGY	Distributed
CENTER STAGE SWITCH DESIGN	NA
LEVEL OF DUPLICATION (CHECK ALL THAT APPLY)	
CENTER STAGE SWITCH	NA
LOCAL SWITCH/BUS	YES
INTERCABINET SWITCH LINKS	YES
TRAFFIC CAPACITIES	
CCS RATING/STATION @MAXIMUM CAPACITY	36 CCS (Non-blocking)
# TALK SLOTS/SYSTEM	Non-blocking
# TALK SLOTS/LOCAL TDM BUS	Non-blocking
# SIMULTANEOUS 2-PARTY CONVERSATIONS/SYSTEM	5,000
MAXIMUM # 3-PARTY CONFERENCES	3,333
PROCESSING SYSTEM DESIGN	
TOPOLOGY	Distributed
MAIN CPU (INDICATE MAKE/MODEL)	MIPS
MAIN CPU OPERATING SYSTEM (INDICATE PLATFORM)	VXWorks
EXTERNAL APPLICATION SERVERS (INDICATED STD OR OPT)	
PROCESSOR TYPE(S) (INDICATE MAKE/MODEL)	Pentium
OPERATING SYSTEM PLATFORM	Windows 2000
FULLY DUPLICATED ELEMENTS (CHECK ALL THAT APPLY)	YES
MAIN CPU (STD OR OPT)	NA
LOCAL PROCESSOR (STD OR OPT)	NA
INTERCABINET PROCESSOR LINKS (STD OR OPT)	NA
MEMORY ELEMENTS	
MAIN MEMORY PLATFORM (INDICATE RAM, EPROM, ET AL)	FLASH, RAM
MAIN SYSTEM MEMORY (# MB)	48M
DUPLICATED MAIN MEMORY (STD OR OPT)	STD (Distributed)
CUSTOMER DATABASE MEMORY (INDICATE DISK, DRIVE, ET AL)	Disk
CUSTOMER DATABASE MEMORY (#MB)	Unlimited
DUPLICATED CUSTOMER DB MEMORY (STD OR OPT)	OPT
SYSTEM RELOAD TIME (SPECIFY #MINUTES)	30 seconds
BATTERY BACKUP (SPECIFY TIME)	Unlimited
CALL PROCESSING RATING (BUSY HOUR CALL COMPLETIONS)	
MAX BHCC RATING	100,000
LOADED BHCC RATING (ACD/CTI CONFIGURATION)	Configuration Dependent
PORT CABINET DESIGN	
CONTROL/PORT CABINET DESIGN	
WALL MOUNTABLE	Rack Mounted
FLOOR-BASED	
NO PORT CABINET REQUIRED	
CABINET TYPE (CHECK ALL THAT APPLY PER MODEL)	
STACKABLE CARRIERS	YES
MULTICARRIER FRAME	
SERVER CABINET W/PCI CARD SLOTS	
# PORT CARD SLOTS	Embedded Interface
STACKABLE CARRIER SHELF	
MULTICARRIER CABINET SHELF	
SERVER CABINET W/PCI CARD SLOTS	
# PORT CARRIERS PER MULTICARRIER CABINET /CABINET STACK	
UNIVERSAL PORT CARD SLOTS	
SYSTEM POWER REQUIREMENTS (CHECK ALL OPTIONS)	
AC	YES
DC	
CIRCUIT CARD DENSITY (INDICATE # PORT TERMINATIONS)	
ANALOG STATION	4, 12, 24
DIGITAL STATION	NA
ISDN BRI STATION	NA
IP STATION (# Gateways)	NA
SOFTPHONE CTI-BASED STATION	NO
ATTENDANT CONSOLE	NA
DATA-ONLY STATION	NA
WIDEBAND (N x DS0) STATION	
GS/LS CO TRUNK	4, 8, 12
DID TRUNK	4, 8, 12
E&M TIE LINE TRUNK	NA
IP TRUNK	NO
DIGITAL TRUNK (VOICE GRADE T1-CARRIER)	24 channels/interface
DIGITAL TRUNK (ISDN PRI)	23 channels/inteface
AUXILIARY TRUNK	NO
MODEM POOL CARD	NO
DTMF (# TONE DETECTORS)	Integrated
SERIAL DATA INTERFACE (# RS-232C CONNECTIONS)	NA

Appendix D

SUPPLIER SYSTEM ARCHITECTURE DESIGN/TECHNOLOGY	SIEMENS HiPath 5500	SIEMENS HiPath 5300
CAPACITIES		
MAX # PORTS	N/A	N/A
MAX # STATIONS	800	300
MAX # ATTENDANT CONSOLES	0	0
MAX # TRUNKS	No Hard Limits	No Hard Limits
MAX # TRUNK GROUPS	No Hard Limits	No Hard Limits
MAX # TRUNKS/GROUP	No Hard Limits	No Hard Limits
CALL TELEPHONY SERVER DESIGN		
SERVER TYPE		
OFF-THE-SHELF SERVER (THIRD PARTY)	YES	
CLOSED CABINET/BUNDLED SOFTWARE (FIRST PARTY)		YES
PROCESSOR TYPE	PENTIUM CLASS	Intel 433 Celeron
SERVER DESIGN TOPOLOGY	CENTRALIZED	CENTRALIZED
REDUNDANT CALL TELEPHONY SERVER OPTION	NO	NO
SERVER OPERATING SYSTEM (INDICATE PLATFORM		EMBEDDED NT
OPTIONAL APPLICATION SERVER(S) (INDICATE MAKE/MODEL)	PENTIUM WINDOWS NT	PENTIUM WINDOWS NT
MAIN MEMORY PLATFORM (INDICATE RAM, EPROM, ET AL)	RAM	RAM
MAIN SYSTEM MEMORY (# MB)	512	512
DUPLICATED MAIN MEMORY (STD OR OPT)	NO	NO
CUSTOMER DATABASE MEMORY (INDICATE DISK, DRIVE, ET AL)	DISK	DISK
CUSTOMER DATABASE MEMORY (#MB)		
DUPLICATED CUSTOMER DB MEMORY (STD OR OPT)	NO	NO
SYSTEM RELOAD TIME (SPECIFY #MINUTES)	NA	NA
BATTERY BACKUP (SPECIFY TIME)	NA	NA
CALL PROCESSING RATING (BUSY HOUR CALL COMPLETIONS [BHCC])		
MAX BHCC RATING	1,800	900
LOADED BHCC RATING (ACD/CTI CONFIGURATION)	750	375
GATEWAY OPTIONS		
CALL TELEPHONY SERVER INTERFACES		YES
DESKTOP	YES	YES
MEDIA GATEWAY MODULE/CARRIER	YES	YES
INTERFACE CARD FOR SWITCH/ROUTER		
TDN/PCM PORT CABINET		

PBX/IP-PBX System Feature and Function Matrices

SUPPLIER SYSTEM ARCHITECTURE DESIGN/TECHNOLOGY	Siemens HiPath 4300	Siemens HiPath 4500	Siemens HiPath 4900
CAPACITIES			
MAX # PORTS	16,752	37632	112896
MAX # STATIONS	2,000	12000	30000
MAX # ATTENDANT CONSOLES	64	64	64
MAX # TRUNKS	10000	10000	30000
MAX # TRUNK GROUPS	2000	2000	6000
MAX # TRUNKS/GROUP	10000	10000	10000
SWITCHING SYSTEM/DESIGN			
TOPOLOGY	Centralized	Centralized	Centralized
CENTER STAGE SWITCH DESIGN	T-S-T	T-S-T	T-S-T
LEVEL OF DUPLICATION (CHECK ALL THAT APPLY)			
CENTER STAGE SWITCH	Optional	Optional	Optional
LOCAL SWITCH/BUS	Optional	Optional	Optional
INTERCABINET SWITCH LINKS	Optional	Optional	Optional
TRAFFIC CAPACITIES			
CCS RATING/STATION @MAXIMUM CAPACITY	36	36	36
# TALK SLOTS/SYSTEM	11,264	25,088	75,264
# TALK SLOTS/LOCAL TDM BUS	128/256	256	256
# SIMULTANEOUS 2-PARTY CONVERSATIONS/SYSTEM	563 2	12544	37632
MAXIMUM # 3-PARTY CONFERENCES	64	64	192
PROCESSING SYSTEM DESIGN			
TOPOLOGY	Dispersed	Dispersed	Dispersed
MAIN CPU (INDICATE MAKE/MODEL)	Pentium+	Pentium+	Pentium+
MAIN CPU OPERATING SYSTEM (INDICATE PLATFORM)	Unix	Unix	Unix
EXTERNAL APPLICATION SERVERS (INDICATED STD OR OPT)			
PROCESSOR TYPE(S) (INDICATE MAKE/MODEL)	Pentium+	Pentium+	Pentium+
OPERATING SYSTEM PLATFORM	Windows NT	Windows NT	Windows NT
FULLY DUPLICATED ELEMENTS (CHECK ALL THAT APPLY)			
MAIN CPU (STD OR OPT)	NO	Optional	Optional
LOCAL PROCESSOR (STD OR OPT)	Optional	Optional	Optional
INTERCABINET PROCESSOR LINKS (STD OR OPT)	Optional	Optional	Optional
MEMORY ELEMENTS			
MAIN MEMORY PLATFORM (INDICATE RAM, EPROM, ET AL)	RAM	RAM	RAM
MAIN SYSTEM MEMORY (# MB)			
DUPLICATED MAIN MEMORY (STD OR OPT)		Optional	Optional
CUSTOMER DATABASE MEMORY (INDICATE DISK, DRIVE, ET AL)	HD/MO	HD/MO	HD/MO
CUSTOMER DATABASE MEMORY (#MB)			
DUPLICATED CUSTOMER DB MEMORY (STD OR OPT)			
SYSTEM RELOAD TIME (SPECIFY #MINUTES)	10 MIN	10 MIN	10 MIN
BATTERY BACKUP (SPECIFY TIME)	NA	NA	NA
CALL PROCESSING RATING (BUSY HOUR CALL COMPLETIONS)			
MAX BHCC RATING	56,000	112500	337500
LOADED BHCC RATING (ACD/CTI CONFIGURATION)			
PORT CABINET DESIGN			
CONTROL/PORT CABINET DESIGN			
WALL MOUNTABLE			
FLOOR-BASED	STD	STD	STD
NO PORT CABINET REQUIRED			
CABINET TYPE (CHECK ALL THAT APPLY PER MODEL)			
STACKABLE CARRIERS	STD	STD	STD
MULTICARRIER FRAME			
SERVER CABINET W/PCI CARD SLOTS	opt	opt	opt
# PORT CARD SLOTS			
STACKABLE CARRIER SHELF			
MULTICARRIER CABINET SHELF	16	16	16
SERVER CABINET W/PCI CARD SLOTS			
# PORT CARRIERS PER MULTICARRIER CABINET /CABINET STACK	3,7	3,7	3,7
UNIVERSAL PORT CARD SLOTS	YES	YES	YES
SYSTEM POWER REQUIREMENTS (CHECK ALL OPTIONS)			
AC	YES	YES	YES
DC	YES	YES	YES
CIRCUIT CARD DENSITY (INDICATE # PORT TERMINATIONS)			
ANALOG STATION	24	24	24
DIGITAL STATION	24	24	24
ISDN BRI STATION			
IP STATION (# Gateways)	30	30	30
SOFTPHONE CTI-BASED STATION	30	30	30
ATTENDANT CONSOLE	24	24	24
DATA-ONLY STATION	16	16	16
WIDEBAND (N x DS0) STATION	na	na	na
GS/LS CO TRUNK			
DID TRUNK	8	8	8
E&M TIE LINE TRUNK	4	4	4
IP TRUNK	30	30	30
DIGITAL TRUNK (VOICE GRADE T1-CARRIER)	24	24	24
DIGITAL TRUNK (ISDN PRI)	24	24	24
AUXILIARY TRUNK	16	16	16
MODEM POOL CARD	16	16	16
DTMF (# TONE DETECTORS)	8x8	8x8	8x8
SERIAL DATA INTERFACE (# RS-232C CONNECTIONS)			

Appendix D

SUPPLIER SYSTEM ARCHITECTURE DESIGN/TECHNOLOGY(Model)	VERTICAL NETWORKS InstantOffice 3000	VERTICAL NETWORKS InstantOffice 3500	VERTICAL NETWORKS InstantOffice 5000	VERTICAL NETWORKS InstantOffice 5500	VERTICAL NETWORKS InstantOffice 6000
CAPACITIES					
MAX # PORTS	106	106	170	170	314
MAX # STATIONS	36	36	84	84	180
MAX # ATTENDANT CONSOLES	16	16	6	6	14
MAX # TRUNKS	70	70	86	86	134
MAX # TRUNK GROUPS	20	20	20	20	20
MAX # TRUNKS/GROUP	70	70	86	86	134
SWITCHING SYSTEM/DESIGN					
TOPOLOGY	Centralized	Centralized	Centralized	Centralized	Centralized
CENTER STAGE SWITCH DESIGN	Switch Matrix	Switch Matrix	Switch Matrix	Switch Matrix	
LEVEL OF DUPLICATION (CHECK ALL THAT APPLY)					
CENTER STAGE SWITCH	NO	NO	NO	NO	NO
LOCAL SWITCH/BUS					
INTERCABINET SWITCH LINKS					
TRAFFIC CAPACITIES					
CCS RATING/STATION @MAXIMUM CAPACITY					
# TALK SLOTS/SYSTEM	256	256	256	256	256
# TALK SLOTS/LOCAL TDM BUS	256	256	256	256	256
# SIMULTANEOUS 2-PARTY CONVERSATIONS/SYSTEM	51	51	85	85	134
MAXIMUM # 3-PARTY CONFERENCES	4	4	4	4	8
PROCESSING SYSTEM DESIGN					
TOPOLOGY	Centralized	Centralized	Centralized	Centralized	Centralized
MAIN CPU (INDICATE MAKE/MODEL)	Pentium III	Pentium III	Pentium III	Pentium III	Pentium III
MAIN CPU OPERATING SYSTEM (INDICATE PLATFORM)	Windows NT 4.0	Windows NT 4.0	Windows NT 4.0	Windows NT 4.0	Windows NT 4.0
EXTERNAL APPLICATION SERVERS (INDICATED STD OR OPT)					
PROCESSOR TYPE(S) (INDICATE MAKE/MODEL)					
OPERATING SYSTEM PLATFORM					
FULLY DUPLICATED ELEMENTS (CHECK ALL THAT APPLY)					
MAIN CPU (STD OR OPT)	NO	NO	NO	NO	NO
LOCAL PROCESSOR (STD OR OPT)					
INTERCABINET PROCESSOR LINKS (STD OR OPT)					
MEMORY ELEMENTS					
MAIN MEMORY PLATFORM (INDICATE RAM, EPROM, ET AL)	DRAM	DRAM	DRAM	DRAM	DRAM
MAIN SYSTEM MEMORY (# MB)	128MB	384MB	256MB	384MB	384MB
DUPLICATED MAIN MEMORY (STD OR OPT)	No	No	No	No	No
CUSTOMER DATABASE MEMORY (INDICATE DISK, DRIVE, ET AL)	Hard Disk Drive	Hard Disk Drive	Hard Disk Drive	Hard Disk Drive	Hard Disk Drive
CUSTOMER DATABASE MEMORY (#MB)	13,000 MB (13 GB)	13,000 MB (13 GB)	13,000 MB (13 GB)	13,000 MB (13 GB)	13,000 MB (13 GB)
DUPLICATED CUSTOMER DB MEMORY (STD OR OPT)	OPT	OPT	OPT	OPT	OPT
SYSTEM RELOAD TIME (SPECIFY #MINUTES)	5	5	8	8	8
BATTERY BACKUP (SPECIFY TIME)	NO	NO	NO	NO	NO
CALL PROCESSING RATING (BUSY HOUR CALL COMPLETIONS)					
MAX BHCC RATING	>5,000	>5,000	>5,000	>5,000	>5,000
LOADED BHCC RATING (ACD/CTI CONFIGURATION)	>5,000	>5,000	>5,000	>5,000	>5,000
PORT CABINET DESIGN					
CONTROL/PORT CABINET DESIGN					
WALL MOUNTABLE	Yes	Yes	No	No	No
FLOOR-BASED	Yes	Yes	Yes	Yes	Yes
NO PORT CABINET REQUIRED	Yes	Yes	Yes	Yes	Yes
CABINET TYPE (CHECK ALL THAT APPLY PER MODEL)					
STACKABLE CARRIERS	Yes	Yes	Yes	Yes	Yes
MULTICARRIER FRAME	No	No	No	No	No
SERVER CABINET W/PCI CARD SLOTS	No	No	No	No	No
# PORT CARD SLOTS					
STACKABLE CARRIER SHELF	7	7	14	14	14
MULTICARRIER CABINET SHELF	n/a	n/a	n/a	n/a	n/a
SERVER CABINET W/PCI CARD SLOTS	n/a	n/a	n/a	n/a	n/a
# PORT CARRIERS PER MULTICARRIER CABINET /CABINET STACK	n/a	n/a	n/a	n/a	n/a
UNIVERSAL PORT CARD SLOTS	No	No	No	No	No
SYSTEM POWER REQUIREMENTS (CHECK ALL OPTIONS)					
AC	Yes	Yes	Yes	Yes	Yes
DC	No	No	No	No	No
CIRCUIT CARD DENSITY (INDICATE # PORT TERMINATIONS)					
ANALOG STATION	12 and 24	12 and 24	12 and 24	12 and 24	12 and 24
DIGITAL STATION	12 and 24	12 and 24	12 and 24	12 and 24	12 and 24
ISDN BRI STATION	n/a	n/a	n/a	n/a	n/a
IP STATION (# Gateways)	n/a	n/a	n/a	n/a	n/a
SOFTPHONE CTI-BASED STATION	integrated	integrated	integrated	integrated	integrated
ATTENDANT CONSOLE	2	2	2	2	2
DATA-ONLY STATION	12 and 24	12 and 24	12 and 24	12 and 24	12 and 24
WIDEBAND (N x DS0) STATION	n/a	n/a	n/a	n/a	n/a
GS/LS CO TRUNK	4 and 8	4 and 8	4 and 8	4 and 8	4 and 8
DID TRUNK	8	8	8	8	8
E&M TIE LINE TRUNK	n/a	n/a	n/a	n/a	n/a
IP TRUNK	4 and 8	4 and 8	4 and 8	4 and 8	4 and 8
DIGITAL TRUNK (VOICE GRADE T1-CARRIER)	2	2	2	2	2
DIGITAL TRUNK (ISDN PRI)	2	2	2	2	4
AUXILIARY TRUNK	n/a	n/a	n/a	n/a	4
MODEM POOL CARD	2	2	2	2	2
DTMF (# TONE DETECTORS)	16	16	16	16	16
SERIAL DATA INTERFACE (# RS-232C CONNECTIONS)	1	1	2	2	2

System Architecture Design/Technology

SYSTEM ARCHITECTURE DESIGN/TECHNOLOGY	CISCO SYSTEMS ICS 7750	CISCO SYSTEMS MCS 7825	CISCO SYSTEMS MCS 7835
CAPACITIES			
MAX # PORTS	694	680	13,000
MAX # STATIONS	214	500	10,000
MAX # ATTENDANT CONSOLES	20	20	96
MAX # TRUNKS	480	180	3000
MAX # TRUNK GROUPS	480	180	3000
MAX # TRUNKS/GROUP	480	180	3000
CALL TELEPHONY SERVER DESIGN			
SERVER TYPE			
OFF-THE-SHELF SERVER (THIRD PARTY)		Optional	Optional
CLOSED CABINET/BUNDLED SOFTWARE (FIRST PARTY)	YES	YES	YES
PROCESSOR TYPE	PENTIUM	PENTIUM	PENTIUM
SERVER DESIGN TOPOLOGY	CENTRALIZED	DISTRIBUTED	DISTRIBUTED
REDUNDANT CALL TELEPHONY SERVER OPTION	OPTIONAL	OPTIONAL	OPTIONAL
SERVER OPERATING SYSTEM (INDICATE PLATFORM	Windows 2000	Windows 2000	Windows 2000
OPTIONAL APPLICATION SERVER(S) (INDICATE MAKE/MODEL)		YES	YES
MAIN MEMORY PLATFORM (INDICATE RAM, EPROM, ET AL)	RAM	RAM	RAM
MAIN SYSTEM MEMORY (# MB)	512 Mbyte	512 Mbyte	1 Gbyte
DUPLICATED MAIN MEMORY (STD OR OPT)	N/A	N/A	N/A
CUSTOMER DATABASE MEMORY (INDICATE DISK, DRIVE, ET AL)	Hard Drive	Hard Drive	Hard Drive (RAID)
CUSTOMER DATABASE MEMORY (#MB)	N/A	N/A	
DUPLICATED CUSTOMER DB MEMORY (STD OR OPT)			STD
SYSTEM RELOAD TIME (SPECIFY #MINUTES)			
BATTERY BACKUP (SPECIFY TIME)	OPT (UPS)	OPT (UPS)	OPT (UPS)
CALL PROCESSING RATING (BUSY HOUR CALL COMPLETIONS (BHCC))			
MAX BHCC RATING	10,000	10,000	150,000
LOADED BHCC RATING (ACD/CTI CONFIGURATION)	10,000	10,000	150,000
GATEWAY OPTIONS			
CALL TELEPHONY SERVER INTERFACES	YES		
DESKTOP		YES	YES
MEDIA GATEWAY MODULE/CARRIER		YES	YES
INTERFACE CARD FOR SWITCH/ROUTER			
TDN/PCM PORT CABINET			

Appendix D

	Mitel Networks 3100 ICP	Mitel Networks 3300 ICP
SYSTEM ARCHITECTURE DESIGN/TECHNOLOGY		
CAPACITIES		
MAX # PORTS	42	1736
MAX # STATIONS	24 IP + 10 Analog =34	1736
MAX # ATTENDANT CONSOLES	NA	16
MAX # TRUNKS	8	384
MAX # TRUNK GROUPS	8	112
MAX # TRUNKS/GROUP	8	175
CALL TELEPHONY SERVER DESIGN		
SERVER TYPE		
OFF-THE-SHELF SERVER (THIRD PARTY)		
CLOSED CABINET/BUNDLED SOFTWARE (FIRST PARTY)	YES	YES
SERVER DESIGN TOPOLOGY	CENTRALIZED	CENTRALIZED
PROCESSOR TYPE	Motorola MPC 860	Motorola 8260
REDUNDANT CALL TELEPHONY SERVER	NO	NO
SERVER OPERATING SYSTEM (INDICATE PLATFORM	VXWorks	VXWorks
OPTIONAL APPLICATION SERVER(S) (INDICATE MAKE/MODEL)	YES (Linux)	YES (Linux)
MAIN MEMORY PLATFORM (INDICATE RAM, EPROM, ET AL)	RAM	RAM/FLASH
MAIN SYSTEM MEMORY (# MB)	128	256MB
DUPLICATED MAIN MEMORY (STD OR OPT)	None	Yes Optional
CUSTOMER DATABASE MEMORY (INDICATE DISK, DRIVE, ET AL)	Disk	Hard Disk
CUSTOMER DATABASE MEMORY (#MB)	TBD	128 MB
DUPLICATED CUSTOMER DB MEMORY (STD OR OPT)	OPT (PC backup)	Yes Optional
SYSTEM RELOAD TIME (SPECIFY #MINUTES)	3 minutes	4 Min
BATTERY BACKUP (SPECIFY TIME)	OPT	OPT
CALL PROCESSING RATING (BUSY HOUR CALL COMPLETIONS [BHCC])		
MAX BHCC RATING	TBD	43346 BHC
LOADED BHCC RATING (ACD/CTI CONFIGURATION)	TBD	43346 BHC
GATEWAY OPTIONS		
CALL TELEPHONY SERVER INTERFACES		YES
DESKTOP		
MEDIA GATEWAY MODULE/CARRIER	YES	YES
INTERFACE CARD FOR SWITCH/ROUTER		
TDN/PCM PORT CABINET		YES

PBX/IP-PBX System Feature and Function Matrices

	Nortel Networks Succession CSE 1000
SYSTEM ARCHITECTURE DESIGN/TECHNOLOGY	
CAPACITIES	
MAX # PORTS	1576
MAX # STATIONS	1256
MAX # ATTENDANT CONSOLES	20
MAX # TRUNKS	576
MAX # TRUNK GROUPS	512
MAX # TRUNKS/GROUP	254
CALL TELEPHONY SERVER DESIGN	
SERVER TYPE	
OFF-THE-SHELF SERVER (THIRD PARTY)	YES
CLOSED CABINET/BUNDLED SOFTWARE (FIRST PARTY)	DISPERSED
SERVER DESIGN TOPOLOGY	Motorola
PROCESSOR TYPE	Optional
REDUNDANT CALL TELEPHONY SERVER	VXWorks
SERVER OPERATING SYSTEM (INDICATE PLATFORM	Pentiium Windows NT
OPTIONAL APPLICATION SERVER(S) (INDICATE MAKE/MODEL)	
MAIN MEMORY PLATFORM (INDICATE RAM, EPROM, ET AL)	RAM
MAIN SYSTEM MEMORY (# MB)	48 MB
DUPLICATED MAIN MEMORY (STD OR OPT)	16M
CUSTOMER DATABASE MEMORY (INDICATE DISK, DRIVE, ET AL)	OPT
CUSTOMER DATABASE MEMORY (#MB)	16 MB
DUPLICATED CUSTOMER DB MEMORY (STD OR OPT)	Optional
SYSTEM RELOAD TIME (SPECIFY #MINUTES)	1-2
BATTERY BACKUP (SPECIFY TIME)	Variable
CALL PROCESSING RATING (BUSY HOUR CALL COMPLETIONS [BHCC])	
MAX BHCC RATING	58K
LOADED BHCC RATING (ACD/CTI CONFIGURATION)	Configuration Dependent
GATEWAY OPTIONS	
CALL TELEPHONY SERVER INTERFACES	
DESKTOP	
MEDIA GATEWAY MODULE/CARRIER	YES
INTERFACE CARD FOR SWITCH/ROUTER	
TDN/PCM PORT CABINET	

INDEX

Note: Boldface numbers indicate illustrations.

A law, 49
AC powered systems, 129–131
access control, converged IP–PBX and, 216
account codes, 388
accounting and billing, 178, 368
adapter or add-on modules for telephones and, 321
adaptive differential pulse code modulation (ADPCM), 18, 225
add-on button module for telephones and, 323
add-on conference, 373
address translation, H.323 protocol and, 182
addresses, 362, 363
  H.323 vs. SIP protocols in, 202
  session initiation protocol (SIP) and, 199, 202
adjunct application servers, 108, 109, 122, 128–129, 133–134
administration, maintenance, and management, 2, 30–31, 128, 359–371
  address assignment, 363
  ARS tables in, 365
  attendant consoles and, 364
  client/server PBX and, 360
  customer database and, 362
  design options in, 360–361, **361**
  diagnostics and maintenance in, 369–371
  dial plans and FACs in, 362–363
  error logs in, 371
  group assignments in, 364–365
  H.323 protocol for, 178
  IP PBX vs. client/server PBX in, 172
  moves additions and changes (MACs) in, 361
  network connection channels in, 364
  order for data entry into switches, 362
  performance management in, 365–369
  recordkeeping in, 361
  sequence of administration duties in, 361–365
  system administration in, 361
  system administration terminal (SAT) for, 360
  translation data in, 361–362
  Web browsers and, 360
administrator's duties, 360
Agent Anywhere option n, 352–353
alarms, 366–367, 370–371
Alcatel, 36, 42, 54, 60, 68, 69, 153, 211, 218, 223, 232, 233
  IP telephones and telephony, 306
  operating systems, 99
  Qsig and, 354–357
  remote port cabinet options for, 337–340
Alcatel Crystal Technology (ACT) system, 69
alphanumeric displays on telephones and, 321
alternate voice data (AVD), 141
Altigen, 43, 44
American National Standards Institute (ANSI) standards, 278

AMISA, 249
analog communications device adapter for telephone, 322, **322**
analog conversion to digital, 46
analog line cards, 139
analog PBX, 16, 148
analog telephones, 148–149, 298–299
annexes to H.323, 184
announcer board, 138
announcers, 128
ANSI/TIA/EIA 568 Commercial Building Cabling Standard for, 278, 283–291
ANSI/TIA/EIA 569 Commercial Building Standard for Telecommunications Pathways and Spaces in, 278, 291–293
ANSI/TIA/EIA 606 Administration Standard for Telecommunications Infrastructure Commercial Bldg., 278
answer detection, 388
answering calls, 8
application cabinet/carriers, 128–129
applications programming interface (API), for computer telephony integration (CTI), 230
application feature buttons in telephones and, 320
Apropos, 242
Archangel expansion link (AEL), 64
architectures for PBX (*See* topologies)
area codes, 363
ARPAnet, 191
Ascend Communications systems, 222, 340
assured forward, DiffServe, 270–271
asynchronous transfer mode (ATM), 54
  ATM trunk card, 142
  center stage switch in 59–60
  H.323 protocol for, 177, 201
  PBX and, 36
AT&T, 20, 28, 34, 35, 37, 38, 42, 43, 105, 164, 341, 349
  cellular phones, 314
  operating system selection, 99
ATM trunk card, 142
attendants/attendant consoles, 2, 3, 138, 389
  administration and management and, 364
  auto manual splitting, 384
  auto start/don't split, 385
  backup alerting, 385
  call waiting, 385
  conference, 385
  control of trunk group access, 385
  delay announcement, 385
  digital attendant console card, 139
  direct station selection with busy lamp field, 385
  direct trunk group selection, 385
  display, 386
  features for, 384–388

attendants/attendant consoles (*continued*)
  intercept treatment, 386
  interposition call and transfer, 386
  intrusion, 386
  lockout, 382
  overflow, 386
  override of diversion, 386
  paging system access, 386
  priority queue, 387
  recall, 387
  release loop operation, 387
  serial calling, 387
  telephony call server in client/server IP–PBX and, 365–366
  through dialing, 387
  trunk group busy/warning indication, 387–388
  trunk identification, 388
  trunk-to-trunk transfer, 387
attenuation to crosstalk (ACR) ratio, 282
audio codecs, 178, 184, 186, **187**, 318
audio transmission, 252
Audix VMS, 43
authorization codes, 388
auto start/don't split, 385
automated attendants (*See* attendants/attendant consoles)
automatic alternate routing (AAR) in, 345, 389
automatic call distributor (ACD), 2, 3, 5, 8–9, 17, 36–38, 84, 138, 389
  announcer board and, 138
  Busy Hour Call (BHC) and, 118–119
  cabinet/carriers, 133–134
  centralized PBX and, 109–110
  IP PBX vs. client/server PBX in, 165–166
  networking with PBX and, 352–353
  voice messaging system (VMS) and, 10
automatic callback, 3, 373
automatic camp on, 389
automatic circuit assurance (ACA) in, 346, 389
automatic intercom, 374
automatic number identification (AIN) in, 29, 329–330, 389
automatic recall, 390
automatic route selection (ARS), 3, 137, 165, 390
  administration and management and, 365
  converged IP–PBX and, 217
  networking with PBX and, 326–328
automatic speech recognition (ASR), 3
automatic transmission measurement system, 390
auxiliary cabinet/carriers, 128
auxiliary trunk card, 142–143
availability, in LAN design, 256
Avaya (*See also* Definity family of systems), 24, 44, 54, 60, 61, 62, 65, 68, 70, 72, 76, 77, 108, 115, 123, 133, 153, 162, 209, 212, 218, 221, 222, 232, 279, 336, 338, 340
  digital telephones and, **300**
  IP telephones and telephony, **305–309, 310, 311,** 312
  mobile phones, 335
  operating system selection, 99
AVVID IP Telephony System, 152, 230, 337

B channel (*See also* ISDN), 329
backbone cabling, 284–285, 292
backplane, cabinet/carriers, 132
backup alerting, 385
backup power, 131
backup servers in, 234
backups, 337
bandwidth management, 49–51, **51**, 252, 253
  cabling and, 278
  converged IP–PBX and, 219
  H.323 protocol and, 180, 182, 185
  in IP PBX vs. client/server PBX, 171–172
  in LAN design, 257–260, **258, 259**
baseline measurement, in LAN design, 256–257
basic rate interface (BRI) (*See also* ISDN), 329
battery power, 131, 136
Bell System, 20
Bell, Alexander G., 278
best effort service, 252, 275
billing systems (*See* accounting and billing)
blocking vs. nonblocking systems, 54, 75–77, **76**, 80–82
  Busy Hour, 82–83
  centum call seconds (CCS), 83–88
  grade of service (GoS), 82–83
  increased traffic on, 133
border elements, H.323 protocol and, 180, 183
BranchHub Sphericall Enterprise Softswitch, converged IP–PBX and, 224
bridged call appearance, 374
broadband
  PBX and, 32–33
  TDM bus using, 56–57
broadcasting, H.323 protocol and, 179–180
buffering, 138, 264
  cards and printed circuit boards, 134
  quality of service (QoS) and, 262
bus topology, 279–280
bus, system processing, 106–107
Busy Hour, 82–88, 366
Busy Hour Call (BHC), 98, 116–119, **118**
Busy Hour Call Attempts (BHCAs), 116–119
Busy Hour Call Completions (BHCCs), 116–119
busy signal, 80, 87–88
busy verification of terminals and trunks, 388

cabinet/carriers, 122, 123–126, **124**
  adjunct application servers and, 133–134
  application cabinet/carriers, 128–129
  auxiliary cabinet/carriers, 128
  backplane, 132
  call processing and, 133
  control cabinet, 125, 126
  expansion requirements and, 125, 132–134
  local controllers in, 105–106
  new application requirements and, 133–134
  port capacity and, 133
  port carrier, 125, 127, **128**
  power cabinet/carriers, 129–131
  remote port cabinets in, 337–340
  switching, 126–127, **127**
  system, 123–126

Index

cabinet/carriers (*continued*)
  telephony over IP (ToIP), 152–153
cabling, 277–293, 369
  American National Standards Institute (ANSI) standards for, 278
  ANSI/TIA/EIA 568 Commercial Building Cabling Standard for, 278, 283–291
  ANSI/TIA/EIA 569 Commercial Building Standard for Telecommunications Pathways and Spaces in, 278, 291–293
  ANSI/TIA/EIA 606 Administration Standard for Telecommunications Infrastructure Commercial Bldg., 278
  backbone, 284–285, 292
  bandwidth and, 278
  Category classification of, 280, 287–288, **287, 288**
  color codes in, 289–290, 292–293
  costs of, IP PBX vs. client/server PBX in, 167
  crossconnects, 286
  crosstalk in, 282
  data transmission and, 278
  digital switching and, 278
  Electronic Industries Association (EIA) standards for, 278
  EN 50173 cabling standard for, 279
  entrance facility, 284, 292
  equipment room, 284, 292
  Ethernet and TCP/IP topologies and, 279–280, 282, 286–289
  flood type, 281
  fundamentals of, 279–281
  home run, 286
  horizontal type, 286–289, **288**, 292
  interbuilding, 292
  interference and noise in, 281–282
  International Standards Organization (ISO) standards for, 278
  intrabuilding, 292
  ISO/IEC IS 11801 standard for, 278–279
  labeling of, 292
  in LAN design, 257
  optical fiber in, 285
  patch panel, 286
  performance losses in, 282
  PowerSum calculation method for, 282
  shielded twisted pair (STP) in, 285, 289
  single pair, 278
  standards for, 278–279
  switching technology and, 278
  telecommunications closet, 285–286, 292
  Telecommunications Industries Association (TIA) standards for, 278
  telecommunications outlet, 289–290, **290**
  topologies and, 279–280
  unshielded twisted pair (UTP) in, 280, 281, 285, 289
  work area, 290–292, **291**
call accounting systems, 2, 128
call admission control (CAC), 203
call back last internal caller, 374
call blocking, 54
call-by-call service selection (CBCSS) in, 36, 329, 330–331, 390
call center, 3, 31, **31**, 128, 241–242
call classifier, 138
call control protocols, 249, 318
call coverage in, 29, 390
call data/detail reports (CDR), 3, 93, 137, 165, 236, 239, 312–315, **314**, 367–368, 391
call forwarding, 4, 5, 29, 165, 374, 375
call hold, 375
call log, 391
call monitoring, 100
call park, 375
call pick-up, 3, 375
call processing, 96, **96**, 100–103, 133, 149
  Busy Hour Call (BHC) in, 116–119, **118**
  cabinet/carriers, 133
  cards and printed circuit boards, limitations on, 144
  client/server IP–PBX and, 229, 249
  legacy PBX, 45–77
  redundancy and duplication in, 112–116
  session initiation protocol (SIP) and, 199–201
  topologies for (centralized, dispersed, distributed), 107–112, **107**
call progress tones, 102–103
call prompting, 138
call registers, 101
call routing (*See also* route selection), H.323 protocol and, 181–182
call sequencing, 100
call set-up, H.323 and, 196–197, **196, 197**
call signaling
  H.225.0 RAS, 189
  H.323 protocol and, 184
call transfer, 376
call vectoring, 37
call waiting, 376, 385
caller ID (CLID), 29, 298, 330
CallManager, redundancy and backup in, 234–238, **235**, 244
CallPilot, 241
capacity planning, in LAN design, 259–260
card slot requirements, 144–145
cards and printed circuit boards, 122, 126, 127, 134–143
  cabinet/carriers backplane and, 132
  common control complex, 97
  control, 134, 136
  diagnostics, 137
  digital line card, 135, **135**
  local loop interfaces, 137
  port, 134, 139–143
  provisioning issues for, 143–145
  service, 134, 137–139
categories of cabling, 280, 287–288, **287, 288**
Category 1 cabling, 289
Category 2 cabling, 289
Category 3 cabling, 280, 281, 282, 287–288, **287, 288**
Category 4 cabling, 282, 287–288, **287, 288**
Category 5 cabling, 280, 281, 282, 283, 287–288, **287, 288**
Category 5e, 287–288, **287, 288**
Category 6 cabling, 280, 287–288, **287, 288**
Category 7 cabling, 289

CBX II 9000, 26, 55, 112
CCIS, 349
CELP, 261
center stage switch, 54–56
   blocking vs. nonblocking systems, 80–82
   control buffer card and, 138
   IP PBX vs. client/server PBX in, 163
   redundancy/duplication in, 71–73
   switch network card and, 136
   switching carrier and, 127
central office (CO), CO, 93, 137, 141
central processing unit (CPU), 136
centralized application servers, networking with PBX and, 353
centralized attendant service (CAS), 141, 334, 391
centralized PBX design, 21, 64–67, 107–112, **107**, 335–337, 343
   IP PBX vs. client/server PBX in, 156–157, 172
   telephony over IP (ToIP), 153
Centrex, IP telephones and telephony, 305
centum call seconds (CCS), 83–88
channel associated signaling (CAS), telephony gateways and, 244
Chorus operating system, 99
Cingular cellular, 314
circuit breakers, 130
circuit emulation service (CES), 142
circuit switched PBX, 46–60, 148
   asynchronous transfer mode (ATM) center stage switch in 59–60
   center stage switching, 54–56
   client/server IP–PBX and, 229, 231
   converged IP–PBX and, 208
   converged IP–PBX and, upgraded, 220–223
   IP PBX vs. client/server PBX in upgrade to, 161–162
   multistage circuit switch matrix in, 58–59, **58**
   pulse code modulation (PCM) and, 47–49, **47, 48**
   single-stage circuit switch matrix in, 57–58
   Superloop and, 58, 62–64, 72–73, 77
   TDM bus, bandwidth and capacity of, 49–51, **51**
   time division multiplexing (TDM) and, 47–48, **47**
   time space time (TST) switch networks in, 59–59
Cisco Systems family of PBX systems, 24, 152, 204, 230, 234, 240, 247, 336, 337
   client/server IP–PBX and, 240, 241, 247
   IP telephones and telephony, 302, 305–306, 308
   remote port cabinet options for, 337–340
   telephony gateways and, 244
CLAN cards, 153
class of restriction, 391
class of service (CoS), 249, 268, 391–392
   network (NCoS), 345–346
client/server IP PBX, 122, 149, 169–173, **170**, 227–249
   administration and management and, 360
   applications layer in, 229
   backup servers in, 234
   cabinet/carrier design for, 124–125
   call centers and, 241–242
   call control protocols for, 249
   call data recording (CDR) in, 236
   call processing layer in, 229, 249

client/server IP PBX (*continued*)
   CallManager, redundancy and backup in, 234–238, **235**, 244
   channel associated signaling (CAS) in, 244
   circuit switching in, 229, 231
   client layer in, 229
   clients or telephone terminals in, 228
   clustered servers in, 236
   common control complex of, 228
   computer telephony integration (CTI) in, 230
   converged IP–PBX vs., 208, 226
   cost of, 166–169
   database synchronization in, 239
   design issues in, 247–249
   dual tone multifrequency (DTMF) in, 243
   Ethernet and TCP/IP in, 229, 240, 243
   features of, 248
   G.711 in, 237
   G.729 in, 237
   gatekeeper in, 228
   gateways in, 230, 242–249
   generic program in, 230
   H.323 vs. SIP protocols in, 201–202, 231, 237, 243, 244, 249
   hardware layer elements of, 229
   integrated communications system (ICS) in, 240–241
   IP PBX vs., 154–174
   IP telephony and, 228, 242–249
   LAN/WAN design in, 228, 229
   layered architecture in, 247–248, **248**
   media gateway (MG) in, 238–239
   media gateway control Protocol (MGCP), 244
   memory in, 231
   mesh topology in, 236
   messaging applications and, 241, 249
   multifunction server design in, 239–242
   multiple active servers in, 234
   operating system in, 99–100, 231, 232–233, 240
   packet switching in, 229
   port capacity of, 249
   power supplies in, 240
   processors in, 231
   quality of service (QoS) in, 249
   redundancy and duplication in, 233–239, 249
   reliability and survivability of, 248–249
   request for proposal (RFP) example for, 407–443
   simple network automated processing (SNAP) in, 238
   SIP in, 243
   survivable remote telephony (SRS) in, 238
   TDM bus in, 228, 229
   telephony call server for, 230–242
   telephony features in, 241
   telephony gateways in, 242–249
   telephony over IP (ToIP), 151–153
   third-party solutions in, 248
   traffic engineering/traffic handling in, 249
   voice communications and voice codecs in, 249
clock rates, 50
clocks, 136, 137
clustered servers, for client/server IP–PBX and, 236
CO trunk card, 141

Index

code calling, 128, 381
codecs, 257
   H.323 protocol and, 178, 179, 184–186, **187**
   quality of service (QoS) and, 261–262
codes, voice messaging system (VMS) and, 10
COHub Sphericall Enterprise Softswitch, converged IP–PBX and, 224
cold standby duplicated control system, 114
color coding
   cabling, 289–290, 292–293
   cards and printed circuit boards, 134
command format, H.323 vs. SIP protocols in, 202
Common Channel Signaling System 7 (CCSS7) in, 329
common control complex, 96–97, **97**
   centralized PBX and, 107–112, **107**
   client/server IP–PBX and, 228
   converged IP–PBX and, 209
   cost of, 300–401
   CTI and, 108, 109
   dispersed PBX and, 110–112, **107**
   distributed PBX and, 107–112, **107**
   IP PBX vs. client/server PBX in, 162
   redundancy and duplication in, 113–114
   telephony over IP (ToIP) and, 150–153
common equipment
   converged IP–PBX, 220
   distributed PBX, 335–337
   legacy PBX, 121–145, **122**
Compaq, 22
complex instruction set computing (CISC), 109
compression/decompression
   H.323 protocol and, 179
   IP telephones and telephony, 304, 307–308
   quality of service (QoS) and, 273–274
Computer Services Telephony Applications (CSTA), 39
computer stored program control (SPC) (*See* stored program control)
computer technology and PBX, 21–22
computer telephony integration (CTI), 3, 17, 21, 27, 29, 38–40, **39**, 149, 239, 368
   adjunct server control for, 108, 109
   API module for telephone using, 321–322
   applications programming interface (API) for, 230
   cards and printed circuit boards, 138–139
   centralized PBX and, 109–110
   client/server IP–PBX and, 230
   common control complex and, 97
   converged IP–PBX and, 225
   data module for telephone using, 321
   digital telephones and, 300
   IP PBX vs. client/server PBX in, 156
   networking with PBX and, 353
   softphones in, 315–317, **316**
   telephony over IP (ToIP), 150–153
   universal serial bus (USB) telephone for, 300–301
computerized branch exchange (CBE), 148
conference calling, 5, 53, 381, 385
   add-on, 373
   analog communications device adapter for, 322, **322**
   analog phones and, 298–299
   cards and printed circuit boards, 137

conference calling (*continued*)
   converged IP–PBX and, 213
   H.323 protocol and, 177, 179
   MGCP and MEGACO in, 204
consecutive speed dialing, 376
constant bit rate (CBR), 253
consultation (broker) hold, 376
Contact Center Solution, 241
control cards and printed circuit boards, 134, 136
control buffer card, 138
control cabinet/carrier, 125, 126
control information, H.323 protocol and, 186–188
control of trunk group access, 385
controlled private calls, 392
controllers, H.323 protocol and, 181–182
converged circuit/packet switched systems, 151
converged IP–PBX, 207–226
   access control and security in, 216, 222
   ARS and, 217
   bandwidth and, 219
   circuit switching in, 208
   client/server PBX vs., 208, 226
   common control complex in, 209
   conference calling, 213
   CTI and, 225
   delay in, 213–214, 217–218, 219
   digital compressors in, 210–211
   digital signal processing (DSP) in, 214
   dispersed common equipment over LAN/WAN infrastructure in, 220
   distributed Internet voice architecture (DIVA) in, 224–225
   distributed modular design in, 223–226
   dual tone multifrequency (DTMF) and, 213
   echo cancellation in, 213, 225
   Ethernet and TCP/IP in, 209, 214, 217, 221–223, 225
   fiber loop exchange (FLEX) in, 222
   G.711 in, 210, 212, 218, 221, 225
   G.723 in, 218
   G.726 in, 221
   G.729 in, 211, 212, 218, 219, 221, 225
   gatekeeper and gateways in, 209–214, 218–219, 223
   H.323 and, 210, 217
   INT–IP card in, 223
   IP station ports and interface cards in, 208–214
   IP telephony and, 208, 212–216, 221
   IP trunk ports in, 216–219
   ISDN and, 217, 218, 221, 222
   ITG line in, 209, 211
   jitter buffering in, 213–214
   jitter in, 221
   LAN/WAN design and, 208, 209, 215, 216–221
   media gateway (MG) and, 210
   media gateway control (MGC) and, 210
   null capability in, 216
   outgoing IP trunk calls in, 217
   peer-to-peer connection in, 209
   port capacity in, 210
   private networking and, 218, 219
   proxy server and, 209
   quality of service (QoS) in, 212–219, 221, 225, 226
   remote peripheral equipment (RPE) in, 221

Index

converged IP–PBX (*continued*)
  silence suppression in, 213, 225
  T1/E1 and, 208, 218
  TDM bus and, 210, 211, 213, 217, 221, 223
  telephony over IP (ToIP) and, 208
  time division multiplexing (TDM) in, 215–217
  transcoding in, 210–212
  upgraded circuit switched PBX in, 220–223
  voice activation detection in, 213
  voice communications and voice codecs in, 210–216, 221, 225
  voice over IP (VoIP) and, 208, 216, 217
converged PBX, 158–169, 170–171
convergence of KTS/hybrid and PBX systems, 11–14
CorNet, 349, 353
cost and pricing issues, 13, 149, 399–406
  client/server PBX, 166–169
  common control cabinet and software, 300–401
  common equipment, 122
  expansion port cabinet/carrier, 401–402
  H.323 vs. SIP protocols in, 202
  installation and cabling, 404, 445
  IP PBX vs. client/server PBX, 154–155, 166–169
  IP PBX, 404–405
  port circuit cards, 402, 445
  price curves for, 406, 445
  system management, 403–404
  telephones and, 296–297, **296**, 317
  voice terminals and telephones, 402–403, 445
CPU bus, 99–100
crossconnects, 286
crossover arbitration, redundancy and duplication in, 114
crosstalk, 282
customer database, 103–104, 136
  administration and management and, 362
  cards and printed circuit boards, 136
  redundancy and duplication in, 113, 239
  synchronization in, 239
customer relationship management (CRM), 240
customer station rearrangement, 376

D channel (*See also* ISDN), 329
D channel backup, 329
D channel handler card, 142
DASSII, 244
data communications, PBX and, 18–19, **19**, 31–33, 170–171
  cabling for, 278
  serial data interface for, 137
  system processing bus and, 106–107
data communications equipment (DCE), 140
data line card, 140
data modules for telephones and, 321
data terminal equipment (DTE), 140
database synchronization, 239
datagrams, 191
DC powered systems, 129–131
DEC, 39
defense Metropolitan Area Telecommunications system (DMATS), 349

Definity family of systems, 33–34, 43, 55, 60–62, 65, 68, 70–73, 133
  blocking vs. nonblocking systems, 81–82
  cabinet/carrier design for, 123–124
  centralized PBX and, 108–110
  converged IP–PBX and, 209, 221–222
  fiber optics and, 338–340
  IP PBX vs. client/server PBX in upgrade, 162
  mobile phones, 335
  operating system for, 99, 232
  port carrier in, 127, **128**
  remote options, 340
  switching carrier in, 127, **127**
Definity Port Networks, 64
Definity ProLogix, 66
delay, 213–214, 217–219, 260–262, 273
delay announcement, 385
delayed ringing, 392
Dell, 22
delta modulation (DM) in, 18
demarcation point, 284
design of PBX system (*See also* networking with PBX)
  distributed common equipment design in, 335–337
  private networks and, multiple system, 341–343
  request for proposal (RFP) for client/server PBX, 407–443
  single system network, on net multilocation support in, 332–335
diagnostics and maintenance in, 126, 127, 137, 163, 369–371
dial by name, 377
dial plan, 362–363, 392
  uniform dial plan (UDP) in, 344–345, 398
dial tone, 100
dialed number identification service, 392
dialing and connection process, 100–101, **101**
DID call waiting, 393
DID trunk card, 141
differentiated services (DiffServ), 203, 249, 268, 270–271, 274–276, 318
differentiated services code points (DSCP), 270, 275
digit reception/analysis, 100, 101–102
digital attendant console card, 139
digital communications device adapter for, 323
digital compressors, converged IP–PBX and, 210–211
digital desktop concept, 22–24
Digital Equipment Corporation (DEC), 21
digital line card, 135, **135**, 139
digital PBX, 46, 125, 148
digital signal processing (DSP), 261
  converged IP–PBX and, 214
  IP telephones and telephony, 304
  latency and, 263, 264
digital subscriber line (DSL), 334
digital switching, 46, 278
Digital Telephone Systems, 17
digital telephones, 23–24, 46, 76, 148–149, 166, 299–302, **300**, **302**
  Busy Hour Call (BHC) in, 117–119
  IP telephones in (*See* IP telephones), 301–317, **302**, **303**
  ISDN BRI type, 301

Index

digital telephones (*continued*)
  proprietary, 299–300
  SuperSet, 242–243
  universal serial bus (USB) type, 300–301
digital trunk interface (DTI), 144
digital wireless office system (DWOS), 42
Dimension PBX, 20, 35, 38, 349
direct department calling, 392–393
direct inward dialing (DID), 3, 93, 102, 137, 393
direct inward system access, 393
Direct Inward System Access (DISA), 102
direct inward termination, 393
direct outward dialing (DOD), 393
direct station selection with busy lamp field, 385
direct trunk group selection, 385
directories, 327, 369
discrete call observing, 376–377
dispersed PBX, 21, 65, 67–70, 107–112, **107**, 335–337
  converged IP–PBX and, 220
  redundancy and duplication in, 116
  telephony over IP (ToIP), 153
display, 386
distance extender module for telephones and, 323
distinctive ringing, 377
distributed communications system (DCS) in, 35, 341, 349
distributed Internet voice architecture (DIVA), 224–225
distributed PBX, 21, 46, 65, 67–70, **68**, 107–112, **107**, 335–337
  converged IP–PBX and, 223–226
  IP PBX vs. client/server PBX in, 172–173
  telephony over IP (ToIP), 153
DLSW, 256
DMS 100, 26, 72
do not disturb, 377
DPNSS, 244
DS1 digital trunk card, 141
dual tone multifrequency (DTMF), 5, 11, 137, 298–299
  analog line cards and, 139
  analog phones and, 298–299
  client/server IP–PBX and, 243
  converged IP–PBX and, 213
  IP telephones and telephony, 304
  PBX and, 23
duplication (*See* redundancy and duplication)
Dynamic Host Configuration Protocol (DHCP), 157, 318, 363, 364

E model for quality of service (QoS) and, 272–276
E&M interface in, 343
E&M tie trunk, 141, 218
E1, 141, 142, 208
early adopters of new technology, 149
echo and echo cancellation in, 213, 225, 272, 318
ECS, IP PBX vs. client/server PBX in upgrade, 162
800 lines, 93
elapsed call timer, 377
electromagnetic compatibility (EMC), 281–282
electromagnetic interference (EMI), cabling, 281–282
Electronic Industries Association (EIA) standards for cabling, 278

electronic programmable read only memory (EPROM), 104
electronic tandem network (ETN), 34–35, 349
electronic telephones, 148
electrostatic damage to cards and printed circuit boards, 134
email, 239, 311
emergency access to attendant, 377
EN 50173 cabling standard, 279
Enhanced Private Switched Communications Service (EPSCS), 341
endpoints
  H.225.0 RAS, 188–189
  H.245 media control, 190
  H.323 protocol and, 180, 182
entrance facility cabling, 284, 292
EPN, 341
equipment room cabling, 284, 292
Ericsson, 25, 41, 64, 112, 133, 211, 243, 284, 315, 336
  remote port cabinet options for, 337–340, 337
Erlang B queuing model in, 88–90, **90**, 93–94
Erlangs, 365
error logs, 371
errors, 370
ESN, 341
Ethernet and TCP/IP, 19–20, 27, 31–33, 106–107, 128, 138, 152, 170
  10Basexx and, 280
  application cabinet/carriers and, 128–129
  cabling and, 279–280, 282, 286–289
  cards and printed circuit boards for, 138–139
  client/server IP–PBX and, 229, 240, 243
  converged IP–PBX, 209, 214, 217, 221–225
  H.323 protocol for, 176–188
  IP PBX vs. client/server PBX in, 157
  IP telephones and telephony using, 304–307
  latency in, 266
  networking with PBX and, 351
  quality of service (QoS) and, 261
  telephones and, 318
  telephony over IP (ToIP), 150–153
  topologies for, 279–280
  voice messaging system (VMS) and, 10
European Computer Manufacturers Association (ECMA), 39, 354
executive access override, 378
executive busy override, 377
executive calling, 378
expansion, IP PBX vs. client/server PBX in, 155–156, 160–162
expansion cabinets, 125
expansion interface boards/cards, 126, 138
expansion port cabinet/carrier, cost of, 401–402
expansion requirements, cabinet/carrier, 132–134
expedited forwarding (EF), DiffServe, 270–271, 275, 276
expert agent selection (EAS), centralized PBX and, 109–110
explicit QoS, 253
extended trunk access, 393
EXTender gateway, 335
extensions, H.323 vs. SIP protocols, 203
external paging with meet me, 378

F9600 PBX, 32, 61, 66, 72, 111, 145
facility busy indication, 378
facility restriction levels, 394
facility test calls, 394
FACs, 362–363
far end crosstalk (FEXT), 282
Fast Connect/Fast Start, 185
Fast Ethernet, 20, 222, 282
fault tolerant systems, IP PBX vs. client/server PBX in, 163
fax, 2, 184
FeatureLinks, 64
features and functions, 3–4, 16–17, 28–44, 100, 149, 367
  analog phone, 298
  attendant, 384–388
  call processing system, 96, **96**
  client/server IP–PBX and, 248
  H.323 protocol for, 178
  IP PBX vs. client/server PBX in, 155–158, 165–166
  IP telephones and telephony, 305–317
  matrices for, 445–467
  networking with PBX and, 351–353
  station user, 373–384
  suppliers, 445–464
  system, 388–398
  telephones and, 296–297, 305–317, 319–320
  terminals, 446–454
  transparency of, 164
Fiber loop exchange (FLEX), converged IP–PBX and, 222
find me feature, 30
FIPN, 349
firewalls, 203, 222, 266, 267
first-digit/second-digit tables, 363
fixed delay, 213–214
Flash memory, 104, 231, 304
flexible night service, 4
flood wiring, 281
floppy disk drives, 104, 136
flow control, H.245 media control and, 190
follow me feature, 374
forced account code, 394
forward error correction (FEC) in, 260–262
forwarding, in DiffServ, 270–271, 275–276
frequency of operation, TDM bus, 50
frequency of wireless PBX, 42
FRL and, 345–346
Fujitsu, 32, 61, 62, 66, 72, 111, 145, 349
fuses, 130
Fusion CCS, 349, 352, 353
FX, 93, 137, 141

G.107, 272
G.711, 184, 186, 249, 257, 261, 264, 307
  client/server IP–PBX and, 237
  converged IP–PBX and, 210, 212, 218, 221, 225
G.722, 184
G.723, 184, 186, 218, 264, 268, 307
G.726, 221
G.728, 184, 186
G.729, 184, 186, 249, 257, 261, 264, 307
  client/server IP–PBX and, 237
  converged IP–PBX and, 211, 212, 218, 219, 221, 225

G.732, 249
Galaxy ACD, 37
gatekeepers
  client/server IP–PBX and, 228
  converged IP–PBX and, 209–214, 218–219
  H.225.0 RAS, 188–189
  H.323 protocol and, 180, 181–182, 196
  telephony over IP (ToIP), 150–153, 153, 161
gateways
  client/server IP–PBX and, 230, 242–249
  converged IP–PBX and, 209–214, 218–219, 223
  H.323 protocol and, 176, 180, 181
  IP telephones and telephony, 304–305
  in LAN design, 255
  latency and, 263
  telephony over IP (ToIP), 150–153, 161
  voice over IP (VoIP), 367
generic software program, 99, 103–104
  cards and printed circuit boards, 136
  client/server IP–PBX and, 230, 234
  IP PBX vs. client/server PBX in, 161, 162
  redundancy and duplication in, 113
Gigabit Ethernet, 282
global system for mobile (GSM), 249, 314
grade of service (GoS), 82–83, 92
graphical user interfaces (GUIs), 30–31
green field installations, IP PBX vs. client/server PBX in, 158, 167, 168
group assignments, 364–365
group listening feature, 378

H.225 call signaling protocol, 184, 187, 305
  RAS, 188–189, **189**
  real time transport protocol (RTP) and, 192
H.235, 184
H.245 control protocol in, 184, 187, 189–190, **191**
  IP telephones and telephony, 305
H.248 (See MGC Protocol)
H.261 protocol, 185
H.263 protocol, 185
H.320 protocol, 177
H.321 protocol, 177
H.323 protocol, 142, 176–188
  accounting and billing using, 178
  address translation and, 182
  administration and management using, 178
  annexes to, 184
  architecture of, 180
  audio codecs for, 178, 184, 186, **187**
  bandwidth and, 180, 182, 185
  benefits of using, 179–180
  border elements in, 180, 183
  call routing in, 181–182
  call set up and, 196–197, **196, 197**
  call signaling and control, 184
  client/server IP–PBX and, 231, 237, 243, 244, 249
  codecs in, 185–186
  compression/decompression and, 179
  control and signaling mechanisms in, 186–188
  converged IP–PBX and, 210, 217
  endpoints in, 180, 182

Index

H.323 protocol (*continued*)
  gatekeepers for, 180, 181–182, 196
  gateways for, 180, 181
  IP telephones and telephony, 304–305
  layered model of, 187
  master/multipoint control unit (MCU), 180, 182–183
  multi-, uni-, and broadcasting in, 179–180, 183
  protocols and procedures in, 183–186, 187, **187**
  quality of service (QoS) and 176–177, 180, 185, 203
  real time transport protocol (RTP) and, 192
  route selection, 182
  security using, 178
  session initiation protocol (SIP) vs., 201
  telephony gateways and, 243, 244
  terminals for, 180, 181
  versions of, 185–186
  video codecs for, 178, 185–186
H.324 protocol, 177
H.450, protocol, 184
H.GCP protocol, 185
hands-free answer intercom (HFAI) in telephones and, 320
hands-free dialing, 378
hands-free intercom, 378
hard QoS, 253
Harris Corporation/Digital, 17
HCX5000, 50
header for RTP, 195, **195**
header for UDP, 195, **195**
header of real time transport control protocol (RTCP), 194, **194**
header of real time transport protocol (RTP) and, 193, **193**
headers, IP, 258
headset interface for telephones and, 320
help/information key, 378–379
Hicom family of systems, 50, 63, 66, 72, 73, 353
  cabinet/carrier design for, 124, **124**
  call center and call center servers in, 241
  converged IP–PBX and, 222
  data transmission rate in, 106–107
  dispersed PBX and, 111
  IP PBX vs. client/server PBX in upgrade, 162
  messaging applications and, 241
  redundancy and duplication in, 116
Highway bus
  blocking and nonblocking design in, 75–77, **76**
  blocking vs. nonblocking systems, 80–82
  centralized PBX and, 66–67
  dispersed PBX and, 70
  local switching network design and, 60–64
  redundancy/duplication in, 71–73
HiPath family of systems, 129
  call center and call center servers in, 241
  client/server IP–PBX and, 231, 232
  converged IP–PBX and, 222–223
  IP PBX vs. client/server PBX in upgrades, 164
  messaging applications and, 241
  mobile phones, 335
  telephony gateways and, 242–244
Hitachi, 50
hold, 5

home run wiring, 286
horizontal cabling, 286–289, **288**, 292
hot line, 379
hot standby duplicated common control, 113–114, 162
hoteling in, 29, 394
house phone, 394
HSRP, 256
HTTP, 184, 200
hunting, 394–395
hybrid systems, 7–8

I/O interfaces, 97
i2050 softphone, 215, **216**
i9150 remote system, 340
IBM, 17, 22, 39, 171, 242
IBX systems, 18–19, 24–25, 38, 148
IEEE 802.3af standard for IP telephones and telephony, 306–307
IMAP, 249
implicit QoS, 253
in-band vs. out-of-band signaling in, 23, 35
incoming call display, 379
individual attendant access, 379
input/output (I/O), 5
installation and cabling, cost of, 404, 445
Intecom family of systems, 18–19, 23, 24, 31, 32, 38, 39, 299, 336
integrated communications system (ICS), 240–241
integrated directory, 395
integrated port interfaces, IP telephones and telephony, 305–306
Intel processors, 98, 109, 231
intelligent feature transparent network (IFTN) in, 349–357
intelligent peripheral equipment module (IPEM), 62
intelligent private networks, 327
intelligent signaling channels in, 23–24, 35
interactive voice response (IVR), 3, 5, 11, 37, 85, 93, 138, 139
interbuilding cabling, 292
intercabinet links, redundancy and duplication in, 116
intercept treatment, 386
intercom, 5, 374, 378, 379, 380
intercom dial, 379
interfaces, 97, 126, 127
interference, in cabling, 281–282
interleaving and, 260–262
intermediate crossconnect, 286
International Standards Organization (ISO) cabling standards, 278
International Telecommunication Union (ITU), 354
Internet, 171, 191, 252
Internet Engineering Task Force (IETF), 305
internetworking, H.323 protocol for, 177
interposition call and transfer, 386
INT–IP card, converged IP–PBX and, 223
intrabuilding cabling, 292
intranets, 267
intrusion, 386
IntuityAudix, 43
inventories, 369
invitation, session initiation protocol (SIP) and, 200–201

Index

IP adapter module, telephones and, 323
IP addresses, 255, 363
IP line card, 140–141
IP PBX, 147–173, **168**
  administration of, 172
  applications across enterprise using, 156–157
  bandwidth in, 171–172
  benefits and advantages of, 154–158
  cabling costs in, 167
  capital, network, operating expenses in, 154–155
  centralized PBX and, 156–157, 172
  client/server type, 169–173, **170**
  computer telephony integration (CTI) and, 156
  converged (*See* converged IP PBX system design)
  converged systems and, 158–169
  cost and pricing of, 166–169, 404–405
  distributed PBX and, 172–173
  features, feature requirements in, 157–158, 165–166
  green field installation of, 167, 168, 169
  investment protection in, 160–162
  MGCP and MEGACO in, 205
  new technology and applications, deployment of, 172
  port capacity of, 166–167
  power supplies for, 166
  private network compatibility with, 163–164
  quality of service (QoS) and, 159–160, 168, 171
  redundancy and duplication in, 162–163, 167
  reliability of, 162–163
  scalability of, 173
  standards and, 156
  telephones for, 166
  telephony over IP (ToIP), 150–153, 159–160
  trunk circuits for, 166
  universality of IP transport and, 171
  upgrades and expansions in, 155–156, 160–162, 168–169
IP precedence, 249
IP station ports and interface cards, converged IP–PBX and, 208–214
IP telephones and telephony, 12, 24, 49, 91, 166, 214–216, 301–317, **302, 303**
  blocking and nonblocking design in, 76–77
  cabinet/carrier design for, 124–125
  cards and printed circuit boards, 140–141
  Centrex and, 305
  client/server IP–PBX and, 228, 242–249
  compression in, 304, 307–308
  converged IP–PBX and, 208, 212–213, 221
  design basics in, 303–305, **303**
  digital signal processing (DSP) in, 304
  dual tone multifrequency (DTMF) and, 304
  email station in, 311
  Ethernet and TCP/IP in, 304–307
  features, 305–317
  gateway subsystem in, 304–305
  H.225 call signaling protocol in, 305
  H.245 Control protocol in, 305
  H.323 and, 304–305
  IEEE 802.3af standard for, 306–307
  integrated port interfaces in, 305–306
  IP line card for, 140–141

IP telephones and telephony (*continued*)
  IP trunk cards, 142
  mobile and cellular phones in, 312–315, **314**
  network interface in, 304
  null capability in, 216
  operating systems in, 308–309
  PBX and, 16, 36
  PDA docking stations and, 306
  power supplies, 306–307, 324
  quality of service (QoS) and, 252, 305–306
  RAS protocol in, 305
  real time transport control protocol (RTCP) in, 305
  sampling rates in, 301–302, **302, 303**
  screenphone type, 309, **309, 310, 311**
  silence suppression in, 308
  SIP in, 305
  softphones and, 40, 315–317, **316**
  software for, 304
  telephony gateways and, 244
  universal serial bus (USB) type, 304, 306
  user interface for, 303
  voice activation detection (VAD) in, 308
  voice interface for, 303–304
  voice over IP (VoIP) and, 175–205, 302
  Web browsers and, 302, 308–311, **309, 310, 311**
IP trunk cards, 142
IPDC, 204
Ipera 3000, 99
IPX, 256
ISDN, 12, 24, 30, 32–34, 36, 93, 244, 368
  blocking and nonblocking design in, 75–77, **76**
  Busy Hour Call (BHC) in, 117–119
  cabling, 290
  call-by-call service selection (CBCSS) in, 330–331
  converged IP–PBX and, 217, 218
  converged IP–PBX and, 221–222
  digital telephones and, 299, 301
  H.323 protocol for, 177, 201
  ISDN BRI line card for, 140
  ISDN BRI trunk cards for, 142
  ISDN PRI trunk cards for, 141–142
  networking with PBX and, 328–329, 350
  Qsig and, 355
  telephones and, 323
  video communications and, 33–34
ISDN BRI, IP PBX vs. client/server PBX in H.323, 156
ISDN BRI line card, 140
ISDN BRI trunk cards, 142
ISDN PRI trunk cards for, 141–142
ISDN Private Network System (IPNS) Forum, 35
ISO/IEC IS 11801 cabling standard for, 278–279
Itecom, 148
ITG line, converged IP–PBX and, 209, 211
IVS2 system, cabinet/carrier design, 126

jitter and jitter buffering, 191, 249
  converged IP–PBX and, 213–214, 221
  quality of service (QoS) and, 264, 265–266

key service unit (KSU), 5–6, 7
key telephone system (KTS), 2–3, 5–6, 10, 22

Index

KTS/Hybrid systems, 2, 3, 7–8
  automated call distribution (ACD) and, 9
  cost of, 13
  digital telephones and, 299
  IP PBX vs. client/server PBX in, 169
  PBX convergence with, 11–14
  scaling of, 13
  voice messaging system (VMS) and, 10

labeling
  cabling, 292
  cards and printed circuit boards, 134
last number redialed, 379
latency, 262–268, **266, 267**
LDAP, 249
least cost routing (LCR) (*See also* automatic route selection), 34, 326
legacy PBX
  call processing design, 45–77
  common equipment, 121–145, **122**
  IP PBX vs. client/server PBX in and, 154
  switch network design, 79–94
  traffic engineering in, 95–119
Lexar, 18
life cycle of PBX, 160–161
line interface module (LIM), 25, 64, 70, 112, 133
line lockout, 379–380
line trunk unit controller (LTUC), 63
line trunk unit (LTU), 61
Linux operating system, 99, 232–233
local area networks (LANs), 16, 19–20, 24, 27, 32–33, 149
  availability in, 256
  bandwidth management in, 257–260, **258, 259**
  baseline measurement in, 256–257
  cabling for, 257
  capacity planning for, 259–260
  client/server IP–PBX and, 228, 229
  converged IP–PBX and, 208, 209, 215, 220, 221
  gateways in, 255
  H.323 protocol for, 176–188
  IP addresses in, 255
  IP PBX and, 147–173
  IP PBX vs. client/server PBX in, 154–158, 170
  networking with PBX and, 336–337
  planning guidelines for, by Cisco Systems, 254–260
  protocols for, 256
  quality of service (QoS) in, 252–254, 257, 260–276
  scalability in, 256
  servers in, 255
  telephony over IP (ToIP), 150–153
  topologies for, 254–255
  voice messaging system (VMS) and, 10
  voice over IP (VoIP) and, 251–276
local controllers, 105–106
  redundancy and duplication in, 113, 116
local loop interface card, 137
local processors, 104
local switching network design and, 60–64
logical channels, H.245 media control, 190
look-ahead routing, 346
loop length limitations, for telephones and, 317

loudspeaker paging access, 380
loudspeakers, 128
low latency queuing (LLQ), 271
Lucent Technologies, 33, 36, 99

MAC addresses, 363
Magellan Passport Asynchronous Transfer Mode (ATM) system, 36
malicious call trace, 380
main crossconnect, 286
main switches in, 344
main system assembly, 122
main system memory, 96–97, **97**, 103–104
main system processor, 96–98, **97**, 105, 106
  cards and printed circuit boards, 134–135
  centralized PBX and, 108–110
  dispersed PBX and, 110–111
  distributed PBX, 111–112
  redundancy and duplication in, 113–116
maintenance/diagnostic card, 137
major alarm, 371
management (*See* administration and management)
management information system (MIS), 2, 12, 37–38
manual intercom, 380
manual originating line service, 380
manual signaling, 380
March Networks (*See* Mitel Networks)
master control unit (MCU), 25
master/multipoint control unit (MCU), H.323 protocol and, 180, 182–183
MCK Communications, 334, 335
MCS 7835, client/server IP–PBX and, 230
MD-110 PBX system (Ericsson), 25, **26**, 41, 64, 133, 284, 336
  converged IP–PBX and, 211
  distributed PBX, 112
  IP telephones and telephony, wireless, 315
mean opinion score (MOS), to judge latency, 262–263
mean time between failures (MTBF), 162
mean time between outages (MTBO), 162
media gateway (MG), 210, 238–239
media gateway control (MGC), 210
media gateway control Protocol (MGCP), 204–205, 244, 249
media processor interface (prowler board), 212
meet-me feature, 378
MEGACO (H.248), 204–205, 249
memory, 96–97, **97**, 103–104, 126, 149
  cards and printed circuit boards, 136
  client/server IP–PBX and, 231
  IP telephones and telephony, 304
Meridian family of systems, 36, 43, 60, 62, 63, 66, 77, 85–87, 133
  blocking vs. nonblocking systems, 82
  cabinet/carrier design for, 124
  call center and call center servers in, 241
  cards and printed circuit boards, 144
  converged IP–PBX and, 209, 221
  data transmission rate in, 106–107
  dispersed PBX and, 110–111
  fiber optics and, 338–340

Meridian family of systems (*continued*)
    IP PBX vs. client/server PBX in upgrade, 162
    mobile phones, 335
    operating system for, 99
    power supply for, 130, **130**, **131**
    redundancy and duplication in, 115–116, **115**
    telephony gateways and, 247
    telephony over IP (ToIP), 153
Meridian Mail, 43
mesh topology, 177, 236
message waiting, 380–381
messaging applications/systems (*See also* voice messaging systems), 2, 17, 42–44, 128
    client/server IP–PBX and, 241, 249
    H.323 vs. SIP protocols in, 202
microprocessors (*See* main system processors; processors)
Microsoft, 39
minor alarm, 371
MIPS3000, 109
mirroring, disk, client/server IP–PBX and, 233–234
MITAI, 231
Mitel Networks and MNxxx family of systems, 99, 152, 231, 232, 240, 242, 244, 247
    client/server IP–PBX and, 231–232, 240, 241
    IP telephones and telephony, 305, 306, 313
    telephones and, 322, **322**
    telephony gateways in, 242–244
mobile and cellular phones (*See also* wireless systems), 17, 41–42, 312–315, **314**, 335
mobile users of PBX, 29–30
MobileOffice, 241
modem pooling, 395
modems, 19
    pooled modem cards for, 143
modular PBX design, 16, 24–27, 223–226
most economical route selection (MERS) (*See also* automatic route selection), 34, 326
Motorola processors, 98, 109, 231
mouth to ear latency, 262
moves additions and changes (MACs), 361
Mu law, 49
multicarrier port cabinets, telephony over IP (ToIP), 153
multicasting
    H.323 protocol and, 179–180, 183
    real time transport protocol (RTP) and, 192
multifrequency (MF), 137
multifunction server design, in client/server IP–PBX, 239–242
multimedia, 157, 170
    H.245 media control, 189–190
    H.323 protocol and, 177, 179–180, 183–186
    H.323 protocol for, 177
    MGCP and MEGACO in, 204
multimedia call handler (MMCH) in, 33–34
multimedia communications exchange (MMCX), 33–34
multiparty conferencing, 381
multiple active servers, 234
multiple listed directory numbers, 395
multiple TDM bus design, 52–54, **53**
multiplexing, H.323 protocol and, 184

multipoint communications, H.323 protocol for, 177
multipoint controller (MC), H.323 protocol and, 183
multiprotocol label switching (MPLP), 267, 271
multistage circuit switch matrix in, 58–59, **58**
multitasking operating systems, 98–100
music on hold, 128, 381, 395
mute buttons on telephones and, 320

National ISDN (NISDN) specifications, digital telephones and, 301
NBX family of systems, 43, 152
    call center applications in, 241–242
    client/server IP–PBX and, 230–232, 239–240
    telephony gateways in, 242
near end crosstalk (NEXT), 282
NEAX family of systems, 27, 38, 63, 70
    cabinet/carrier design for, 124–126
    cards and printed circuit boards, 143–144
    converged IP–PBX and, 209
    dispersed PBX and, 111
    IP PBX vs. client/server PBX in, upgrades, 162, 164
    networking with PBX and, 352
NEC, 27, 36, 38, 39, 42, 63, 70, 111, 115, 124, 125, 162, 209, 212, 312, 341, 349
NetVision Phone, wireless, 313–314
network class of service (NCoS) in, 345–346
network connection channels, 364
network demarcation point, 284
network interface, IP telephones and telephony, 304
network interface cards (NICs), 19
network latency, 266
networking with PBX (*See also* private networks; virtual private networks), 34–36, 64–70, 325–357
    Agent Anywhere option n, 352–353
    automatic alternate routing (AAR) in, 345
    automatic call direction (ACD) in, 352–353
    automatic circuit assurance (ACA) in, 346
    automatic number identification (ANI) in, 329–330
    automatic route selection (ARS) in, 326–328
    call-by-call service selection (CBCSS) in, 330–331
    caller ID (CLID) and, 330
    centralized application servers in, 353
    centralized attendant service (CAS), 334
    Common Channel Signaling System 7 (CCSS7) in, 329
    computer telephony integration (CTI) and, 353
    defense Metropolitan Area Telecommunications system (DMATS), 349
    directories in, 327
    distributed common equipment design in, 335–337
    distributed communications system (DCS), 349
    E&M interface in, 343
    Ethernet and TCP/IP in, 351
    features of, 351–353
    FRL and, 345–346
    intelligent feature transparent network (IFTN) in, 349–357
    intelligent private networks and, 327
    ISDN features for, 328–329, 350
    LAN/WAN structure and, 336–337
    network class of service (NCoS) in, 345–346
    nonfacility associated signaling (NFAS), 331

Index

networking with PBX (*continued*)
  off-premises extensions (OPX) and, 333–335
  off- vs. on-net routing in, 326–327
  optical fiber, 336–337
  private networks and, 327, 332, 341–349
  Qsig and, 354–357
  redundancy and duplication in, 340
  remote port cabinet options for, 337–340
  single system network, on net multilocation support in, 332–335
  small office/home office (SOHO) support, 333–335
  station identification (SID) in, 329–330
  switching technologies and, 344
  tail end hop off-(TEHO) in, 327
  tandem tie trunk networks (TTTNs) in, 343–344
  telephony over IP (ToIP) and, 340
  tie trunks in, 343
  uniform dial plan (UDP) in, 344–345
  virtual private networks (VPNs) in, 346–349
  wireless phones and, 335
night service, 395
911 calls, 363
no progress tone, 101
noise, in cabling, 281–282
nonblocking switch networks (*See* blocking vs. nonblocking systems)
nonbroadcast media access (NBMA), 256
nonfacility associated signaling (NFAS), 329, 331
Nortel Networks (Northern Telecom), 17, 18, 20, 24–26, 36, 42–43, 60, 62–63, 66, 72, 77, 82, 85–87, 99, 106, 110, 115, 124, 133, 153, 162, 211, 221, 222, 231, 238, 241, 244, 245, 340, 341
  IP telephones and telephony, 309, 312, 313, 315
  mobile phones, 335
  remote port cabinet options for, 337–340
  telephones and, 319
North American Dialing Plan (NADP), 363
Novell, 39
null capability, in converged IP–PBX, 216
number analysis, 100, 102

off-hook, 80
off-hook alarm, 381, 395
off-premises call forwarding, 375
off-premises extensions (OPX), 30, 93, 137, 333–335
off-premises station, 396
off- vs. on-net routing in, 326–327
OmniPCX system, 60, 68, 69, 99
  converged IP–PBX and, 211, 218–219, 223
  operating system in, 232, 233
on-hook dialing feature, telephones and, 320
1B+D protocol, 140
Open Applications Interface (OAI), 38–39
open design standards, 122, 179, 247
open system speed dial, 396
operating memory, 103–104
operating systems, 98–100
  client/server IP–PBX and, 231–233, 240
  IP telephones and telephony, 308–309
operations, administration, maintenance (OA&M), 155

optical fiber, 285, 336–337
  cabinet/carriers backplane and, 132
  converged IP–PBX and, 222
  distance limitations for, 132
  fiber loop exchange (FLEX), 222
  interference in, 281–282
  remote port cabinet options for, 338–340
optiKeyboard, softphone, 215, **216**
optiSet digital telephones, 76
OPX line card, 139
order for data entry into switches, 362
Oryx/Pecos, 99
overflow, 386
override of diversion, 386

Pacific Telephone and Telegraph, 20
packet-based switching, 2
packet loss, 260–262, 274
packet of real time transport protocol (RTP) and, 193
packet switching, 148
  client/server IP–PBX and, 229
  H.323 protocol for, 177
  telephony over IP (ToIP), 151–153
PacketCable, 205
packetization, quality of service (QoS) and, 264
padlock, 381
paging system access, 386
paging systems, 2, 5, 128, 381
parking calls, 375
patch panel cabling, 286
payload of real time transport protocol (RTP) and, 192
PCM bus, latency and, 263
peer-to-peer converged IP–PBX and, 209
Pentium processors, 98, 109
per hop behaviors (PHBs), 270, 275
peripherals
  IP PBX vs. client/server PBX in, 157–158
  telephony over IP (ToIP), 150–153
personal digital assistant (PDA), IP telephones and telephony, docking stations, 306
personal speed dialing, 381
personalized ringing, 381
PhoneHub Sphericall Enterprise Softswitch, converged IP–PBX and, 224
plain old telephone service (POTS), 122
planning for system, request for proposal (RFP) for client/server PBX, 407–443
point-to-point communications, H.323 protocol for, 177
Poisson distribution, 82–83, 87–88
pooled modem cards, 143
port cards and printed circuit boards, 134, 139–143
port addresses, 362
port cabinets, 125
port capacity, 122
  blocking vs. nonblocking systems, 80–82
  cabinet/carrier design for, 123–124, 133
  centralized PBX and, 109–110
  centum call seconds (CCS) and, 85–88
  client/server IP–PBX and, 249
  converged IP–PBX and, 210
  dispersed PBX and, 110–111

port capacity (*continued*)
  distributed PBX, 111–112
  expansion requirements and, 132–134
  IP PBX vs. client/server PBX in, 166–167
  port carrier and, 127, **128**
  port circuit card microcontroller and, 104–105
  power cabinet/carriers, 129
  Superloop, 85, **86**, 87
  telephony over IP (ToIP), 152–153
port carrier, 127, **128**
port circuit card microcontrollers, 104–105
port circuit cards, cost of, 402, 445
port interface modules (PIMs), 27
port interfaces, IP telephones and telephony, 305–306
port numbers, IP, 191
port processor interface, 27
port seizure, 100–101, **101**
port-to-port communications, single TDM bus for, 51–52
portable operating system (POSIX), 233
power cabinet/carriers, 129
power failure transfer, 396
power failure transfer card, 140
power supplies, 129–131
  application cabinet/carriers, 128–129
  auxiliary cabinet/carriers, 128
  cabinet/carriers, 129–131
  cards and printed circuit boards, 144
  client/server IP–PBX and, 240
  IEEE 802.3af standard for, 306–307
  IP PBX vs. client/server PBX in, 166
  IP telephones and telephony, 324
  IP telephones and telephony, Ethernet, 306–307
  port carrier and, 127, **128**
  power cabinet/carriers, 129
  power failure transfer card, 140
  power system monitor card for, 136
  switching carrier and, 127
  telephones and, 324
  uninterruptible (UPS), 130, 131
power system monitor card, 136
PowerDsine, 306
PowerSum, 282
price curves for cost, 406, 445
primary rate interface (PRI) (*See also* ISDN), 329
printed circuit boards (*See* cards and printed circuit boards)
priority calling, 382
priority queue, 387
privacy, 382
private line, 382
private networking, 332, 343–349
  automatic alternate routing (AAR) in, 345
  automatic circuit assurance (ACA) in, 346
  converged IP–PBX and, 218, 219
  E&M interface in, 343
  FRL and, 345–346
  intelligent, 327
  IP PBX vs. client/server PBX in compatibility, 163–164
  multiple system, 341–343
  network class of service (NCoS) in, 345–346
  tandem tie trunk networks (TTTNs) in, 343–344

private networking (*continued*)
  tie trunks in, 343
  uniform dial plan (UDP) in, 344–345
  virtual private networks (VPNs) in, 346–349
probability calculation, grade of service (GoS), 82–83
processor, 96–98, **97**, 104, 106, 126, 136
  centralized PBX and, 108–110
  port circuit, 104–105
  redundancy and duplication in, 116
processor cards, 136
processor occupancy, 366
prompting, call, 138
property management system interface, 396
proprietary digital phones, 299–300
proprietary vs. open design standards, 122, 247, 299–300
protocols, in LAN design, 256
prowler board, 212
proxy servers, 203, 209, 266, 267
public switched telephone network (PSTN), 177, 201
Pulse, 20
pulse code modulation (PCM), 18, 47–49, **47, 48**, 186

Q signaling (Qsig), 33, 35–36, 164, 184, 244, 329, 350, 354–357
  H.225.0 RAS, 189
  H.323 protocol and, 187
Q.931, 184, 187, 189, 329
quality of service (QoS), 252–254, 363
  802.1p/Q tagging in, 268–269
  best effort in, 275
  buffers in, 262, 264
  class of service (CoS) and, 268
  client/server IP–PBX and, 249
  codecs in, 261–262
  concealment of errors and, 261–262
  controls for, 268–271
  converged IP–PBX and, 212–219, 221, 225, 226
  delay and, 260–262, 273
  differentiated services (DiffServ) and, 268, 270–271, 274–276
  digital signal processing (DSP) and, 261, 263, 264
  E model for, 272–276
  echo in, 272
  forward error correction (FEC) in, 260–262
  H.323 protocol and, 176–177, 180, 185, 203
  hard vs. soft, 253
  implicit vs. explicit, 253
  interleaving and, 260–262, 260
  IP PBX vs. client/server PBX in, 159–160, 168, 171
  IP telephones and telephony, 305–306
  jitter buffering and, 264, 265–266
  in LAN design, 257, 260–276
  latency and, 262–268, **266, 267**
  low latency queuing (LLQ) in, 271
  multiprotocol label switching (MPLP) in, 267, 271
  packet loss and, 260–262, 274
  packetization and, 264
  real time transport protocol/real time transport control protocol (RTP/RTCP) in, 192
  service level agreements (SLAs) and, 271

Index

quality of service (*continued*)
  session initiation protocol (SIP), 203
  speech compression and, 273–274
  subnet bandwidth management (SBM) in, 271
  summary of control of, 271–272
  tandeming and, 274
  telephones and, 318
  telephony over IP (ToIP), 159–160
  TIA recommendations for, 272–276
  transcoding and, 274
  transmission rating (R) in, 272–273
  type of service (ToS) prioritization in, 268, 269–270
  virtual LANs (VLANs) and, 271
  vocoding and, 263–268, **264**
queuing calls, 8
queuing models, 87–90, **90**

R300 Remote Office Communicator, 221–222, 340
random access memory (RAM), 104, 304
reach line card, 221
read only memory (ROM), 304
Real Time Application Interface (RTAI), Linux, 233
Real Time Linux (RTL), 233
real time operating systems (RTOS), 98–100, 232
real time streaming protocol (RTSP), 198
real time transport control protocol (RTCP), 187, 190–197, 305
real time transport protocol (RTP), 187, 190–194, 198, 258
real time voice, H.323 protocol for, 177
real time, soft, 233
recall signaling, 382
recent change history, 4, 396
recorders, 128
recordkeeping, 361
reduced instruction set computing (RISC), 109
redundancy and duplication, 71–73, 112–116
  backup servers in, 234
  CallManager system and, 234–238, **235**
  client/server IP–PBX and, 233–239, 249
  database synchronization in, 239
  IP PBX vs. client/server PBX in, 162–163, 167
  local controller, 106
  media gateway (MG) in, 238–239
  mirroring in, 233–234
  multiple active servers in, 234
  networking with PBX and, 340
  RAID in, 233–234
  simple network automated processing (SNAP) in, 238
  survivable remote telephony (SRS) in, 238
  switching, 71–73
  trivial file transfer protocol (TFTP) in, 235
redundant array of independent disks (RAID), 233–234
registration, admission and status (RAS), 187, 188–189, 305
release link trunks (RLT), 141
release loop operation, 387
reliability levels, 113–116, 149
  client/server IP–PBX and, 248–249
  H.323 vs. SIP protocols and, 201
  IP PBX vs. client/server PBX, 162–163
remote cabinet option, 25–26
remote designs, distributed PBX, 111–112
remote office unit, 221
remote peripheral equipment (RPE), 221
remote switch center (RSC) option, 26
request for proposal (RFP) for client/server PBX, 407–443
reserve power, 131
reserving bandwidth, 253
resource management, 100
resource reservation protocol (RSVP), 198, 249
restricted incoming station, 396
restriction features, 396
ring topology, 279–280
ringer cutoff, 382
ringing call forwarding, 375
ringing tone control, 382
RJ11, USB bus and, 301
Rockwell, 37
Rolm, 17, 18, 26, 43, 55, 112, 148, 341
Rolm CBX, 18, 37
RolmNet, 341
round-trip latency, 262
route advance, 397
route optimization (*See also* automatic route selection), 326
route selection (*See also* automatic route selection), 34, 100
  Busy Hour Call (BHC) and, 118–119
  H.323 protocol and, 182
routers and routing (*See also* automatic route selection), 8, 252, 256
  analysis of, 100
  automatic alternate routing (AAR) in, 345
  automatic route selection (ARS) in, 326–328
  latency in, 266–267
  look ahead, 346
  off- vs. on-net routing in, 326–327
  tail end hop off-(TEHO) in, 327
RS-232C, 32, 132, 266, 360
RS-449, 32

S/T interface, 140
sampling, 18, 48–49, **48**, 186, 301
satellite switches, 344
save and redial, 382
scalability
  H.323 vs. SIP protocols in, 203
  IP PBX vs. client/server PBX in, 173
  in LAN design, 256
screening calls, 8
screenphone, 309, **309, 310, 311**
second communication line in, 23–24
secondary extension feature activation, 383
security
  converged IP–PBX and, 216, 222
  H.323 protocol for, 178
security module for telephones and, 323
segmentation, time slot access and, 73–75, **74**
selective multiple line call forwarding, 375
Selsius System, 230, 244
send all calls, 383
serial calling, 387
serial data interface, 137

servers, 14
  in LAN design, 255
  PBX and, 22
  telephony call server for client/server IP–PBX, 230–242
service cards and printed circuit boards, 134, 137–139
service level agreements (SLAs), 82–83, 271
session announcement protocol (SAP), 198
session description protocol (SDP), 198
session initiation protocol (SIP), 156, 197–199, **198**
  client/server IP–PBX and, 243, 249
  IP telephones and telephony, 305
  telephony gateways and, 243
SG-1, 20
shadowing, redundancy and duplication in, 114
shared tenant service, 397
shielded twisted pair (STP), 285, 289
Shoreline/ShoreGear converged IP–PBX and, 224–226
Siemens, 17, 36, 50, 63, 66, 72, 73, 76, 111, 115, 124, 129, 135, 148, 162, 222, 231, 241, 242, 244, 247, 349, 353, 407, 417
  IP telephones and telephony, 306, 308, 312, 315
  mobile phones, 335
  Qsig and, 354–357
signaling mechanisms, H.323 protocol and, 186–188
signaling system 7 (SS7), 200, 329
silence suppression, 213, 225, 308, 318
simple endpoint terminal (SET), H.323 protocol and, 184
simple network automated processing (SNAP) in, 238
single pair cabling, 278
single-stage circuit switch matrix, 57–58, 57
skinny gateway control protocol (SGCP), 204, 249
skinny protocol, 204
SL-1 family of systems, 18, 25, 26, 72, 337–340
small office/home office (SOHO) support, 333–335
SMDI, 230
soft QoS, 253
soft real time, 233
softkey buttons for telephones and, 319
softphones, 40, 157, 297–298, 315–317, **316**, 335
Softswitch Consortium, 205
software code, main system processor, 98
software defined network (SDN), 164
software options for PBX and, 27
spanning tree configuration, 256
speaker verification, 11
speakerphones, 320
Spectrum 24 wireless phone, 312–313
speech codecs, H.323 protocol and, 184
speech compression, quality of service (QoS) and, 273–274
speech synthesizers, 137
speed dialing, 376, 381, 383, 397
Sphere Communications, 336, 337–340
Sphere Sphericall Enterprise Softswitch, 224–226
splitting calls, 384
spread spectrum communications, 312
Sprint cellular, 314
stackable cabinet design, 26–27
standardization efforts, 35–36, 46
  cabling, 278–279
  IP PBX vs. client/server PBX in, 156
star topology, 177, 279–280, 286

static electricity (*See* electrostatic damage)
station identification (SID) in, 329–330
station user features, 373–384
stations, performance management in, 366
step call, 383
store redial, 383
stored program control (SPC), 2, 20–22, 278
straightforward outward completion, 388
subnet bandwidth management (SBM), 271
Succession family of systems, 99
  call center and call center servers in, 241
  client/server IP–PBX and, 231, 238
  telephony gateways and, 245–246
Superloop, 58, 62–64, 72–73, 77
  port capacity of, 85, **86**, 87
SuperSet digital telephones, 242–243
SuperStack 3 (SS3), 152, 231
supervisor/assistant calling, 383
supervisor/assistant speed dial, 383
suppliers, features provided by, 445–464
survivable remote telephony (SRS), 238
switch network cards, 136
switching carrier, 126–127, **127**
switching technologies, 2, 3, 17–20, 46–60, 149
  blocking in, 75–77, **76**
  cabling and, 278
  call processing system, 96, **96**
  centralized PBX and, 65–67, **65**
  circuit switched, 148, 208, 229, 231
  client/server IP–PBX and, 229
  computer stored program control (SPC) and, 20–22
  converged circuit/packet switched systems, 151
  digital, 17–20
  dispersed PBX, 67–70, **70**
  distributed PBX, 67–70, **68**
  duplication in, 71
  Highway bus in, 60–64
  legacy, 79–94
  local switching network design and, 60–64
  network design using, 64–70, 344
  nonblocking, 75–77, **76**
  order for data entry into switches, 362
  packet, 148, 229
  redundancy in, 71–73
  Superloop in, 58, 62–64, 72–73, 77
  switch network cards in, 136
  switching carrier and, 126–127, **127**
  TDM bus in, 47–64
  time slot access and segmentation in, 73–75, **74**
  time slot availability in (*See also* blocking; nonblocking), 75–77, **76**
SX 2000 system, 152, 243, 338
Sybase, 242
Symbol Technologies, IP telephones and telephony, wireless, 312–313
Symposium Call Center Server, 241
synchronization in distributed PBX, 112
System 75, 105
System 85, 28, 109–110, 109
system administration (*See also* administration and management), 361

Index

system administration (*continued*)
　cost of, 403–404
system administration terminal (SAT), 360
system cabinets (*See* cabinets/carriers)
system control interfaces, 97
system processing bus, 106–107
system speed dial, 397
Systems Network Architecture (SNA), 171
SYSTIMAX guidebook, cabling, 279
system processing engine (SPE), 234

T.120, 185
T1/E1, 141, 142, 152
　centralized PBX and, 67
　client/server IP–PBX and, 232
　converged IP–PBX and, 208, 218, 221, 222
　fiber optics and, 338–339
　PBX and, 35
Tadiran, 42
tagged frames, 802.1p/Q tagging in, 268–269
tail end hop off-(TEHO), 327
talk slot assignment
　blocking vs. nonblocking systems, 81–82
　cards and printed circuit boards, 143–144
tandem tie trunk networks (TTTNs) in, 343–344
tandeming, for quality of service (QoS) and, 274
tape drives, 136
TCP/IP (*See* Ethernet and TCP/IP)
TDM bus, 47–48
　analog line cards for, 139
　asynchronous transfer mode (ATM) center stage switch in 59–60
　bandwidth and capacity of, 49–51, **51**
　blocking and nonblocking design in, 75–77, **76**
　blocking vs. nonblocking systems, 80–82
　broadband, 56–57
　Busy Hour Call (BHC) and, 118–119
　cards and printed circuit boards, 134 143–144
　center stage switch complex and, 55–56
　centralized PBX and, 65–67, **65**
　centum call seconds (CCS) and, 87
　client/server IP–PBX and, 228, 229
　control buffer card and, 138
　converged IP–PBX and, 210, 211, 213, 217, 221, 223
　digital attendant console card, 139
　digital line card, 139
　dispersed PBX, 67–70, **70**
　distributed PBX, 67–70, **68**
　local controllers, cabinet/carrier shelf, 106
　local switching network design and, 60–64
　multiple, 52–54, **53**
　multistage circuit switch matrix in, 58–59, **58**
　port capacity of, 87
　port circuit card microcontroller and, 105
　port to port communications over, 51–52
　redundancy/duplication in, 71–73
　single-stage circuit switch matrix in, 57–58
　Superloop and, 58, 62–64, 72–73, 77
　switch network card and, 136
　switching carrier and, 127
　system control interface and, 97

TDM bus (*continued*)
　system processing bus and, 106–107
　talk slot assignment and switching, 100
　time slot access and segmentation in, 73–75, **74**
　time space time (TST) switch networks in, 59–59
　tone clock for, 137
TDM/PCM, 151–153, 176
telecommunications closet cabling, 285–286
Telecommunications Industries Association (TIA) standards for cabling, 278
telecommunications outlet cabling, 289–290, **290**
telephones and voice terminals, 148, 295–324
　adapter or add-on modules for, 321
　add-on button module for, 323
　alphanumeric displays on, 321
　analog, 148–149, 298–299
　analog communications device adapter for, 322, **322**
　application feature buttons in, 320
　audio codecs in, 318
　call control in, 318
　compression in, 307–308, 307
　computer telephony integration (CTI) API module for, 321–322
　cost and price ranges for, 296–297, **296**, 317, 402–403
　data modules for, 321
　design basics in, 303–305, **303**
　digital communications device adapter for, 323
　digital, 148–149, 299–302, **300, 302**
　distance extender module for, 323
　dual tone multifrequency (DTMF) and, 298–299, 304
　dynamic host configuration protocol (DHCP) and, 318
　echo suppression in, 318
　email station in, 311
　Ethernet and TCP/IP in, 304–307, 318
　features of, 296–297, 305–317, 319–320
　fixed feature buttons, 319
　gateway subsystem in, 304–305
　H.225 call signaling protocol in, 305
　H.245 Control protocol in, 305
　H.323 and, 304–305, 304
　hands-free answer intercom (HFAI) in, 320
　headset interface for, 320
　IEEE 802.3af standard for, 306–307
　integrated port interfaces in, 305–306
　IP adapter module for, 323
　IP PBX vs. client/server PBX in, 166
　IP telephones in (*See also* IP telephones), 297–298, 301–317, **302, 303**
　ISDN BRI adapter module for, 323
　ISDN compatible, 301
　loop length limitations in, 317
　mobile and cellular phones in, 312–315, **314**, 335
　mute buttons on, 320
　network interface in, 304
　on-hook dialing feature, 320
　operating systems in, 308–309
　PDA docking stations and, 306
　performance vs. price in, 317
　power supplies and, 306–307, 324
　programmable line/feature buttons, 319
　quality of service (QoS) and, 305–306, 318

telephones and voice terminals (*continued*)
  RAS protocol in, 305
  real time transport control protocol (RTCP) in, 305
  screenphone type, 309, **309, 310**, **311**
  security module for, 323
  silence suppression in, 308, 318
  SIP in, 305
  softkey buttons in, 319
  soft-, 297–298, 315–317, **316**, 335
  software for, 304
  speakerphones in, 320
  telephony over IP (ToIP), 152–153
  universal serial bus (USB) type, 300–301, 304, 306
  user interface for, 303
  voice activity detection (VAD) in, 308, 318
  voice interface for, 303–304
  voice over IP (VoIP) and, 302
  Web browsers and, 302, 308–311, **309, 310, 311**
  wiring pair numbers in, 318
Telephony Applications Programming Interface (TAPI), 39
telephony call server for client/server IP–PBX, 230–242
  administration and management in, 230
  call centers and, 241–242
  call data recording (CDR) in, 236
  call processing in, 230
  CallManager, redundancy and backup in, 234–238, **235**
  cluster topology in, 236
  database synchronization in, 239
  gatekeeper in, 230
  integrated communications system (ICS), 240–241
  media gateway (MG) in, 238–239
  mesh topology in, 236
  messaging applications and, 241
  multifunction server design in, 239–242
  operating system in, 232–233
  proprietary type, 230
  redundancy and duplication in, 233–239
  simple network automated processing (SNAP) in, 238
  survivable remote telephony (SRS) in, 238
  telephony features in, 241
  third party type, 230
telephony gateways
  CallManager, 244
  channel associated signaling (CAS) in, 244
  client/server IP–PBX and, 242–249, 242
  dual tone multifrequency (DTMF) in, 243
  Ethernet and TCP/IP in, 243
  H.323 in, 243, 244
  media gateway control Protocol (MGCP), 244
  proprietary vs. open design in, 247
  SIP in, 243
telephony over IP (ToIP), 150–153, 159–161
  converged IP–PBX and, 208
  H.323 protocol and vs., 176
  networking with PBX and, 340
telephony services application programming interface (TSAPI), 39
teleworkers, 30, 333–335
Teltone, 334
10Basexx Ethernet, 280

terminal handling, 100
terminals
  features for, 446–454
  H.323 protocol and, 180, 181
text messages, 383–384
3Com, 43, 44, 152, 230, 232, 239, 241, 242, 306
threshold alarms, 366–367
through dialing, 387
tie trunks in, 343
time-based switching, 2
time division multiple access (TDMA), 314
time division multiplexing (TDM) (*See also* TDM bus), 12, 16, 18, 47–48, **47**, 215–217
time slot access, 73–75, **74**
time slot assignment, 132, 143–144
time slot interchanger (TSI), 27, 56
time space time (TST) switch networks in, 59–59
timed queue, 384
timed reminder, 397
token ring topology, 279–280
tone clock, 126, 137
tone receiver, 137
topologies, 64–70, 107–112, **107**, 286, 465–467
  cabling, 279–280
  H.323 protocol for, 177
  LAN, 254–255
Toshiba IP telephones and telephony, 312
tracing calls, 380
traffic analysis, 46
traffic distribution, 366
traffic engineering, 95–119
  blocking vs. nonblocking systems, 80–82
  Busy Hour, 82–88, 366
  Busy Hour Call (BHC) in, 116–119, **118**
  Busy Hour Call Attempts (BHCAs), 116–119
  Busy Hour Call Completions (BHCCs), 116–119
  call data reports (CDRs) for, 93
  centum call seconds (CCS), 83–88
  client/server IP–PBX and, 249
  Erlang B queuing model in, 88–90, **90**
  Erlang B, 93–94
  grade of service (GoS), 82–83, 92
  legacy PBX, 95–119
  models for, 91, **91**
  performance management and, 365–369
  Poisson distribution in, 87–88
  queuing models in, 87–90, **90**
  trunk circuit usage/analysis, 88–94, **91, 92**, 365
transaction processing, session initiation protocol (SIP) and, 200
transcoding, 210–212, 274
transfer call, 5
translation data in, 361–362
transmission control protocol (TCP) (*See also* Ethernet and TCP/IP), 191–192
transmission rating (R) in quality of service (QoS) and, 272–273
transparency of features, 164
traveling class mark (TCM), 346
tributary switches in, 344
trivial file transfer protocol (TFTP), 235

Index

trunk answer any station, 397
trunk callback queuing, 397
trunk circuits, 122
   ATM trunk card, 142
   auxiliary trunk card for, 142–143
   Busy Hour Call (BHC) in, 117–119
   CO trunk card for, 141
   converged IP–PBX and, 216–219
   DID trunk card for, 141
   digital trunk interface (DTI), 144
   DS1 digital trunk card for, 141
   E&M tie trunk card for, 141
   Erlang B queuing models in, 88–90, **90**
   grade of service (GoS), 92
   inbound vs. outbound, 93–94
   IP PBX vs. client/server PBX in, 166
   IP trunk cards, 142
   ISDN BRI trunk cards for, 142
   ISDN PRI trunk cards for, 141–142
   number required, 92–94
   release link trunks (RLT) in, 141
   telephony over IP (ToIP), 151–153
   traffic engineering for, 88–91, **91, 92**
   usage and traffic analysis, 365
trunk flash, 384
trunk group busy/warning indication, 387–388
trunk identification, 388
trunk-to-trunk connection, 384
trunk-to-trunk transfer, 387
trunk usage, 365
2B+D protocol, 139, 140, 142, 299
type of service (ToS) prioritization in, 249, 268, 269–270

U interface, 140
unicasting, 179–180, 183, 192
unified messaging systems (UMSs), 10, 43, 241
uniform call distribution (UCD) in, 37, 398
uniform dial plan (UDP) in, 344–345, 398
uninterruptible power supplies (UPS), 130, 131
universal port circuit card slot designs, 143
universal serial bus (USB) telephone, 300–301, 304, 306
UNIX, 99, 232, 233
unshielded twisted pair (UTP), 280, 285, 289
upgrades, IP PBX vs. client/server PBX in, 155–156, 160–162, 168–169
URLs, session initiation protocol (SIP), 199, 202
user agent client/server (UAC/S), in SIP, 199, **199**
user datagram protocol (UDP), 191–192, 195, **195**, 258

variable delay, 213–214
VAX minicomputer, 21
video codecs, 178, 185–186
video transmission, 17, 33–34, 252
videoconferencing, 177, 323
videophones, 33
VIM Sphericall Enterprise Softswitch, 224
virtual circuits, 252
virtual connections, 191
virtual extensions, 398
virtual LANs (VLANs), 256, 271

virtual private networks (VPNs), 164, 222, 346–349
VLCBX, distributed PBX, 112
vocoding and, 263–268, **264**, 263
voice activity detection (VAD) in, 213, 258–259, 308, 318
voice announcers, 2
voice communications and voice codecs, 28–31, 257
   client/server IP–PBX and, 249
   converged IP–PBX and, 210–216, 221, 225
   vocoding and, 264–268
voice interface, for IP telephones and telephony, 303–304
voice mail, 29, 239
voice messaging system (VMS), 2, 5, 9–10, 42–43, 84, 138, 139, 241, 335, 398
voice over IP (VoIP), 155, 175–205
   converged IP–PBX and, 208, 216, 217
   gateways in, 367
   H.323 protocol for, 176–188
   IP telephones and telephony, 302
   LAN/WAN design guidelines for, 251–276
   media gateway control Protocol (MGCP), 204–205
   MEGACO (H.248) in, 204–205
   protocols for, 176
   real time transport protocol/real time transport control protocol (RTP/RTCP) in, 191
   session initiation protocol (SIP) for, 176, 197–199, **198**
   telephony over IP (ToIP) vs., 176
voice terminals (*See* telephones and voice terminals)
VoiceXML, 240
VPIM, 249
VTP, 256
VX Works operating system, 14, 99, 231, 232, 240, 308–309

warning alarm, 371
WATS, 137, 141
Web browsers, 302, 308–311, **309, 310, 311**, 360
Webswitch, 243
Wescom 580, 37
wide area networks (WANs), 16
   client/server IP–PBX and, 228, 229
   converged IP–PBX and, 208, 209, 216– 221
   H.323 protocol and, 176
   IP PBX and, 147–173
   IP PBX vs. client/server PBX in, 154–158
   networking with PBX and, 336–337
   telephony over IP (ToIP), 150–153
   voice over IP (VoIP) and, 251–276
Windows/Windows NT, 14, 22, 98, 99, 231, 232, 233, 247, 353
wireless PBX systems, 139, 149, 335
   frequency of operation for, 42
   H.323 protocol for, 177
   IP telephones and telephony, 312–315, **314**
   PBX and, 41–42, **41**
wiring pair numbers in telephones and, 318
work area cabling, 290–292, **291**
working memory (*See* operating memory)

Xpressions server, 241

ABOUT THE AUTHOR

ALLAN SULKIN is the founder and president of TEQConsult Group, which advises enterprise communications providers on product development, marketing, and competitive analysis. He is widely acknowledged as the foremost expert in PBX (Private Branch Exchange), Key Telephone/Hybrid, and ACD (Automatic Call Distribution) communications systems today. His clients include Cisco Systems, Avaya, Siemens, NEC, and Alcatel. Sulkin is also a contributing editor to B*usiness Communications Review*, and the designer/presenter of BCR's PBX training seminar. His company has its offices in Hackensack, New Jersey.